Lecture Notes in Physics

Edited by J. Ehlers, München K. Hepp, Zürich
R. Kippenhahn, München H. A. Weidenmüller, Heidelberg
and J. Zittartz, Köln

156

Resonances in Heavy Ion Reactions

Proceedings of the Symposium
Held at the Physikzentrum
Bad Honnef, October 12-15, 1981

Edited by K.A. Eberhard

Springer-Verlag
Berlin Heidelberg New York 1982

Editor

Klaus A. Eberhard
Sektion Physik der Universität München
Am Coulombwall 1, D-8046 Garching

ISBN 3-540-11487-4 Springer-Verlag Berlin Heidelberg New York
ISBN 0-387-11487-4 Springer-Verlag New York Heidelberg Berlin

This work is subject to copyright. All rights are reserved, whether the whole or part of the material is concerned, specifically those of translation, reprinting, re-use of illustrations, broadcasting, reproduction by photocopying machine or similar means, and storage in data banks. Under § 54 of the German Copyright Law where copies are made for other than private use, a fee is payable to "Verwertungsgesellschaft Wort", Munich.

© by Springer-Verlag Berlin Heidelberg 1982
Printed in Germany

Printing and binding: Beltz Offsetdruck, Hemsbach/Bergstr.
2153/3140-543210

CONTENTS

I

INTRODUCTORY TALK

Resonances in Heavy Ion Reactions
  D.Allan Bromley 3

II

RESONANCES VERSUS FLUCTUATIONS

Extended Critical Analysis of Structures in the
   Excitation Functions for $^9Be+^{12}C$
  L.Jarcyk 37

Use of the Deviation Function in a Search for
   Resonances in the System $^{12}C+^{16}O$
  M.Hugi, J.Lang, R.Müller, J.Sromicki, L.Jarcyk
  A.Strzalkowski, H.Witala, and K.A.Eberhard 51

Nucleon Decay of $^{12}C+^{12}C$ and $^{12}C+^{16}O$:
   Resonances or Statistical Fluctuations?
  D.Evers 53

Resonant Structures in the $^{16}O+^{16}O$-System
   near the Coulomb Barrier
  G.Gaul, W.Bickel, W.Lahmer, and R.Santo 72

Partial Coherence in Heavy-Ion Reactions
  K.M.Hartmann, W.Dünnweber, and W.E.Frahn 74

RESONANCE STUDIES IN PARTICULAR REACTIONS - CARBON-OXYGEN MASS REGION AND LIGTHER

Search for Resonances in Light Heavy Ion Systems
    H.Fröhlich, P.Dück, W.Treu, and H.Voit 79

$^{12}C+^{12}C$ Resonances Studied in the Elastic,
    Inelastic, and Transfer Channels
    T.M.Cormier 95

The Spins and Spectroscopy of $^{12}C+^{12}C$ Intermediate
    Structure Resonances
    E.R.Cosman, R.J.Ledoux, M.J.Bechara,
    C.E.Ordonez, and H.A. Al-Juwair 112

Direct Observation of $^{12}C-^{12}C$ Configuration
    States in Their ^{12}C Decay
    K.Katori, K.Furuno, J.Schimizu, Y.Nagashima,
    and M.Sato 129

Elastic and Inelastic Scattering of $^{14}C+^{14}C$ and $^{12}C+^{14}C$
    D.Konnerth, K.G.Bernhardt, K.A.Eberhard, R.Singh,
    A.Strzalkowski, W.Trautmann, and W.Trombik 131

Microscopic Investigation of the $^{14}C+^{14}C$ Interaction
    D.Baye and P.-H.Heenen 133

$^{10}B+^{14}N$ and $^{12}C+^{12}C$ Reaction Data near Molecular
    Resonances
    W.Hoppe, E.Klauß, D.Sprengel, J.Drevermann,
    R.Isenbügel, H.v.Buttlar, and N.Marquardt 135

High-Resolution Excitation Functions of $^{14}N+^{14}N$
    Reactions Near Resonances in ^{28}Si
    M.Treichel, R.Isenbügel, H.v.Buttlar,
    and N.Marquardt 137

Intermediate and Fine Structure Studies in the
System $^{16}O + ^{12}C$
  P.Braun-Munzinger and H.W.Wilschut 139

Correlation Measurement Searching for Resonant
$^{12}C-^{12}C$ States Induced by the $^{12}C(^{16}O,\alpha)$ Reaction
  K.Katori, T.Shimoda, T.Fukuda, H.Ogata,
  I.Miura, and M.Tanaka 152

Resonances in ^{26}Al, ^{29}Si and ^{30}Si: Are They Entrance
  Channel Dependent?
  S.T.Thornton 154

Gross Structure in Mismatched Channels
  Peter Paul 161

IV

RESONANCE STUDIES IN PARTICULAR REACTIONS —
sd-SHELL NUCLEI AND HEAVIER

Resonances in sd-Shell Nuclei
  J.P.Schiffer 177

High Angular Momentum Resonances in $^{28}Si + ^{28}Si$ Scattering
  R.R.Betts 185

Micoscopic Study of Elastic $^{28}Si - ^{28}Si$ Scattering
  K.Langanke and R.Stademann 199

Search for Intermediate Structure in ^{36}Ar Via the
$^{24}Mg(^{12}C,\alpha)^{32}S$ Reaction
  R.Čaplar, G.Vourvopoulos, X.Aslanoglou,
  and D.Počanić 202

Molecular Structure in $^{12}C + ^{12}C$, Orbiting in $^{12}C + ^{28}Si$,
  and first Studies of the $^{60}Ni + ^{60}Ni$ Interaction
  K.A.Erb, J.L.C.Ford,Jr., R.Novotny, and
  D.Shapira 204

Phase Shift Analysis and Heavy Ion Scattering
   C.Marty 216

Local-Potential Description of the Bound, Quasi-Bound and
       Scattering States of the α-Nucleus System
   R.Ceuleneer, F.Michel, and G.Reidemeister 227

Calculation of the Internal and Barrier Wave Contributions
       to Heavy Ion Elastic Scattering Made Simple
   J.Albinski and F.Michel 229

V

RESONANCE PHENOMENA IN FUSION AND
TOTAL REACTION CROSS SECTIONS

Unitarity of the S-Matrix and Resonance Phenomena in Nuclear
       Reaction Cross Sections
   I.Rotter 233

Characteristic Resonances and the Limits to Fusion
       in Light Heavy-Ion Systems
   J.J.Kolata 256

Fusion Resonances in $^{12}C(^{16}O,\gamma)^{28}Si$
   A.M.Sandorfi and M.T.Collins 264

Total Reaction Cross Section of $^{12}C+^{16}O$
       Near the Coulomb Barrier
   E.C.Schloemer, M.Gai, A.C.Hayes, J.M.Manoyan,
   S.M.Sterbenz, H.Voit, and D.A.Bromley 266

Structure in Heavy Ion Reactions Involving ^{14}C
   R.M.Freeman 268

Structure in Symmetric Light Heavy-Ion Fusion Cross Sections
   N.Rowley, N.Poffé, and R.Lindsay 279

VI

SEARCH FOR DIRECT γ-DECAY OF RESONANCES

Search for γ-Rays from the Quasimolecular $^{12}C+^{12}C$ System
  V. Metag, A. Lazzarini, K. Lesko, and R. Vandenbosch 283

Search for Direct γ-Transitions in $^{12}C+^{12}C$
  R.L. McGrath, D. Abriola, J. Karp, T. Renner
  and S.Y. Zhu 290

VII

SPIN ALIGNMENT AND POLARIZATION MEASUREMENTS

Measurement of Spin Alignment in $^{12}C+^{12}C$
    Inelastic Scattering
  W. Trombik 297

DWBA Analyses of Resonance Structure in the
    $^{16}O(^{16}O,^{12}C)^{20}Ne$ Reaction
  Yosio Kondō and Taro Tamura 314

VIII

MODELS AND SYSTEMATICS

Theory of Nuclear Molecular States
  D. Hahn, W. Scheid, and J.Y. Park 337

Structure and Formation of Molecules
  U. Mosel 358

Weak Coupling Model Approach to Heavy Ion
    Molecular Resonances
  O. Tanimura 372

Multistep Transfer of Nucleons and the
    Formation of Molecular Oribitals
        W.von Oertzen and B.Imanishi 388

Validity of the Adiabatic Molecular Orbital Concept
    in the Interaction of Heavy Ions
        B.Imanishi and W.von Oertzen 405

Resonances in $^{16}O+^{16}O$ and the Systematic Occurence of
    $J^{\pi}=8^{+}$ Resonances in Heavy Ion Resonant Systems
        M.Gai, E.C.Schloemer, J.E.Freedman, A.C.Hayes,
        S.K.Korotky, J.M.Manoyan, B.Shivakumar, S.M.Sterbenz,
        H.Voit, S.J.Willett, and D.A.Bromley 407

Schematic Models of Resonances: Predictions and Comparison
        N.Cindro and D.Počanić 409

On the Structural Similarity of Nuclear Molecules
        N.Marquardt 411

IX

SUMMARY OF THE CONFERENCE

    W.Greiner 415

LIST OF PARTICIPANTS

INTERNATIONAL WORKSHOP ON RESONANCES IN HEAVY ION COLLISIONS
Bad Honnef, Oct. 12-15, 1981

W. Assmann, Munich
D. Baye, Brussels
R. Betts, Argonne
P. Braun-Munzinger, Stony Brook
D. A. Bromley, Yale
H. v. Buttlar, Bochum
R. Caplar, Zagreb
R. Ceuleneer, Mons
N. Cindro, Zagreb
T. M. Cormier, Rochester
E. Cosman, MIT
W. Dünnweber, Munich
K. A. Eberhard, Munich
K. A. Erb, Oak Ridge
D. Evers, Munich
R. M. Freeman, Strasbourg
H. Fröhlich, Erlangen
M. Gai, Yale
G. Gaul, Münster
W. Greiner, Frankfurt
D. Hahn, Giessen
K. M. Hartmann, Berlin
P.-H. Heenen, Brussels
B. Imanishi, Tokyo
L. Jarcyk, Cracow
E. Klauß, Bochum
J. J. Kolata, Notre Dame
R. Könnecke, Frankfurt

Y. Kondo, Austin
D. Konnerth, Munich
J. Lang, Zürich
A. Lazzarini, Seattle
N. Marquardt, Bochum
C. Marty, Orsay
R. L. McGrath, Stony Brook
V. Metag, Heidelberg
F. Michel, Mons
U. Mosel, Giessen
W. v. Oertzen, Berlin
P. Paul, Stony Brook
N. Poffe, Oxford
I. Rotter, Rossendorf
N. Rowley, Daresbury
A. Sandorfy, Brookhaven
R. Santo, Münster
W. Scheid, Giessen
J. P. Schiffer, Argonne
P. Sperr, Munich
A. Strzalkowski, Cracow
O. Tanimura, Giessen
A. Thiel, Frankfurt
S. T. Thornton, Charlottesville
W. Trautmann, Munich
M. Treichel, Bochum
W. Trombik, Munich
H. Voit, Erlangen

FOREWORD

An international workshop on "Resonances in Heavy Ion Collisions" was held from Oct. 12 to 15, 1981 at the Physikzentrum at Bad Honnef. The purpose of the workshop was to review the current status of the experimental and theoretical aspects of the field. Through the financial support of the Volkswagenstiftung it became possible to have an internationally representative meeting with nearly all the experts actively working in the field.

The workshop was attended by 60 delegates from ten different countries. Every effort was made to provide enough time for informal discussions, including the time during the afternoon walk through the Siebengebirge near Bad Honnef.

The participants felt that some basic expressions and concepts widely used in the field should be agreed upon in a more rigorous way in order to avoid unnecessary confusion in discussions and in the literature. In an evening session chaired by John Schiffer most of the participants - following an extended discussion - agreed to suggest the following convention concerning the definition of a molecular resonance, the calculation of a reduced width and the notation of deduced spin assignments:

Definition of "Molecular Resonance"
(1) Unique spin and parity, defined separately
(2) Partial width larger than 2% of Wigner limit in at least one channel with both nuclei heavier than α-particles
(3) Appearance in the angle-integrated cross section $\sigma_{cc'}$ of at least two exit channels

Calculation of Reduced Widths
(1) Use R-matrix theory (one-channel case)
(2) Use penetrability (not transmission) coefficients
(3) Adopt radius $R = 1.5 \, (A_1^{1/3} + A_2^{1/3})$
(4) Quote in units of Wigner limit

Notation for Spin Measurement of a Resonance
(1) Angular distribution measurements at $N > 4 \, L_{max}$ angles at more than 5 energies over the resonance; statistics < 3% at maxima; χ^2 are "reasonable"
NOTATION: J^π

(2) Angular range $\geq 90°$ measured at 3 energies; otherwise same as for (1)
   NOTATION: (J^π)
(3) Some angular distributions measured
   NOTATION: $(J^\pi?)$
   Equivalent to (1): Use quantum theory of angular momentum (Racah coefficients) in model-independent way

Finally, I would like to take this opportunity to thank the Volkswagen Stiftung for their financial support, and Dr. Debrus from the Physikzentrum for his friendly help throughout all stages of organizing this workshop. Thanks go also to him and to his crew at the Physikzentrum for the pleasant atmosphere during our meeting.

K.A. Eberhard

I. INTRODUCTORY TALK

RESONANCES IN HEAVY ION COLLISIONS
by
D. ALLAN BROMLEY
A. W. Wright Nuclear Structure Laboratory, Yale University, New Haven, Conn.

INTRODUCTION

Over the past twenty years the study of resonant phenomena in heavy ion interactions has grown, almost exponentially, from study of what many considered a curious isolated result in the $^{12}C + ^{12}C$ system to what is now recognized as a very general feature of nucleus-nucleus interactions. This growth reflects both the availability of more powerful instrumentation and data handling facilities and the greatly increased understanding of nuclear interaction mechanisms generally.

In opening this international workshop I want to emphasize what we do not yet know even more than what we do know. It will clearly be impossible to even attempt an exhaustive review of all the very high quality experiment and theory appearing in the recent literature. I shall instead select a number of topics that span this field and that highlight some of my major points and questions in the hope that some of these questions, at least, will be answered before we leave Bad Honnef. Perhaps not surprisingly I shall draw a large fraction of my illustrations from work, both experimental and theoretical, that we have underway in the Wright Laboratory at Yale. I am most familiar with it. And my apologies, in advance, go to those many other researchers, around the world, whose excellent data and calculations I have not included.

I believe that in our study of resonant phenomena -- of nuclear molecular phenomena -- we at long last are at the end of the beginning. We can now hope to understand this fascinating new kind of nuclear behavior from a fundamental viewpoint -- and as a fundamental, ubiquitous mode of nuclear behavior.

I shall begin with a brief review of the experimental situation and then turn to the theoretical attempts at understanding of this situation.

EXPERIMENTAL STUDIES

ELASTIC SCATTERING:

Figure 1 shows typical excitation functions for identical particle scattering in three different systems[1] and illustrates three of the four characteristic, energy, dependences of scattering cross sections. Very rapid, statistical cross section fluctuations have been extensively studied,[2] are now well understood, and do not appear on the scale to which this figure has been drawn.

In the Coulomb barrier regions, particularly in the $^{12}C + ^{12}C$ system, where the insert shows this structure in greater detail, the cross section is characterized by sharp ($\Gamma \leq 300$ keV) structure, now well characterized in this system in terms of angular momentum and parity, and molecular character (through determination of partial decay widths and appearance in many correlated exit channels primarily). As we shall see, below, much less information is available concerning systems such as $^{12}C + ^{16}O$ and $^{16}O + ^{16}O$ beyond the fact that such structures exist; it will be very important to complete their study -- an activity we have underway at Yale -- and to search with greater precision for their appearance elsewhere -- particularly in heavier systems.

Above the Coulomb barriers the average cross section drops by between one and two orders of magnitude below the Rutherford value. Superposed on this average is broad, relatively regular structure first observed,[3] and most clearly evident in the $^{16}O + ^{16}O$

system; typical widths are $\Gamma \sim 3$ to 4 MeV. These broad structures are further fragmented into peaks of intermediate width, $\Gamma \sim 500$ keV.

Finally, at some characteristic energy well above the Coulomb barrier (just barely reached in the ^{28}Si + ^{28}Si data shown) the average cross section drops by at least another order of magnitude, the superposed structure becomes much wider, $\Gamma \sim 10$ MeV, and the fragmentation of this broad structure is very much reduced.

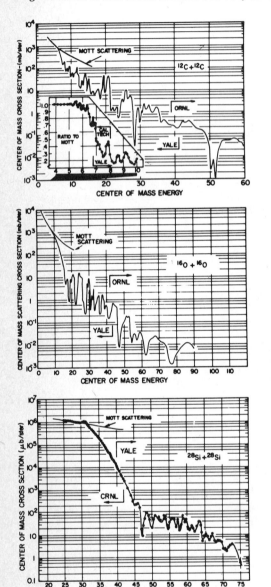

FIG I

As Figure 1 illustrates, these appear to be common features of all the identical particle systems yet studied. These features are not yet understood although I shall review some of the more successful attempts at such understanding below.

TOTAL CROSS SECTIONS

Figure 2 illustrates one of the earliest, and continuing puzzles -- as well as the quality of the data now available in this field; the ^{12}C + ^{12}C and the ^{16}O + ^{16}O systems are strikingly different. A few years ago we attempted to explain this difference in terms of the very much greater level density reached in the ^{32}S compound system as compared to ^{24}Mg. As we shall see below this explanation was premature.

Figure 3 presents a summary of the salient features of Figure 1, just discussed, together with a sketch of an interaction potential that might plausibly produce such behavior. It is characterized by a second minimum very reminiscent of that now well known in the fission of heavy nuclei and the imaginary part of the potential has a weak long-range tail i.e. the nuclear surface is a largely transparent one. All these features are consistent with all the data now available. Unfortunately, as we shall see, all attempts to produce such a potential from more fundamental or microscopic bases have been less than fully successful. In particular it has been difficult, if not, indeed, impossible, to obtain an outer potential minimum at adequately large radius or with adequate depth.

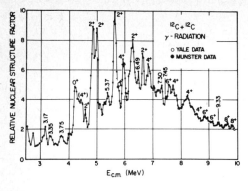

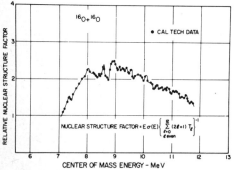

FIG 2

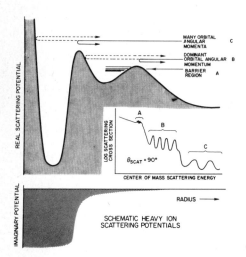

FIG 3

ELASTIC COLLISIONS IN ATOMIC SYSTEMS

Figure 4 reminds us that molecular phenomena have long been studied in the atomic domain and that the different classes of excitation structure referred to above have possible counterparts in atomic physics. What is plotted here[4] are excitation functions for the elastic scattering of hydrogen and deuterium atoms from mercury vapor; the energy scale is in atomic units, a.u. where 1 a.u. = 27.21165 eV and the cross section scale in atomic units a_0^2 where $1 a_0^2 = 0.2800285 \text{ Å}^2$. The solid curve is that calculated with a Rydberg-Klein-Rees potential and the dashed curve that for a Morse type potential. The quantum-numbers (ν, J) of the vibration-rotation states corresponding to these resonances (orbiting resonances in atomic parlance) are given below the curves and the indices N of the glory oscillations are given above. Bernstein[5] has studied these situations in detail and has noted that according to Levinson's theorem[6] the number of the latter broad maxima must correspond to the number of quasi-bound vibrational states of the molecular complex. Thus the number of sharp resonances should be matched by the subsequent number of broad maxima. Further examination of the extent to which the well developed concepts of atomic physics can be applied to the nuclear situation would appear to hold promise.

THE EFFECT OF VALENCE NUCLEONS

In heavy ion resonance studies it has, for some time, been an article of faith that valence neutrons tended to damp or smear resonant structure very strongly in heavy ion interactions[7] Figure 5 is a striking illustration of the error in this belief. The $^{14}C + ^{14}C$ system, only recently accessible to experimental study[8] shows striking intermediate width structure (class B of Figure 3) although sufficiently precise studies have not been done, as yet, to resolve class A structure, if present, nor have measurements been carried out at high enough energies to delineate the

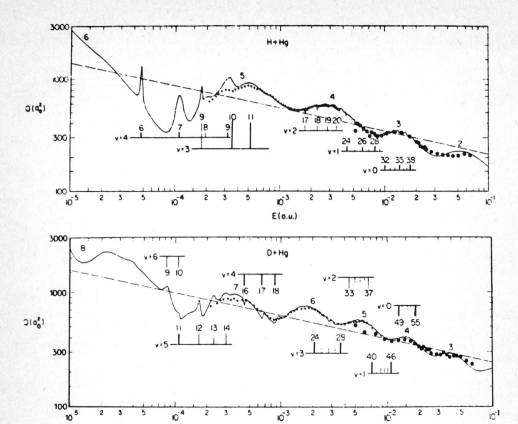

FIG 4

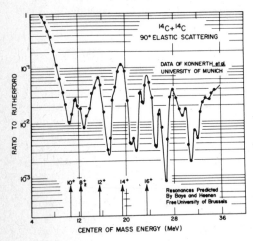

FIG 5

transition to class C structure if present (by analogy with the data of Figure 1).

Also shown on this figure are the angular momenta predicted by Baye and Heenen[9] for these energies; it will be of substantial interest to determine these angular momenta experimentally to test these predictions. It, perhaps, bears emphasis that data thus far available support the assumption that each of the class B structures is dominated by a single angular momentum while in contrast, the class C structures show no such simplicity.

STRUCTURE IN TRANSFER EXCITATION FUNCTIONS

Again demonstrating that resonant phenomena survive the presence of valence nucleons, Figure 6 shows energy-angle cross section surfaces as measured

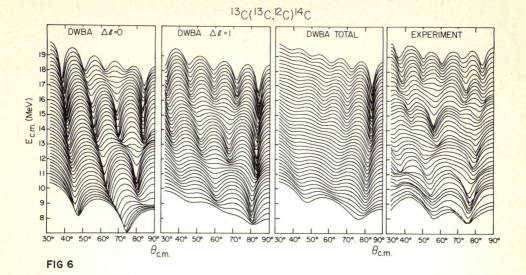

FIG 6

and calculated by Korotky et al.[10] in the $^{13}C(^{13}C,^{12}C)^{14}C$ neutron transfer reaction. Although a judicious mixture of $\Delta L = 0$ and $\Delta L = 1$ transfer does reproduce the gross trends of the experimental data when calculated within a DWBA framework, quite striking resonances are observed superimposed upon these general trends.

Moving to more complex transfer situations, Figure 7 illustrates the very strong energy dependence of elastic and inelastic cross sections at extreme forward and backward angles in the $(^{16}O, ^{12}C)$ quartet transfer reactions on ^{24}Mg[11] and ^{28}Si.[12] As illustrated, at the peaks of structure in the excitation functions remarkably pure $|P_L|^2$ angular distributions are observed -- over limited angular ranges. It bears emphasis that over these ranges $|P_L|^2$ angular distributions are remarkably similar to the $|j_L|^2$ ones that wave optics predicts for an observation along the axis of a uniformly illuminated ring source.[13] Further detailed study of the extent to which such an explanation, reflecting scattering of the incident beam from an effectively opaque sphere with an equatorial illuminated ring would be worthwhile.

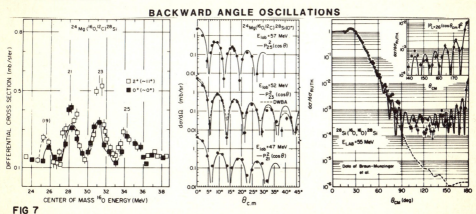

FIG 7

Clearly nuclear phenomena, including resonances, would be expected to surface first at very large angles in elastic scattering inasmuch as the $\csc^4 \theta/2$ term in Rutherford scattering dominates at forward angles.

But that these phenomena may well be resonant is illustrated in Figure 8 where new data obtained with the pre-ATLAS superconducting linac at Argonne[14] are combined with lower energy tandem data. Here we show direct forward angle data. A DWBA calculation in this energy region, assuming a direct alpha particle transfer, shows no structure whatever, although it can produce angular distributions very much like those observed.

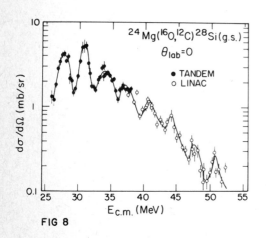

FIG 8

As yet, data on these heavier systems is rather sparse but already the evidence appears rather conclusive for resonant behavior. Higher energies, beyond those available with most current tandems will be essential to these studies.

Current data acquisition and analysis techniques make possible the search for more subtle nuclear structure dependence in heavy ion interactions than has been the case. Ascuitto[15] has suggested a search for resonances in the $^{17}O + ^{48}Ca$ interaction leading to the $2s_{1/2}$ state in ^{17}O. He notes that the degeneracy of this level with the $2p_{1/2}$ state of ^{49}Ca could lead to resonant neutron transfer between these levels in analogy to the well known resonant electron transfer results obtained by Everhart[16] in atomic scattering. This search has not yet been made.

ENERGY DEPENDENT STRUCTURE IN MORE COMPLEX INTERACTIONS

As will be illustrated in Figure 25, below, energy dependent structure is now well known in fusion excitation functions and the structure itself has been shown to provide a sensitive measure of the angular momentum to be associated with each of the type B structures of Figure 3; moreover these data demonstrate rather clearly that each of the maxima in type B structure is characterized by a single partial wave as noted above.

It may well be possible to extend such measurements to even more complex interactions involving deep inelastic scattering and fragmentation -- and, in going to higher energies, hence higher angular momenta, to regions where fusion is no longer possible.

Figure 9 shows results of recent studies[17] of 143 MeV ^{32}S ions incident on ^{40}Ca and ^{48}Ca. The left panel shows the yield of various fragments (summed over energy) and the even-odd mass staggering is much more pronounced for the ^{40}Ca than the ^{48}Ca target. This is not understood. The right panel shows the energy dependence of the Si yield from these interactions. Again these shapes are not understood; in particular the glitch in the yield from a ^{48}Ca target is experimentally real but difficult to understand. Such new data pose a real challenge -- both experimentally and theoretically.

THE $^{28}Si + ^{28}Si$ SYSTEM

In Figure 10 we return to the $^{28}Si + ^{28}Si$ data of Figure 1. Obviously the angular momenta involved in the structure here is of interest; Figure 11 shows angular distribu-

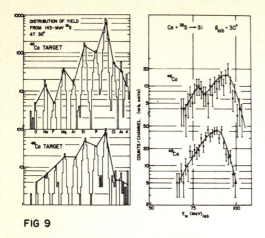

FIG 9

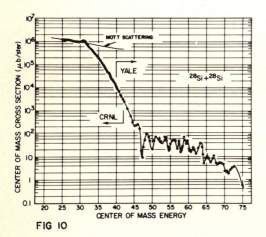

FIG 10

tions[18] at selected energies. The L = 42 $\hbar$ data here represent the highest angular momentum yet characterized uniquely in nuclear science. The fact that the angular distribution is so clearly characterized by a single partial wave, and that this wave changes systematically as shown here is difficult to reconcile with the known fact that ^{28}Si is oblate and thus in random orientations should appear as a central core with a relatively diffuse surface penumbra in its density distribution.

In a series of very elegant measurements Betts et al.[19] have obtained much higher resolution data some of which are shown in Figure 12 (top panel). Clearly, as in the ^{16}O + ^{16}O data of Figure 1, but to a much greater extent, the gross structure peaks of Figure 10 are fragmented into sharp components. The top panel shows the total elastic and inelastic scattering cross section and the lower two panels the deviation and correlation functions defined in the usual way for the elastic and 3 inelastic channels. For the energy averaged C(E), these authors find < C(E) > = 0.18 which is to be compared to the value 0.0 ± 0.04 calculated on the assumption of statistical fluctuations. From Figure 11 it follows that the three broad structures encompassed in Figure 12 are characterized by 36, 38 and 40 $\hbar$ respectively, and it is known that the angular distribution shapes remain essentially constant in traversing the broad structure. Betts et al.[19] conclude that they are indeed observing compound system resonances in ^{56}Ni; if so these must be unusual states indeed inasmuch as one can readily calculate that at an excitation energy of 64 MeV in ^{56}Ni the density of J = 36 $\hbar$ states is ~ 10^5 per MeV and of J = 40 $\hbar$ is ~ 5 x 10^3 at 70 MeV, and these observed resonances stand out from these very dense continua. A natural assumption is that the observed states are those having very strong overlap with the ^{28}Si + ^{28}Si entrance channel, i.e. to a dinuclear molecular configuration, and thus very weak matrix elements coupling them to the complex structure of the underlying continuum members. Such sharp states at ~ 40 $\hbar$ and E_x ~ 70 MeV represent qualitatively new aspects of nuclear structure.

Obviously great interest now attaches to the demanding but very important experiments that can provide additional information concerning these states in ^{56}Ni and to the search for similar phenomena in other heavy ion systems.

HEAVIER SYSTEMS

Figure 13 is the calculated[20] excitation curve for the ^{60}Ni + ^{60}Ni system analogous

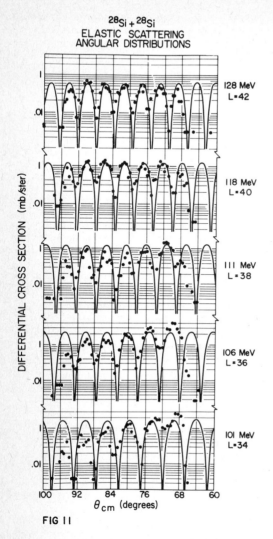

FIG 11

to that of Figures 1 and 10 here plotted however as ratio to Rutherford. As illustrated, the standard MP tandem does not produce ^{60}Ni beams of high enough energy to be of much interest -- while an ESTU tandem (for which we await the necessary federal support at Yale) spans a very interesting region; Erb et al.[21] are currently studying this system using the newly available Holifield tandem facility at the Oak Ridge National Laboratory. Already striking structure reminiscent of Figure 13 has been found, in both elastic and inelastic scattering, persisting to the highest energies available.

One of the most important measurements not yet performed is that of determining the extent to which this structure in the ^{60}Ni + ^{60}Ni case is fragmented as in Figure 12 -- and then of determining the angular momentum of the fragments.

FRAGMENTATION OF MOLECULAR RESONANCES

Feshbach has suggested[22] that true molecular resonances should be considered as particularly simple doorway states whose strength would at least partially dissolve into those of the underlying continuum states <u>of the same angular momentum</u> that happen to have significant coupling matrix elements.

Were this true we would expect to find groupings of resonances all having the same angular momentum -- and with entrance channel partial widths that, in principle, should sum to the single particle value of the archtypical molecular doorway. Erb et al.[23] indeed have shown that in the ^{12}C + ^{12}C barrier region this expectation is fulfilled. Figure 14 shows work at higher energy by Cosman et al.[24] in which a multilevel Breit-Wigner fit to the 90° excitation function yields rather good agreement assuming the individual resonance angular momenta and reduced widths shown. Here, too, there is an apparent clumping of strength for a given angular momentum consistent with the doorway concept. It must be emphasized, however, that this fitting approach is a highly speculative one given the number of free parameters.

^{16}O + ^{16}O REVISITED; A NEW CLASS OF RESONANCES

It also bears emphasis that despite heroic efforts by many of the world's experimentalists working with heavy ions, an enormous amount of work remains to be done on even the most carefully studied systems. I remarked earlier that the ^{16}O + ^{16}O system was noteworthy because of the apparent absence of sharp resonances -- see Figure 4. Recently, however, we have had occasion to go back to this system to take a more careful

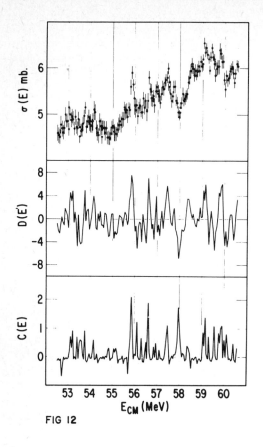

FIG 12

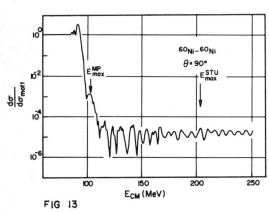

FIG 13

look than we had before. Figure 15 shows the results obtained by Gai et al. in this study. We have very clearly identified two $J^\pi = 8^+$ and one $J^\pi = 10^+$ resonances in the vicinity of $E_{cm} \simeq 16$ MeV and have been able to extract all the pertinent resonance and background S matrix elements from our data. The background we find to be dominated by a narrow angular momentum window centering upon the L = 10 grazing angular momentum. We have taken advantage of measurements made at zeros of the pertinent Legendre Polynomials to make unique and unambiguous J^π assignments.

With these assignments in hand we have been led to a systematic look at the 8^+ resonances observed in heavy ion systems at energies above the Coulomb barrier. This is illustrated in Figure 16 from Gai et al.[26] where we show elastic scattering excitation functions for six heavy ion systems where we note that in each system an 8^+ resonance, frequently the most prominent in its energy region, appears at $E_{cm} = 4 \times 2.8 + N \times 2.4$ MeV where $0 \leq N \leq 4$ and N is integral. In other words, in moving from one of these alpha particle systems to the next over the compound system range from ^{24}Mg to ^{44}Ti we see that these resonances show a systematic behavior that might suggest addition of a characteristic energy increment of 2.4 MeV carrying no angular momentum i.e. $\Delta E_{cm} = 2.4$ MeV; $\Delta J^\pi = 0^+$.

If we recall the long known Ikeda rule[27] to the effect that the energy penalty, in moving from a 4F fermion system to a molecular (4F-4) $\otimes$ 4 system, is 2.4 MeV in the p and sd shells, the above-noted systematic behavior rather strongly suggests that the alpha particle substructure of the nuclei involved is playing an active role. In such an event, these molecular resonances would not be of the single two-nuclear character considered thus far in this field (except for a very few rare exceptions) but rather would be of polynuclear character with the alpha particle as an active participant. This, of course, was raised as a conjecture by Michaud and Vogt[28] a long time ago. What we here appear to be seeing is the systematic presence of two quite distinct kinds of molecular resonances; two nuclear and polynuclear.

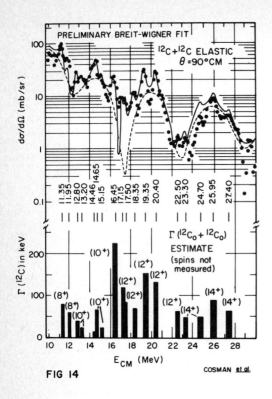

FIG 14

Further evidence that would support such a picture has been adduced by Gai et al.[26] but the data are still fragmentary and much less convincing; this concerns the systematic occurrence of a second set of resonances at higher energies, at $E_{cm} = N_\alpha + 2.8$ MeV, $J^\pi = (2 \times N_\alpha)^+$ with $6 \le N_\alpha \le 11$ in these same systems i.e. with $\Delta E_{cm} = 2.8$ MeV and $\Delta J^\pi = 2^+$. The 8^+ set shown in Figure 16 are conspicuous by their absence in inelastic scattering while this second set are pronounced. It will be very important to pursue these systematic studies across systems to search for signatures of new resonant structures and possible simplicity in their interpretation.

Figure 17 is a further very preliminary but suggestive proposal of Gai et al.[29] who have noted that the inelastic scattering excitation functions for different systems, here illustrated by $^{16}O + ^{12}C$ and $^{16}O + ^{20}Ne$ show a tantalizing similarity if the energy scales, E_{cm}, are scaled according to the number of quartets (alpha particles ?) in the compound system i.e. by the factor 7/9 in this case. This again is a suggestion that will bear much more study, both experimental and theoretical.

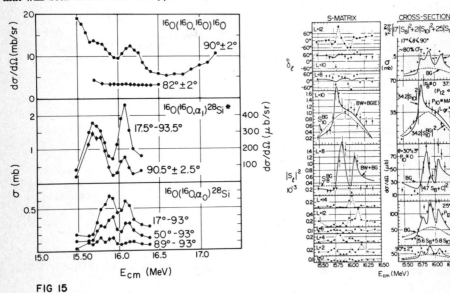

FIG 15

Again, as an illustration of quite striking structure in $^{16}O + ^{16}O$ reactions, Figure 19 shows data of Rossner et al.[30] on the $^{16}O(^{16}O,^{12}C)^{20}Ne^*(4_1^+)$ channel. The dashed and dotted curves are calculations made by Kondo and Tamura[31] using an EFR-DWBA code and the Vandenbosch[32] and Gobbi[33] potentials respectively. Obviously the former provide a better reproduction of the data illustrating the possible role of band crossing in enhancing the potential structure in such data.

RESONANCES IN RADIATIVE CAPTURE

Before leaving this brief review of the experimental situation I must include Figure 20 taken from some beautiful work of Sandorfi[34] on the radiative capture of ^{12}C by ^{12}C; shown here are the $\theta = 45°$ excitation functions measured for the radiative capture transitions populating the four lowest states of ^{24}Mg

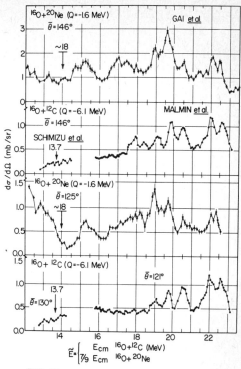

FIG 17

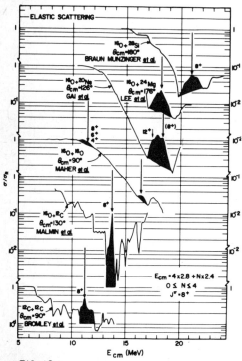

FIG 16

together with statistical model calculations of these same functions. Sandorfi concludes that the ^{12}C partial widths of the capture resonances are substantially enhanced over statistical predictions, a signature for molecular configurations. What is striking, however, is that all the previously known resonances in the $^{12}C + ^{12}C$ system in the energy region studied here appear in the radiative capture data but only with the strengths expected from statistical arguments; the dominant radiative capture resonances are new ones and are enhanced above statistical predictions. This suggests that in the radiative capture situation we are dealing with correlations between the E2 and the $^{12}C + ^{12}C(2^+)$ strength functions reflecting an unusual configuration strongly coupled both to the entrance channel and to the low states of Mg^{24}. As such it provides very convincing evidence for a close linkage between the $^{12}C + ^{12}C$ molecular states and the structure of ^{24}Mg. We still do not understand the detailed correspondence between the resonances that dominate radiative capture and those that appear in other channels.

Figure 21 is taken from earlier data of Litherland et al.[35] in which the electron induced fission of ^{24}Mg was studied, and in the lower panel shows very clearly the characteristic $\sin^2 2\theta$ pattern that is the signature for the E2 decay of a 2^+ state in ^{24}Mg to the ground state.

It is gratifying to see these data on the electromagnetic resonance interactions. For a long time it had appeared that the ^{7}Li(^{16}O, γ)^{23}Na resonance observed by Feldman and Heikkinen[36] was an isolated example. But considerable work remains to be done, both experimentally and theoretically, in order to answer the detailed questions that these radiative capture data raise.

HEAVIER SYSTEMS, HIGHER ENERGIES

The greatest experimental challenge in this field is currently that of establishing the extent to which molecular phenomena remain present in going to heavier systems and to higher excitations in systems where they are already well known. As we have seen, above, rather compelling evidence exists for such structure at some 70 MeV

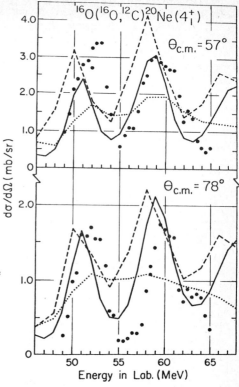

FIG 19

of excitation in ^{56}Ni, the heaviest system yet studied with adequate resolution to delineate the molecular structure. This provides a strong argument in support of new accelerator systems that are capable to reaching new, higher energy regimes while retaining the resolution essential to study of such new and striking phenomena.

THEORETICAL STUDIES

REVIEW OF EARLY MODELS

The number of theoretical studies of nuclear molecular phenomena is now legion beginning with the simple, rather intuitive, models of Vogt and McManus[37] and of Davis,[38] through the seminal work of Nogami and Imanishi[39] on two nuclear configurations -- bound as a consequence of temporary transfer of energy from relative motion to internal excitation -- and the work of Michaud and Vogt[28] on polynuclear configurations involving alpha particles, to modern two-center shell models, constrained time-dependent Hartree Fock and generator coordinate approaches. Obviously I cannot hope to even mention all of this work. Let me instead select a few examples that will lead naturally into some of the most recent model studies and that will illustrate some of the very real difficulties that have been encountered along the way.

Figure 22 illustrates some pioneering work of Greiner and his co-workers[40] on the so-called double resonance variant of the Nogami-Imanishi mechanism in the ^{16}O + ^{16}O system. The left panel shows the real part of the interaction potential assumed, without

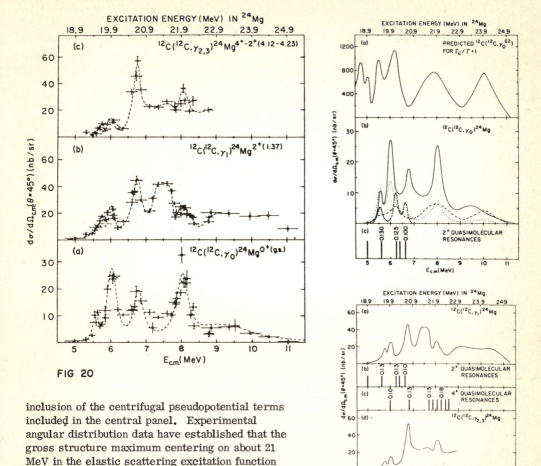

FIG 20

inclusion of the centrifugal pseudopotential terms included in the central panel. Experimental angular distribution data have established that the gross structure maximum centering on about 21 MeV in the elastic scattering excitation function was dominated by an L = 12 orbiting resonance in the entrance channel; beyond that it was known that the peak was fragmented into three components as shown in the right panel. Greiner et al. noted that coupling the angular momentum of the orbiting resonance to that of the intrinsic (3⁻) state, inelastically excited during the collision, led to the required fragmentation in the model prediction; a further check on the realism of this approach was obtained by calculating the inelastic scattering cross section itself as shown in the bottom panel.

POSSIBLE RELEVANCE OF A DIFFRACTION MODEL

As more and more evidence became available, from many laboratories, on gross structure in reaction and inelastic scattering excitation functions, and as more and more structure in any excitation function was presented as evidence for molecular structure we became interested in examining the extent to which we could reproduce such structure without invoking explicit resonances. This led Phillips et al.[41] to examine a simple Austern-Blair[42] model wherein the inelastic scattering is determined by the energy derivatives of the S matrix elements required to reproduce the elastic scattering. Gross structure then appears as a consequence of the overlap of these derivatives in entrance and exit channels differing by two units of angular momentum as shown in the left panel of Figure 23.

The right panel shows inelastic scattering data for the $^{12}C + ^{12}C$ and $^{16}O + ^{16}O$ systems as measured by Haas et al. and the predictions of this simple diffractive model. The different curves correspond to slightly different model parameterizations; what appears is that it is possible to obtain the gross structure without explicitly invoking resonance mechanisms. McVoy et al. have emphasized, however, that our success can be reinterpreted in terms of orbiting resonances so the situation is not as clear as it might appear. What is clear is that all structure in excitation curves does not reflect resonances.

COUPLED CHANNEL CALCULATIONS

Returning now to the coupled models illustrated by Figure 22, I show, in Figure 24, rather typical[43] results obtained to date in reproducing measured elastic excitation functions including both single and mutual excitation of the lowest 2^+ state in ^{12}C. While structure reminiscent of the data appears in the calculations no real fits have ever been obtained. In part this can be understood as a consequence of neglect of higher states that would be expected to participate via inelastic excitation -- particularly at higher energies. The fact remains, however, that quantitative fits still elude us.

A special, and rather fully developed variant of the coupled model is the band-crossing model of Abe, Kondo and Matsuse.[44] Figure 25 illustrates the model concept itself in the upper left panel. The aligned excited band crosses the

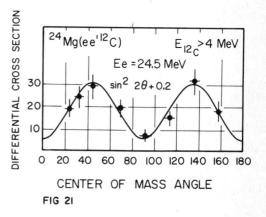

FIG 21

elastic molecular band and in so doing couples the entrance channel to a particular inelastic exit channel. This greatly limits the complexity of the model calculations and it has been rather successful in reproducing a wide range of experimental data as illustrated in the remaining panels of this figure.[45] In particular the lower left panel shows the reproduction of the fusion data of Fernandez et al.,[46] the upper right, of the elastic scattering data of Maher et al.[47] and the lower right the inelastic scattering data of Haas et al.[48] Again, however, while qualitative agreement is present, quantitative fits still are absent.

ALIGNMENT IN INELASTIC SCATTERING

In an effort to resolve some of the ambiguities in the application of these different potential models we undertook to measure a new observable, specifically the alignment of

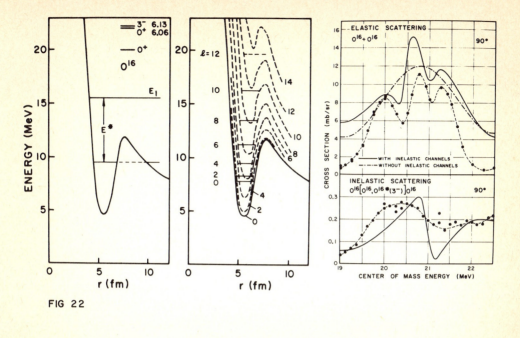

FIG 22

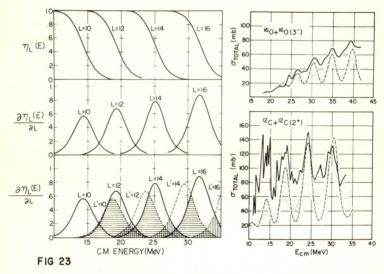

FIG 23

the 2^+ ^{12}C state involved in the inelastic scattering. Figure 26 shows the results obtained by Willett et al.[49] The upper left panel shows the total inelastic excitation function again exhibiting pronounced gross structure and fragmentation of it. The right panel compares the band crossing model calculations with these data for $15 \leq E_{cm} \leq 35$ MeV and with the measured alignment. Although we obtain rough qualitative reproduction of the total cross section, we fail completely to reproduce the alignment data. We do not yet fully understand why. The panel at the lower left compares DWBA and Austern-Blair type

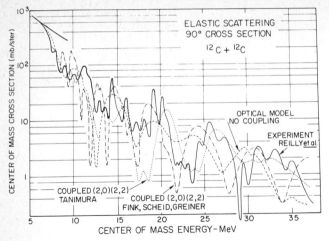

FIG 24

diffraction model predictions with these same measured alignments. Again the DWBA fails completely to reproduce the structure in the alignment data; the diffraction model, however, is rather successful for $E_{cm} \leq 26$ MeV.

We believe that we understand the apparent failure at $E_{cm} > 26$ MeV as reflecting the transition from type B to type C structure of Figure 3; the fact that several partial waves contribute to the higher energy, type C structure washes out the alignment and is not included in the diffraction model.

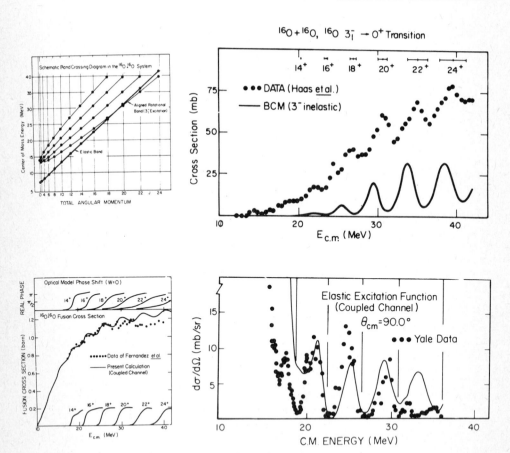

FIG 25

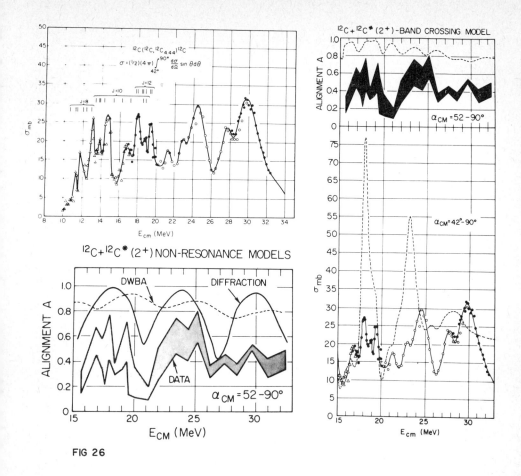

FIG 26

What we find, then, is that DWBA and band crossing models, at least as now constituted, fail even qualitatively in reproducing alignment data. A diffraction model, within its range of validity, on the other hand, fares rather well. It remains unclear why so simple a model should be uniquely successful in this respect.

A GROUP THEORETIC APPROACH

At about the time we were completing this alignment study our discussions with Iachello[50] led us to consider a new examination of the lower-energy, barrier region resonances within a group theoretic format. In the early 1960's when only a few of the resonances were known the apparent simplicity of the excitation function suggested to many investigators that they must reflect some simple underlying mechanism. More recently, as some 38 resonances became established in the $^{12}C + ^{12}C$ barrier region -- as shown in Figure 2, for example, hopes for a simple understanding had essentially vanished; the group theory approach has rekindled these hopes. Figure 27 illustrates three spectrum generating algebras, for elementary particles, and for nuclear structure and reactions respectively. On the left is shown the simplest group chain that suffices for particle physics including flavor, color, strangeness, angular momentum and isospin; its successes in ordering quantumchromodynamic solutions is well known. In nuclear structure, Iachello's work with Arima,[51] Talmi,[52] and others[53] has led to the recognition that almost all earlier nuclear structure models can be encompassed as special cases of the U(6) classification shown here as appropriate to already known s and d nucleon-pair bosons with higher f.g etc. bosons remaining to be discovered at higher energies in analogous fashion to the top, t, quark. The three group chains shown lead respectively to vibrational limits U(5), to rotational limits U(3) and to triaxial limits O(6). This approach has been remarkably successful in ordering a vast body of nuclear structure static and dynamic data and including, on equal footing, those regions of the periodic table that previously fell between regions of validity of the earlier nuclear models.

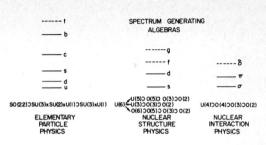

FIG 27

This success has in the main been attained with respect to bound states in the nuclear domain. In turning to unbound states, to nuclear interactions, Iachello recognized that the group structure could be U(4) (a radius vector plus three Euler angles) rather than U(6) and introduced the concept of σ and π bosons. Such a U(4) classification has a unique chain including O(3)--and thus leading to good angular momentum of the states generated--as shown in Figure 27.

Figure 28 illustrates application[54] of this U(4) approach to the simplest 2 nuclear molecular system, ^{8}Be. The right panel shows the complete spectrum predicted for unexcited alpha particles (states of ^{8}Be involving excited alpha particles are at much higher excitations above 20 MeV). The experimental situation shown in the left panel is well matched by this prediction when we recall that the $\nu = 2$, 0^+ state would be expected to have a width well in excess of 10 MeV and thus unobservable in ^{8}Be.

Erb and I[55] undertook to apply the U(4) idea to the ^{12}C + ^{12}C molecular system noting that Iachello had demonstrated[50] that any molecular Hamiltonian having U(4) symmetry necessarily generated a vibration-rotation spectrum of the form:

$$E(\nu, L) = \sum_{m,n} A_{mn} (\nu+\tfrac{1}{2})^m L^n (L+1)^n$$

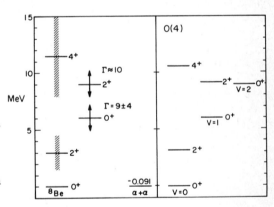

FIG 28

As the lower panel of Figure 29 shows, we were remarkably successful in fitting all the known ^{12}C + ^{12}C resonances to the U(4) pattern. In such a fitting we cannot, of course, establish the vibration quantum number since it is easy to show that the pattern is invariant under change in ν. Two rather important observations emerge from this pattern, however. The fact that the resonances, in general, do not appear as multiplets suggests that if an intrinsic excitation is involved, as in Figure 22, for example, then that excitation may have J=0, rather than the J=2 of the 4.43 MeV state long considered the pertinent excitation since it appeared strongly in inelastic scattering while the 7.6 MeV

0^+ state did not. Furthermore, the pattern of Figure 29 suggests a molecular dissociation energy near 7 MeV. As we shall see below, this new focus on the 0 MeV ^{12}C state at 7.6 MeV has borne interesting new results.

Obviously odd L values are missing from this figure. For this reason one of our students, Schloemer,[56] has undertaken a corresponding study of the $^{16}O + ^{12}C$ system: Figure 30 shows some of his preliminary results. In this system, relatively fewer of the resonances have received assignments and we are currently working to correct this; already it seems clear however that the U(4) classification is useful here too. Both here and in the ^{12}C case of Figure 29 we have found it possible to correlate a large body of data with only 4 fitting parameters--and of these, one, C, could well have been estimated in advance in terms of the moment of inertia appropriate to touching spheres.

Clearly it would be very desirable to obtain experimental evidence supporting the physical reality of this vibration-rotation pattern. The gamma radiation deexcitation selection rule would be $\Delta\nu = 0$ so that in principle enhanced gamma ray deexcitation within one of the rotational bands shown in this figure would provide such a signature. Unfortunately, with even enhanced E2 widths as required, in the identical particle case of Figure 29 we estimate a branching ratio for gamma radiation $\sim 10^{-7}$; despite this we have searched at Yale, without success, for such intraband gamma transitions.

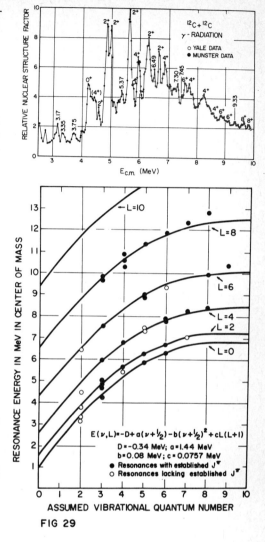

FIG 29

In the $^{12}C + ^{16}O$ system of figure 30, although the odd angular momenta are now present, the E1 matrix elements vanish in first order inasmuch as there is no separation of the centers of mass and charge in such a self-conjugate system. In consequence the branching ratios are still estimated to be less than 10^{-5} for electromagnetic deexcitation; we plan to search for such transitions between well established resonances at Yale in the near future.

Both we and others have also searched for higher energy gamma radiation deexciting type B structure but without success. To the extent that a diffraction model represents the physical situation giving rise to this structure this electromagnetic radiation could not be expected to be as strongly enhanced as has sometimes been suggested and in consequence the failure to observe it experimentally is perhaps not surprising.

I believe that our best opportunity to observe such transitions lies in the region spanned by Figure 30.

In the meantime, however, stimulated by our search for a situation where an E1 matrix element does not vanish in first order we have been studying states in ^{18}O via resonant radiative capture of alpha particles by ^{14}C and we believe that we do now have convincing evidence for strongly enhanced E1 transitions within a rotational band having ^{12}C + α molecular structure (accompanied by 2 valence neutrons) in ^{18}O.

NUCLEUS-NUCLEUS INTERACTION POTENTIALS

The difficulties that we have experienced with the traditional molecular potential approaches, and their elaborations, together with our apparent successes with this U(4) approach has led us, at Yale, to a re-examination of the whole question of our understanding of nuclear molecular phenomena on the basis of nucleus-nucleus potentials. In particular we have focussed on such basic questions as the identification of the correct molecular collective variables, the need to introduce effective masses into the problem, and the general structure of the nucleus-nucleus potentials themselves.

We can already identify problems associated with some of the above mentioned studies. In the band crossing model, for example, use of a phenomenological potential yields qualitative reproduction of type B structure but does not address the well defined type A structure. This model does not account for internal degrees of freedom or the Pauli principle when computing the interaction between asymptotic configurations.

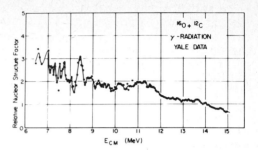

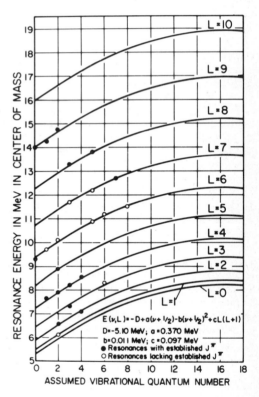

FIG 30

In contrast, the original Nogami-Imanishi model addressed type A structure, but, as shown by Abe, Kondo and Matsuse[57] in a very important piece of work that reinvigorated this entire field, the original calculations did not exploit anything like the potential richness of the model spectrum. Even with a more complete spectrum calculation, however, only isolated quantitative agreement is obtained with the details of the type A spectrum.

One of the most ambitious treatments was the 24 nucleon, two-center shell model treatment of Park, Greiner and Scheid[58]. Inclusion of several asymptotic channels corresponding to excited states of ^{12}C permitted a further qualitative reproduction of the data of Figure 24. This adiabatic model would be expected to be most successful in a weak coupling limit where the structure of the molecular configurations is similar to that of the asymptotic channels. When this is not the case, however, new collective variables appear to be required.

Larsson and Leander[59] have constructed models that discard entirely the dependence on the entrance channel and focus instead on the collective potential energy of the composite system in terms of quadrupole, total angular momentum and other shape variables. Here the technique is to search for stationary shape isomers in the potential surface and then attempt to relate levels in these shape isomers to the observed resonances-- in a situation very analogous to fission as illustrated in Figure 3. Such calculations have two difficulties in that they are unable to address quantitatively the distinctions between the types A, B and C structure of Figure 3 and in that serious problems have been encountered in attempting to obtain appropriate potential parameters for the relatively light systems so far studied.

The type of problem encountered is illustrated in Figure 32 where I plot the Morse potential[60] that would correspond to the spectra of Figure 29. The molecular potential (the second minimum of Figure 3) systematically occurs at too small a radius and is too shallow. This figure also illustrates the effect of assuming different vibrational quantum numbers in Figure 29.

In their calculations Larsson and Leander have used a Nilsson model basis and were unable, in consequence, to separate the system into some of the asymptotic channels of interest. Chandra and Mosel[61] avoided this difficulty by using 2 two-center spherical shell model. Although the results as shown in Figure 33 are suggestive of the kind of potential appearing in Figure 3 no quantitative comparison with data was attempted. In all of these calculations the effective mass was taken to be the reduced mass.

Fink et al.[62] pointed out in 1973 that this is probably a poor assumption inasmuch as when nuclear matter densities overlap the mass parameters for collective motion of the nuclei are no longer independent of the distance between the interacting nuclei and both velocity and coordinate dependence of the effective mass should be expected. In preliminary calculations these authors demonstrated that inclusion of these effects could indeed yield an interaction potential with a secondary minimum like that of Figure 3 but no attempt was made to carry through a detailed calculation.

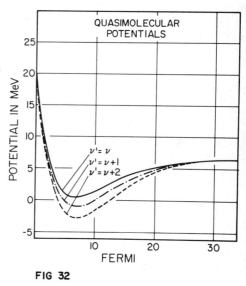

FIG 32

TIME DEPENDENT HARTREE FOCK STUDIES

Moving to even more complex calculations, the work of Cugnon et al.[63] involved calculation of the energy of the 24 nucleon system of $^{12}C + ^{12}C$, using $Z = \sqrt{Q_2}$ as a separation collective variable, for various orientations of the intrinsic state. These authors have used realistic nuclear forces and constrained Hartree-Fock techniques (constrained in the sense of requiring quadrupole deformations); using their channel coupling potentials and a phenomenologic imaginary part they compute elastic, inelastic and mutual inelastic $90°$ excitation functions. Again, as in Figure 24, although rough qualitative reproduction of the data was obtained the model failed in any quantitative description of the type A, B or C structure.

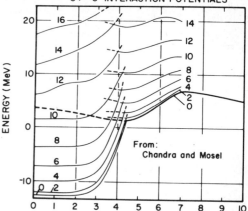

The adiabatic constrained Hartree Fock (ATDHF) technique has been used by many authors--Berger and Cogny,[64] Cusson, Hilko and Kolb[65] and Flocard et al.[66] among them. No repulsive core is allowed by the true adiabatic nature of the calculation and potentials such as that in the lower panel of Figure 34 were obtained by Berger and Cogny[64] for example; the upper panel shows the effective mass $\tilde{M}$ obtained by these same authors as a function of the nuclear separation R. The details of the $\tilde{M}$ dependence on R depend critically on the level crossings (See Figure 33) and are thus very dependent upon the character of the forces used in the calculations.

Additional constrained Hartree-Fock calculations which make a sudden rather than an adiabatic approximation yield rather similar potentials at large R but become strongly repulsive at small R leading to repulsive cores. Such calculations have been done by Brink and Stancu[67], Zint and Mosel[68] and by Strayer et al.[69] Figure 35

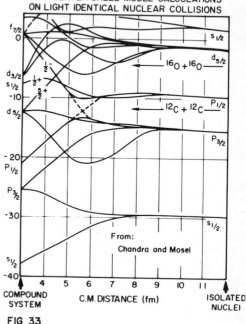

FIG 33

from the latter authors, is typical of what we find. The exchange term is clearly the operative one in leading to the repulsive core.

Almost all of these constrained Hartree-Fock calculations use Q_2 as the collective coordinate. It has become clearer that this cannot yield an adequate description of the type A resonances. We note, for example, that the adiabatic potentials have no repulsive core that could prevent the molecular components from fusing. While the sudden potentials do provide the required separation the sudden approximation itself is inconsistent with the existence of long lived resonant states. And our somewhat preliminary studies would appear to indicate that use of improved effective mass does not alter these conclusions.

I have earlier summarized some of the other difficulties with existing model potentials. The radius is too small to match the moments of inertia required in Figures 29 and 30; the vibrational spacing of about 1 MeV requires a much broader Morse type potential minimum than is found or a very unphysical effective mass. After considerable effort and calculation we have convinced ourselves at Yale that these problems cannot be solved without introducing some fundamental new ingredients into the understanding of nuclear molecular phenomena.

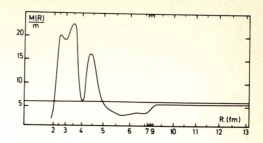

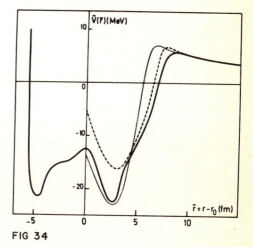

FIG 34

MICROSCOPIC TDHF CALCULATIONS

We have undertaken to try to understand these difficulties within the framework of the microscopic TDHF theory and thus far have been concentrating on the $^{12}C + ^{12}C$ system although our techniques are clearly more general. We are led to this by the recent realization that this theory, in addition to its successes in describing fusion, fission and deep inelastic scattering appears capable of giving a quantitative reproduction of large amplitude nuclear collective phenomena such as the giant multipole resonances.

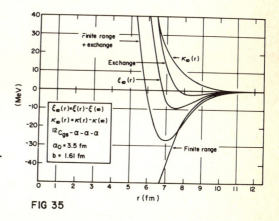

FIG 35

I want to report here some of our very recent results obtained by Strayer et al.[69,70] In the Coulomb barrier region (type A structure) within the framework of this theory we have first taken the results of Cusson et al.[71] shown in the left panel of Figure 36 where we plot the calculated total energy of ^{12}C versus the quadrupole deformation β. The ground state is a relatively soft oblate structure in this model while the first excited 0^+ state at 7.6 MeV is a strongly prolate state having an essentially linear three alpha particle structure as initially postulated many years ago by Morinaga. In the right panel of this figure we show the calculated density contours for the A and B intrinsic states at deformations so indicated by the arrows on the left.

We recognize that there are many inelastic processes that can raise a ^{12}C nucleus from its ground to its first 0^+ state but for simplicity in our calculations we are simply preparing[69] the two nuclei in our study, one in the 0^+-ground state and one in this excited 0^+ state. We do this in view of the suggestion of Erb and Bromley[55] (see Figure 29) that it may well be this state rather than the 2^+ state at 4.43 that is involved in the molecular interactions. As we shall see shortly the fact that it does not appear to be strongly excited in ^{12}C + ^{12}C inelastic scattering has a ready explanation in view of the strong coupling to the ^{8}Be + ^{16}O channel in such interactions.

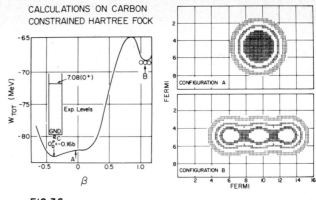

FIG 36

Starting with our two intrinsic states[69] as shown in Figure 36 we have allowed them to collide, in the orientation shown in the upper left of Figure 37 and have then allowed the

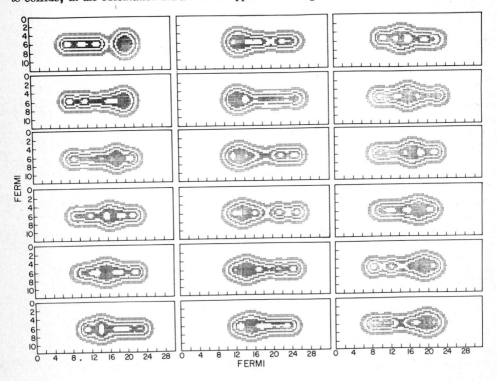

FIG 37

^{12}C + ^{12}C* (7.6 MeV)

system to evolve as the TDHF model requires. From this series of time slices we can readily calculate that the period of the motion wherein these intrinsic states appear to oscillate through one another is 2.3×10^{-21} seconds corresponding to a vibrational energy of ~ 1.79 MeV. The moment of inertia parameter has a value that is very close to the experimental one (see Figure 29).

We note that for a large fraction of the period of the large amplitude oscillation depicted here, the system has a large and obvious overlap with the ^{8}Be + ^{16}O channel and relatively much less with the ^{12}C + ^{12}C*(4.43 MeV) channel consistent with the known experimental situation.

The molecular motion that we show here can be viewed as a strong octupole rather than quadrupole motion. Even more pictorially it can be viewed, in analogy with the well known inversion transition in the ammonia (NH_3) molecule, as reflecting the oscillation of the linear alpha particle chain of the excited state through the equilateral alpha particle triangular structure of the carbon ground state. The coupling[72] to the ^{8}Be + ^{16}O in this picture then simply reflects breakage of the linear chain with capture of a single alpha particle by the ^{12}C ground state to form ^{16}O with simultaneous release of the ^{8}Be fragment.

It is our hope that we can use this view of the interaction[69] as representing multiple sequential cluster exchanges to consider evolution of a fragmentation model from the dominant octupole motion and thus at least attempt to relate the collective molecular wavefunction of our model to the partial widths for its decay into various asymptotic exit channels.

In the meantime it is of interest to us to examine our new data on ^{18}O within such a framework and this work is now in progress. Within the simplest model framework we might assume that our ^{18}O molecular states, apart from the 2 spectator valence nucleons might be amenable to treatment as a single alpha particle oscillating through the ^{12}C system with a frequency somewhere between $\sqrt{3}$ and 3 times higher than that of the three alpha structure of Figure 37. Having a very general TDHF code running on our in-house Laboratory IBM 4341 computer greatly facilitates such calculations but we do not as yet have results on the ^{18}O system.

It bears emphasis that the shapes we encounter in Figure 37 are far from those that we would associate with either the ^{24}Mg ground state or most of the asymptotic channels. This has suggested to us that we should map the intrinsic potential energy surface of ^{24}Mg for large changes in the Q_2 and Q_3 variables; such a map is shown in Figure 38 where we have omitted the symmetric portion of Q_3 for simplicity. The quasimolecular valley appears clearly for large Q_2 and over a range of Q_3 values as does the ^{24}Mg ground state valley at much lower Q_2 and Q_3 values; a few of the prominent decay channels are indicated schematically as well.

Our work thus far with this constrained Hartree-Fock approach has been encouraging and we are currently exploring its potential extensions in this and other nuclear interactions.

NONCENTRAL TDHF CALCULATIONS

Thus far our calculations have only been for head-on collisions. Figure 39, from Flocard, Koonin and Weiss[66] shows similar TDHF time slices of a non-central ^{16}O + ^{16}O collision at 52.5 MeV center-of-mass energy and with 13 $\hbar$ relative angular momentum. The times shown below each slice are in units of 10^{-22} seconds. Computer memory limitations precluded following the collision process further but already well developed vibration-rotation molecular characteristics are evident.

To illustrate that at least calculationally these phenomena persist in heavier systems, Figure 40 shows comparable time slices[66] for a $^{40}Ca + {}^{40}Ca$ collision at E_{cm} = 139 MeV and with relative angular momentum of $70\hbar$. Again the time interval between different slices is 10^{-22} seconds and again well developed vibration-rotation characteristics are evident. It will be of great interest to attempt to find such phenomena experimentally in these heavier systems.

A URANIUM-URANIUM MOLECULE?

Finally, let me move directly to one of the heaviest possible systems $^{238}U + {}^{238}U$ now under study at GSI by a group including my Yale colleague Jack Greenberg.[73] On the upper right of Figure 41 is shown the now-familiar Dirac diagram illustrating that in a collision wherein $Z_1 + Z_2 \gtrsim 170$, the supercritical Coulomb field results in the K-shell binding energy of the transient superatom diving into the Dirac continuum; if the initial stages of the collision should have produced a vacancy in the K shell of one of the participants (a process that Greenberg has shown to have high probability[74]) then the vacancy can be filled by one of the negative energy Dirac electrons with the simultaneous appearance of an unaccompanied positron. This process and its various competing and possibly obscuring ones have been very carefully studied theoretically by Greiner and his collaborators.[75]

On the left I show recent unpublished data of Greenberg et al.[73] The upper figure shows the direct positron spectrum

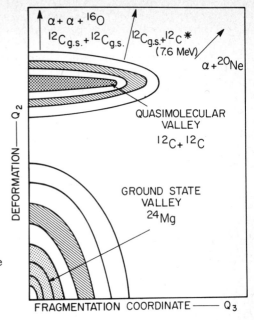

FIG 38

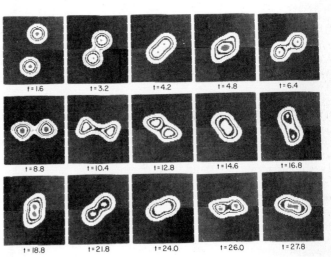

FIG 39

as detected in a long solenoidal spectrometer carefully baffled against electrons. It has the general shape and yield dependence on $Z_1 + Z_2$ that Greiner et al. have predicted.

What is very exciting, however, and not yet understood is what happens when the positrons are examined in coincidence with the scattered ion. Greenberg et al. have found that over very narrow angular and energy ranges, striking peaks appear in the positron spectra -- as shown in the lower left panel. One intriguing suggestion made by Greiner,[76] is that the kinematic restraints imposed by the coincidence measurement select a very special set of collisions where the impact parameter is such that a uranium-uranium molecule forms thus extending the duration of the supercritical field, extending the interval during which the K shell vacancy is submerged at a particular depth in the Dirac continuum, and increasing the probability of emission of a positron of a particular energy. Indeed there are suggestions of secondary peaks that could perhaps be correlated with the existence of excited states of the U + U molecule. All this, however, is very highly speculative as yet but it does suggest that it will be very important to undertake studies dir directed toward the nuclear as opposed to the atomic aspects of the collision. Greiner et al.[76] have already found from an examination of the proximity

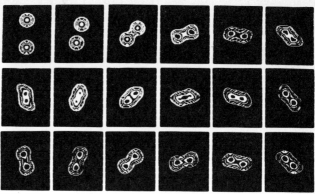

FIG 40

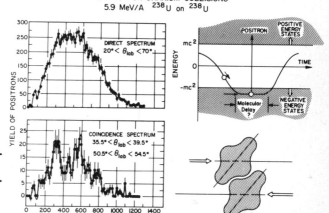

FIG 41

potential for U + U, and including the large static hexadecapole deformation of uranium, that collisions of the kind sketched very schematically on the lower right would be those that would most probably form transient molecular configurations. It would be immensely gratifying to find confirmation that molecular complexes do indeed exist in these heaviest ion collisions. I have suggested previously that superheavy nuclear species may well be observable only as resonances in such collisions. The time has come, I think, to measure a careful excitation function for, say U + U collisions, detecting positrons (as evidence of atomic phenomena) and both elastic and inelastic ions as well as interaction fragments (as evidence for nuclear phenomena). Correlations between structure in these excitation functions would be indicative of mechanisms such as that discussed above.

DIPROTON RESONANCES

And, finally, having touched on very heavy systems as compared to those traditionally studied in our field let me conclude by reminding you of a set of phenomena at very high energies that may have more than token connection to our work [77]. Figure 42, on the right, plots the locations of the proton-proton dibaryon resonances at 2140, 2260 and 2430 MeV against $J(J+1)$. Also shown is the pp "ground state" at 1877 MeV and what might be considered as the ppπ band head some 143 MeV higher following transfer of this energy from relative motion into internal excitation of one proton (to produce a pion at rest in the proton frame). This figure bears a marked resemblance both in appearance and in underlying physics to that in the upper left of Figure 25 for example. The panel on the left of this figure plots $E_{rot} = \hbar^2/2I$ where I is the moment of inertia for a very wide range of rotational bands against the band head energy in GeV (including rest mass in all cases); the open circles are for mesonic examples, the small solid circles for baryons, the square for the diproton, and the large solid circles for light nuclei. This work of MacGregor[78] is tantalizing but, in view of some experimental uncertainty regarding the status of the diproton resonances themselves, has not been widely quoted or extended. It would, again be gratifying to find the Nogami-Imanishi mechanism in the elementary particle domain!

CONCLUSIONS:

Nuclear molecular phenomena have come of age and are increasingly recognized as a ubiquitous aspect of nuclear interactions. We are indeed at the end of the beginning. But only there. After much work, both experimental and theoretical we appear to be on the threshold of microscopic understanding of the molecular configurations; and we have hints that these phenomena span ranges of mass and energy vastly larger than any yet explored.

It bears emphasis that the existence of a sharp state at ~70 MeV excitation in ^{56}Ni, in regions where there are as many as 100 thousand other states of the same spin and parity per MeV, is truly a new and fascinating piece of evidence for the behavior of a

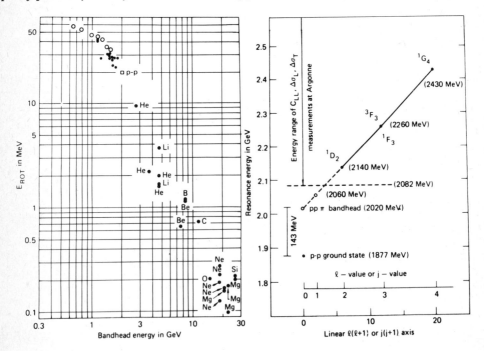

FIG 42

56 body system at high excitation. This is new physics. So also is that in the uranium uranium collisions.

Although we still do not understand these, and a great many other related phenomena, we have reason to be optimistic. And much work remains to be done.

ACKNOWLEDGEMENTS:

I am indebted to my Yale colleagues, Franco Iachello, Karl Erb, Jack Greenberg, Michael Strayer, and Moshe Gai for much discussion concerning these matters and for permission to quote some of their unpublished data and results. Walter Greiner, Ron Cusson, Steve Koonin and Joseph Weneser, as frequent visitors to Yale have been generous with their time and effort; and most of all I am indebted to several generations of Yale graduate students without whom a large fraction of the work I discuss herein would not have been done.

Finally, and as many times before, Mary Anne Schulz, Rita Bonito and Sandy Sicignano have converted my notes and sketches into this manuscript; they have my sincere thanks.

REFERENCES:

1. For $^{12}C+^{12}C$ and $^{16}O+^{16}O$ the data are from:
   D.A. Bromley, J.A. Kuehner, and E. Almqvist, Phys. Rev. Lett. $\underline{4}$, 365 (1969);
   M.L. Halbert, C.B. Fulmer, S. Raman, M.J. Saltmarsh, A.H. Snell, and P.H. Stelson, Bull. Am. Phys. Soc. II $\underline{18}$, 1387 (1973). Paul Stelson, private communication to D.A. Bromley (1975).
   In $^{28}Si+^{28}Si$ the data are from:
   R.R. Betts, J. DiCenzo and J.F. Petersen Phys. Lett. $\underline{100B}$ 117 (1981); A.J. Ferguson, O. Hauser, A.B. McDonald and T.K. Alexander, Argonne National Laboratory report Na ANL-7837, 187 (1971)

2. T.E. Ericson and T. Mayer-Kuckuk, Ann. Rev. of Nucl. Sci. $\underline{16}$, 183 (1966).

3. J.V. Maher, M.W. Sachs, R.H. Siemssen, A. Weidinger, and D.A. Bromley, Phys. Rev. $\underline{188}$, 1665 (1969).

4. W.C. Stwalley, A. Niehaus and D.R. Herschbach, J. Chem. Phys. $\underline{63}$ 3081 (1975).

5. R.B. Bernstein, J. Chem. Phys. $\underline{37}$ 1880 (1962); J. Chem. Phys. $\underline{38}$ 2599, (1963); R.B. Bernstein and R.A. LaBudde, J. Chem. Phys. $\underline{58}$ 1109 (1973).

6. See P. Swan Proc. Roy. Soc. (London) A228 10 (1955).

7. P.T. Debevec, H.J. Korner and J.P. Schiffer, Phys. Rev. Lett., $\underline{31}$, 171 (1973).

8. D.M. Drake, M. Cates, N. Cindro, D. Pocanic, and E. Holub, Phys. Lett. 98B (1981) 36
   D. Konnerth, K.G. Bernhardt, K.A. Eberhard, R. Singh, A. Strzalkowski, W. Trautmann, and W. Trombik, Phys. Rev. Lett. $\underline{45}$ (1980) 1154.

9. D. Baye and P.H. Heenen, Nuc. Phys. A283 176 (1977); D. Baye, Nuc. Phys. $\underline{A272}$ 445 (1976).

10. S. Korotky, S. Willett, R. Phillips, K.A. Erb and D.A. Bromley, Phys. Rev. (to be published) 1982;
    S. Korotky, Ph. D dissertation, Yale University unpublished (1981).

11. W. Henning, D.G. Kovar, R.L. Kozub, C. Olmer, M. Paul, M. Paul, F.W. Prosser, S.J. Sanders, and J.P. Schiffer, to be published.

12. P. Braun-Munzinger, G.M. Berkowitz, M. Gai, C.M. Jachcinski, T.R. Renner, C.D. Uhlhorn, J. Barrette, and M.J. LeVine, to be published in Phys. Rev. C.

13. M. Born and E. Wolf Principles of Optics Pergamon (1959).

14. S.J. Sanders, W. Henning, H. Ernst, D.F. Geesaman, C. Jachcinski, D.G. Kovar, M. Paul and J.P. Schiffer, IEEE Trans. on Nucl. Sci. $\underline{NS-28}$ 1246 (1981).

15. R.G. Ascuitto, Transfer Reactions, a review chapter in Heavy Ion Science edited by D.A. Bromley, Plenum Press (in press) 1982 and private communication (1981).

16. E. Everhart, Phys. Rev. Lett 14 247 (1965); Phys. Rev. A140 175 (1965) A136 674 (1964); 132 2078, 2083 (1963).

17. W. Henning, B. Back, D.F. Geesaman, C.M. Jachcinski, D.G. Kovar, C. Olmer, M. Paul, S.J. Sanders and J.P. Schiffer, BAPS 25 524 (1980).

18. R.R. Betts, S.B. DiCenzo, and J.F. Petersen, Phys. Rev. Lett. 43 (1979) 253; Phys. Lett. 100B (1981) 117;

19. R.R. Betts, B.B. Back, and B.G. Glagola, Phys. Rev. Lett. 47 (1981) 23.

20. W. Greiner, private communication (1980).

21. K.A. Erb, private communication (1981).

22. H. Feshbach, J. Phys. (Paris) Colloq. 37 C5-177 (1976).

23. K.A. Erb, et al. (to be published) 1981.

24. E.R. Cosman, R. Ledoux, M.J. Bechara, C. Ordonez, R. Valicenti, and A. Sperduto, MIT preprint (1981).

25. E.C. Schloemer, M. Gai, J.E. Freedman, A.C. Hayes, S.K. Korotky, J.M. Manoyan, B. Shivakumar, S. Sterbenz, H. Voit, S.J. Willett and D.A. Bromley (to be published) 1981.

26. M. Gai and D.A. Bromley, to be published (1981).

27. K. Ikeda, N. Takigawa and H. Horiuchi, Prog. Theor. Phys. Suppl. Extra Number 464 (1968).

28. G. Michaud and E.W. Vogt, Phys. Letters 30B, 85 (1969).
    G. Michaud and E.W. Vogt, Phys. Rev. C5, 350 (1972). See also G. Michaud, Phys. Rev. C8, 525 (1973).

29. M. Gai and D.A. Bromley, Phys. Rev. Lett. submitted for publication (1981).

30. H.H. Rossner, G. Hinderer, A. Weidinger and K.A. Eberhard, Nuc. Phys. A218 606 (1974).

31. Y. Kondo and T. Tamura, University of Texas preprint (1981).

32. R. Vandenbosch, M.P. Webb and M.S. Zisman, Phys. Rev. Lett. 33 842 (1974).

33. A. Gobbi, R. Wieland, L. Chua, D. Shapira and D.A. Bromley, Phys. Rev. C7 30 (1973).

34. A.M. Sandorfi, M.T. Collins, D.J. Millener, A.M. Nathan and S.F. Lebrun, Phys. Rev. Lett 42 700 (1979).

35. A.E. Litherland, Bull. Am. Phys. Soc. 22, 573 (1977). A.E. Litherland (private communication to D.A. Bromley, 1977). A Sandorfi, L. Kilius, H. Lee and A.E. Litherland, Bull. Am. Phys. Soc. 22, 610 (1977); A.M. Sandorfi, L.R. Kilius, H.W. Lee and A.E. Litherland, Phys. Rev. Lett. 40 1248 (1978); Phys. Rev. Lett. 46 884 (1981).

36. W. Feldman and D.W. Heikkinen, Nuc. Phys. A133 177 (1969).

37. E.W. Vogt and H. McManus, Phys. Rev. Lett 4 518 (1960).

38. R.H. Davis, Phys. Rev. Lett. 4 521 (1960).

39. A. Nogami and B. Imanishi, private communication (1962); B. Imanishi, Phys. Letters 27B, 267 (1968).
    B. Imanishi, Nucl. Phys. A125, 33 (1968).

40. W. Scheid, W. Greiner and R. Lemmer, Phys. Rev. Letters 25, 176 (1970).

41. R.L. Phillips, K.A. Erb, D.A. Bromley and J. Weneser, Phys. Rev. Lett. 42 566 (1979).

42. N. Austern and J. Blair, Ann. Phys. (New York) 33 15 (1965).

43. H.J. Fink, W. Scheid and W. Greiner, Nucl. Phys. A188, 259 (1972).
    W. Reilly, R. Wieland, A. Gobbi, M.W. Sachs, J.V. Maher, R.H. Siemssen, D. Mingay and D.A. Bromley Il Nuovo Cimento 13A, 897 (1973). W. Reilly, R. Wieland, A. Gobbi, M.W. Sachs, J.V. Maher, R.H. Siemssen, D. Mingay, and D.A. Bromley, Il Nuovo Cimento 13A, 913 (1973).
    R. Koennecke, W. Greiner and W. Scheid (to be published) 1981. W. Greiner, private communication (1981).

44. Y. Abe, Proceedings Conference on Nuclear Clustering Phenomena, ed. by D. Goldman, University of Maryland (1975); Y. Abe, Y. Kondo and T. Matsuse Theor. Phys. Suppl. 68 303 (1980) and references therein.

45. Y. Kondo, D.A. Bromley and Y. Abe, Phys. Rev. C22 1068 (1980), Prog. Theor. Phys. 63 722 (1980).

46. B. Fernandez et al. Nucl. Phys. A306 259 (1978).

47. J.V. Maher, Doctoral Dissertation, Yale University unpublished (1969).

48. F. Haas and Y. Abe, Phys. Rev. Lett. 46 (1981) 1667, and references therein; J.J. Kolata et al. Phys. Rev. C16 891 (1977); Phys. Rev. C19 2237 (1979).

49. S.J. Willett, K.A. Erb, S.K. Korotky, R.L. Phillips, and D.A. Bromley, Bull. Am. Phys. Soc. 25 (1980) 591; to be published 1981.

50. F. Iachello, Phys. Rev. C23 2778 (1981).

51. A. Arima and F. Iachello, Annual Review of Nuclear and Particle Science (in press).

52. A. Arima, T. Otsuka, F. Iachello and I. Talmi, Phys. Lett. 66B 205 (1977); Phys. Lett. 76B (1978); Nuc. Phys. A309 (1978).

53. F. Iachello Interacting Bosons in Nuclear Physics Plenum Press (1978).

54. M. Gai, private communication (1981).

55. K.A. Erb and D.A. Bromley, Phys. Rev. C23 2781 (1981).

56. M. Gai, E.C. Schloemer, J.E. Freedman, A.C. Hayes, S.K. Korotky, J.M. Manoyan, B. Shivakumar, S. Sterbenz, H. Voit, S.J. Willett, and D.A. Bromley, to be published.

57. Y. Abe, Proceedings Conference on Nuclear Clustering Phenomena, ed. by D. Goldman, University of Maryland (1975).

58. J.Y. Park, W. Greiner and W. Scheid, Phys. Rev. C16 2276 (1977).

59. G. Leander and S.E. Larsson, Nuc. Phys. A239 93 1975; S.E. Larsson, G. Leander, I Ragnarsson and N.G. Alenius, Nuc. Phys. A261 77 (1976).

60. K.A. Erb, private communication (1981).

61. H. Chandra and U. Mosel, Nuc. Phys. A298 151 (1978).

62. H.J. Fink, W. Scheid and W. Greiner, Proceedings of the International Conference on Reactions between Complex Nuclei, Nashville 1974 edited by R.L. Robinson, Francis K. McGowan and James B. Ball (North Holland, Amsterdam) and Doctoral Dissertation, University of Frankfurt, 1973.

63. J. Cugnon, H. Doubre and H. Flocard, Nuc. Phys. A331 213 (1979).

64. J.F. Berger and D. Cogny, Nuc. Phys. A333 302 (1980).

65. R.Y. Cusson, R. Hilko and D. Kolb, Nuc. Phys. A270 437 (1976).

66. H. Flocard, P.H. Heenen and D. Vautherin, Nuc. Phys. A339 336, (1980).

67. D.M. Brink and F. Stancu, Nuc. Phys. A243 175 (1975).

68. P.G. Zint and U. Mosel, Phys. Lett. 56B 424 (1975).

69. M. Strayer, R.Y. Cusson, J.A. Maruhn, D.A. Bromley and W. Greiner (to be published) 1982.

70. M. Strayer, private communication (1981).

71. R.Y. Cusson, R. Hilko and D. Kolb, Nuc. Phys. A270 437, (1976).

72. N.R. Fletcher, J.D. Fox, G.J. KeKelis, G.R. Morgan and G.A. Norton, Phys. Rev. C13, 1173 (1976).

73. H. Backe, L. Handschung, F. Hessberger, E. Kankeleit, L. Richter, F. Weik, R. Willwater, H. Bokemeyer, P. Vincent, Y. Nakayama, and J.S. Greenberg, Phys. Rev. Lett. 40 1443; J.S. Greenberg, private communication (1981).

74. C.K. Davis and J.S. Greenberg, Phys. Rev. Lett. $\underline{32}$, 1215 (1974).

75. W. Pieper and W. Greiner, Zeits. fur Phys. $\underline{218}$, 327 (1969).
    B. Mueller, J. Rafelski, and W. Greiner, Zeits. fur Phys. $\underline{257}$, 62 (1972).
    K. Smith, H. Peltz, B. Mueller and W. Greiner, Phys. Rev. Lett., $\underline{32}$, 554 (1974).
    B. Mueller, J. Rafelski, and W. Greiner, Zeits. fur Phys. $\underline{257}$ 183 (1972).

76. W. Greiner, private communication (1981).

77. I.P. Auer et al. Phys. Rev. Lett. $\underline{41}$ 1436 1978, Phys. Rev. Lett. $\underline{41}$ 354 (1978); Phys. Lett. $\underline{70B}$ 475 (1977).

78. M.H. MacGregor, Phys. Rev. Lett. $\underline{42}$ 1724 (1979).

II. RESONANCES VERSUS FLUCTUATIONS

EXTENDED CRITICAL ANALYSIS OF STRUCTURES IN THE EXCITATION FUNCTIONS FOR ^{9}Be + ^{12}C

Lucjan Jarczyk

Institute of Physics, Jagellonian University, 30059 Cracow,
Laboratorium für Kernphysik, Eidg. Technische Hochschule,
8093 Zürich

1. Introduction

In the last 20 years, since the discovery of quasi-molecular resonances in the ^{12}C + ^{12}C system [1], many experimental efforts have been made in order to clarify the problem of existence of such configurations in other heavy ion systems. Hanson et al.[2] and also Cindro et al.[3] formulated the conditions favourable for the observation of such quasi-molecular resonances in different heavy ion systems. On the top of a list of the best candidates Hanson places the system ^{9}Be + ^{12}C. Since that time this system has been investigated quite extensively [2,4-7] in different energy regions and in various outgoing channels. A summary of the obtained results based on excitation curves and in some cases also on angular distributions is presented in table I. The conclusions drawn by different authors are not unambiguous, in some cases even contradictory. The structures interpreted as resonances seem to appear at different energies, are not very distinct and not well correlated in different channels.

Table I

Ref.	E_{cm} (MeV)	ΔE_{cm} (keV)	Exit channel observed	Type of measurement	Analysis	Resonances observed
2	2.4 - 6.3	87	p, d, α	σ(E)	$\Sigma \sigma_k$	no
4	10.0 - 15.0	111	α, ^{8}Be	σ(E)	C(E)	11.2, 11.5, 12.6, 13.8, 14.5
5	5.0 - 12.0	143	α	σ(E) σ(θ)	D(E), σ(θ)	6.57, 7.57, 8.1, 8.42, 9.71
6	5.1 - 11.4	114	α	σ(E) σ(θ)	D(E), C(E), σ(θ)	6.3, 7.5, 8.9, 9.7
7	5.9 - 15.4	107	p, d, t, α, ^{8}Be, elast, inelast	σ(E)	D(E), C(E), N(E)	no

Some authors are rather reluctant to recognize the observed peaks in excitation curves as resonances.

These large discrepancies in the interpretation of experimental data have its origin in the complex nature of the reaction mechanism in the system under investigation. Due to the properties of the interacting nuclei, especially the loosely bound structure of ^{9}Be, many direct processes play a significant role, even at not very high energies still close to the Coulomb barrier. It was shown in ref.[8], that in the energy range 6.9 - 15.4 MeV cm the fusion process with compound nucleus formation exhausts only 70% of the total reaction cross section ($\sigma_{fu} = 0.7\ \sigma_r$). Thus a considerable part of the ^{9}Be + ^{12}C reaction (30% of σ_r) proceeds through some direct reaction mechanism, such as neutron and ^{4}He transfer observed in the (^{9}Be,^{8}Be) channel, ^{5}He or ^{8}Be transfer in the α-particle outgoing channel, ^{3}He transfer in elastic scattering at backward angles or more complicated reactions.

In the energy range 5 - 11.5 MeV cm the fusion of ^{9}Be and ^{12}C nuclei to ^{21}Ne leads to a region of excitation, where a strong overlap of individual levels will produce quite large statistical fluctuations in the excitation curves. As the contribution of direct reaction processes enhances the absolute value of fluctuations, they can be increased in some channels (e.g. α and ^{8}Be) by a factor even as great as 10.

This behaviour is illustrated by fig. 1 for synthetic excitation curves[9] calculated for the ^{9}Be + ^{12}C system with two extreme values of direct reaction contribution d=0 and d=0.9. It is clearly visible that the fluctuations increase by approximately the same factor as the mean cross section.

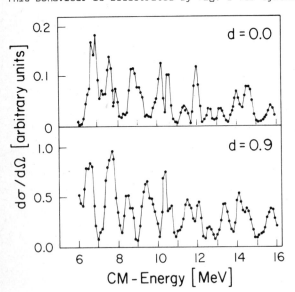

Fig. 1. Synthetic excitation curves for d = 0 and d = 0.9.

The principal problem in the search for intermediate structures in excitation functions is to distinguish them from the usually large fluctuating background. In case of considerable differences between the widths of the fluctuations and possible intermediate structures, the averaging applied to the experimental excitation curves could already reveal the presence of resonances. In such a procedure the high frequency part due to the fluctuations would be much stronger damped leaving the more slowly varying part in the averaged cross section [10].

This however is not the case for the ^{9}Be + ^{12}C system. Here the expected width of quasi-molecular resonances ($\Gamma_{is} \approx 500$ keV [2]) would be of the same order of magnitude as the width of statistical fluctuations of the cross section ($\Gamma_{fl} = 300 - 400$ keV [11]). Thus averaging the excitation curves could not help to visualize the resonances and much more sophisticated statistical methods should be applied.

According to the modus procedendi for indentifying intermediate structures in the experimental data accepted in such investigations [12], one has not only to show a significant deviation from the assumption of the statistical model but also to prove interchannel correlations, to ascribe to the observed structures definite values of angular momentum and parity and to propose a simple entrance configuration for their interpretation.

To meet these requirements the analysis should be based on a large experimental material comprising the excitation curves and angular distributions in many reaction channels measured in small energy steps in a rather broad energy range.

From the experimental data being presently at our disposal those from ref. 4,5 and 6 concern the elastic, α and ^{8}Be outgoing channels, all three very strongly affected by the contribution from direct reactions. The most extended experiment, performed by the Cracow-Zürich group at the ETH Zürich tandem accelerator, contains 266 excitation curves taken in 107 keV cm energy steps in the energy range from 5.9 - 15.4 MeV cm for different emission angles between 5° and 175° [7]. The individual excitation curves concern elastic and inelastic scattering and the emission of p, d, t, α and ^{8}Be, to different excited states of the residual nuclei [7]. Table II gives details of the experimental material while typical examples of experimental excitation curves are presented as points in fig. 2 and 3.

This experimental material was very carefully and critically analysed in order to investigate to what extent the structures appearing in the excitation curves could be attributed to phenomena outside the scope of the statistical reaction model.

Table II

p1	^{20}F:	0.0 (2 +)
p2		1.824 (5 +), 1.873 (2 -), 1.971 (3 -), 2.044 (2 +), 2.219 (3 +)
d1	^{19}F:	0.0 (1/2+), 0.109 (1/2-), 0.197 (5/2+)
d2		1.346 (5/2-), 1.459 (3/2+), 1.554 (3/2+)
d3		2.780 (9/2+)
d4		3.907 (3/2+), 3.999 (7/2+), 4.033 (9/2-)
d5		4.378 (7/2+), 4.555 (5/2+), 4.557 (3/2+), 4.648(13/2+), 4.683 (5/2-)
d6		5.106 (5/2-), 5.337 (1/2+), 5.425 (7/2+), 5.465 (7/2+), 5.500 (3/2+), 5.54 (5/2+), 5.62 (3/2-)
t1	^{18}F:	0.0 (1 +)
t2		0.937 (3 +), 1.042 (0 +), 1.081 (0 -), 1.121 (5 +)
t3		1.701 (1 +)
t4		2.101 (2 -)
t5		2.524 (2 +)
t6		3.060 (2 +), 3.135 (1 -), 3.357 (3 +)
t7		3.734 (1 +), 3.787 (3 -), 3.835 (2 +)
t8		4.119 (3 +), 4.229 (2 -), 4.361 (1 +), 4.402 (4 -)
t9		4.650 (4 +), 4.739 (0 +), 4.849 (1 -), 4.957 (2 +)
α1	^{17}O:	0.0 (5/2+)
α2		0.871 (1/2+)
α3		3.055 (1/2+)
α4		3.841 (5/2-)
^{8}Be1	^{13}C:	0.0 (1/2-)
^{8}Be2		3.086 (1/2+)
^{8}Be3		3.68 (3/2-), 3.85 (5/2+)

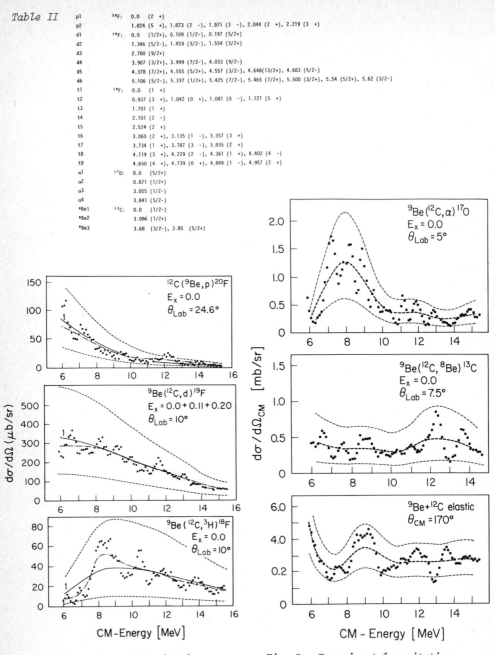

Fig. 2. Experimental excitation curves •, average cross sections —·—, 1% significance limits - - - and the Hauser-Feshbach cross sections ———.

Fig. 3. Experimental excitation curves •, average cross section ———, 1% significance limits - - -.

2. Statistical model (fluctuation) analysis of individual excitation curves

In the first step of analysis the experimental distribution of the cross sections was compared with the theoretical distribution of fluctuations given by the formula [13]:

$$P(y_k) = \left(\frac{n_k}{1-d_k}\right)^{n_k} y_k^{n_k-1} \exp\left(-n_k \frac{y_k+d_k}{1-d_k}\right) \frac{I_{n_k-1}\{2n_k(y_k d_k)^{1/2}/(1-d_k)\}}{\{n_k(y_k d_k)^{1/2}/(1-d_k)\}^{n_k-1}} \quad (1)$$

where $y_k = \frac{d\sigma_k}{<d\sigma_k>}$ is the value of the cross section in the reaction channel k reduced by dividing with its properly energy averaged value; n_k denotes the number of effective channels and $d_k = \frac{d\sigma_k^d}{<d\sigma_k>}$ is the contribution of direct reactions.

As the results of statistical analysis at its different stages depend quite significantly on the averaging of experimental excitation curves, a considerable effort was put into the problem of selection of the most suitable averaging procedure [14]. Both original experimental data as well as especially generated synthetic excitation curves were examined by means of different methods. A repeated running average over 17 points covering the energy interval of 1.8 MeV was adopted as the most suitable [9]. Such a procedure damps the fluctuations of widths up to 0.5 MeV better than by a factor of 10.

The number of effective channels n_k was estimated from a statistical model calculation using a method proposed by Duyras et al.[15]. The contribution of direct reactions was calculated from the expression:

$$d_k = 1 - \frac{d\sigma_k^{H.F.}}{<d\sigma_k>}$$

using the Hauser-Feshbach predictions $d\sigma_k^{H.F.}$ based on an extended analysis of angular distributions corresponding to various reaction channels at different energies [8]. Especially large contributions of direct reactions were found in the alpha-particle, ^8Be, elastic and inelastic channels.

Both experimental and calculated distributions of $d\sigma_k/<d\sigma_k>$ are presented in fig. 4, while figs. 2 and 3 show the energy dependence of the averaged cross section $<d\sigma_k>$ together with the fluctuation band corresponding to a significance level of 1%.

In all cases under investigation the experimental distributions of $d\sigma_k/<d\sigma_k>$ are in quite satisfactory agreement with those given by the fluctuations theory. Almost all extreme values of excitation curves remain inside the 1% significance interval.

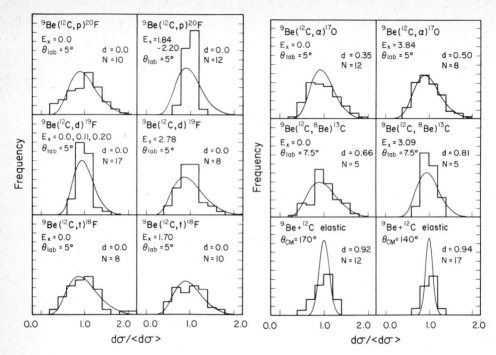

Fig. 4. Histograms of experimental cross section distribution. The curves present the theoretical statistical theory prediction.

The statistical model analysis of the excitation curves provides us also with an estimate of the coherence width Γ_{coh}. The obtained value of 470 ± 60 keV is consistent with that of 470 ± 110 keV obtained from the same data by means of the "counting of maxima" [13,16] method, and both agree quite well with the value Γ_{coh} = 400 keV which follows from systematics of A dependence of experimental and theoretical coherence widths [11].

A rough estimation of the expected cross section for a resonance in one exit channel gives a value of the order of 10 mb/sr. A comparison of this estimate with the absolute values of the cross sections in individual channels and the results of the fluctuation analysis lead to the conclusion that the distribution of cross section is not contradictory with statistical theory prediction and, if the intermediate structures really exist, they are distributed among many decay channels.

3. Statistical search for correlated structures in excitation curves.

Taking into account that an intermediate structure, if any, could appear in many exit channels, in the next step of analysis the excitation curves for different reaction channels were searched for appearance of correlated structures. At this stage of analysis the experimental material was devided into 9 subsets. In each set only those excitation curves were included, which differ in the scattering angle by more than the value of the coherence angle, being according to [17] in our case ca. $30°$. Details concerning the composition of individual sets are given in table III.

Table III

Set	Number of exc. curves					$d\sigma(\theta, E_x)$	
	total	p	d	t	α	^{3}Be	elastic
A	58	10	19	2	19	6	2
B	34	6	10	2	12	3	1
C	20	4	6	1	9		1
Protons		10					
Deuterons			19				
Tritons				14			
Alphas					19		
^{8}Be						6	
Elastic + ^{8}Be						6	3

The correlation analysis contained investigation of:
1. energy dependent deviation function:

$$D(E) = \frac{1}{N} \Sigma \left(\frac{d\sigma_i(E)}{\langle d\sigma_i(E) \rangle} - 1 \right) \quad (2)$$

2. number of maxima N(E) in excitation curves of the ensemble at each energy
3. cross correlation function:

$$C(E) = \frac{2}{N(N-1)} \sum_{i>k} \frac{\left(\frac{d\sigma_i(E)}{\langle d\sigma_i(E) \rangle} - 1\right)\left(\frac{d\sigma_k(E)}{\langle d\sigma_k(E) \rangle} - 1\right)}{\sqrt{R_i R_k}} \quad (3)$$

where R_i is the autocorrelation coefficient for channel i.

The results of this analysis are shown in figs. 5-7. In the case of the deviation function the probability density function derived recently [18] could be used to attribute the statistical significance to the results of the

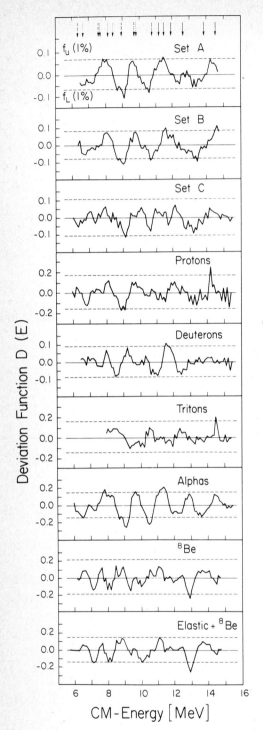

analysis. Fig. 8 presents an example of the theoretical probability density distribution of D as compared with the experimental results. As can be seen the agreement is quite satisfactory without significant discrepancies with the statistical predictions. The maxima appearing in the energy dependent deviation function (figs. 2 and 3) are distributed inside the band corresponding to the 1% level of significance.

Similarly the method of "counting of maxima" does not indicate any correlations significant within the 1% confidence limits derived from the binomial distribution of N(E). No large maxima appear also in the cross correlation curves although in this case the lack of the knowledge of the distribution does not allow to attribute the statistical significance to the results.

Fig. 5. The energy dependence of the deviation function.

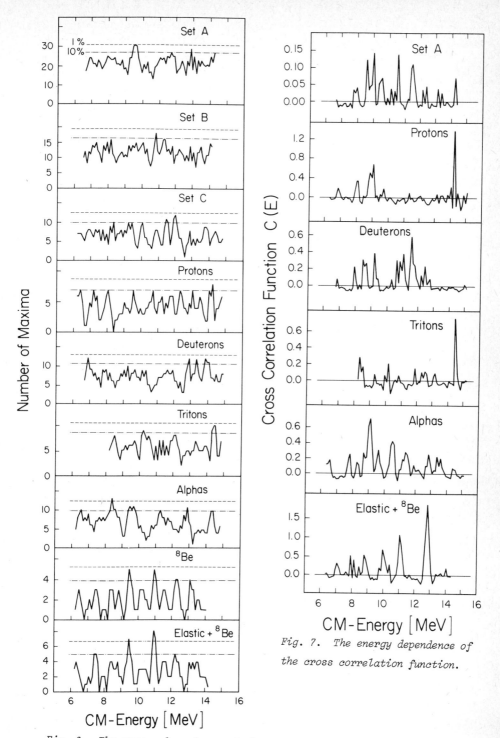

Fig. 6. The energy dependence of the number of maxima.

Fig. 7. The energy dependence of the cross correlation function.

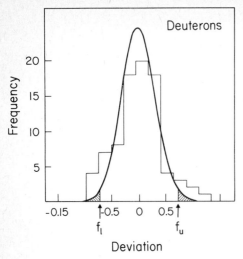

Fig. 8. Histogram of the distribution of deviation for deuterons. The curve presents the statistical theory prediction.

In conclusion we can state that within the 1% significance limits no correlations were found in the experimental excitation curves. In fig. 5 the arrows indicate energies, at which the other authors were inclined to see the resonances. At some of these energies, i.e. at 8.1, 9.7 and 11.2 MeV, a maxima in D(E) reaching the significance limits in the alpha-particle channel indicate a weak inter-channel correlation.

4. The average cross section and direct reaction contribution

As it was mentioned before, there appears a very strong contribution from direct reaction processes in some channels, particularly in the α, ^{8}Be and elastic ones. The Hauser-Feshbach model calculations based on parameters, determined from the analysis of angular distributions for many reaction channels at several energies, (1_{f_u} and level density parameters), follow very closely[8] the average level in the excitation curves for p, d and t channels for all angles and all observed states of the residual nuclei (see fig. 2, curve 1). In the case of α's, ^{8}Be and elastic channels the compound model contribution calculated with the same Hauser-Feshbach parameters is considerably smaller (by approximately one order of magnitude) than the average cross section both in the forward as well as in the backward hemisphere. This proves evidently a strong contribution from the direct reactions in those channels.

The problem arises whether and to what extent the direct reaction calculation can explain the difference between the average experimental and the calculated compound nucleus cross sections. The calculations for different transfer processes in various reaction channels were performed in the DWBA model using the MARS-SATURN code of Tamura and Love[19] with the theoretical values of the spectroscopic factors[20]. In fig. 9 the curves show the sums of com-

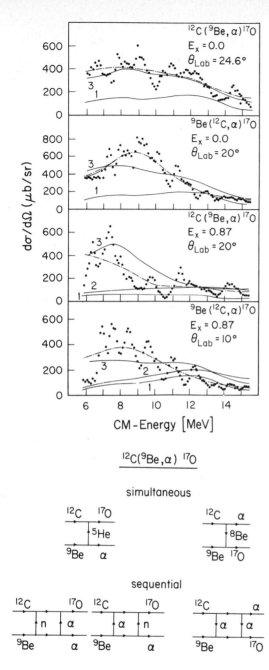

Fig. 9. *Excitation curves: dots - experimental points, 1 - compound nucleus contribution, 2 - compound nucleus + single step transfer contributions, 3 - compound nucleus + single step and sequential transfer contributions, dash-point line - average cross section.*

pound nucleus and direct reaction contributions. Although the general trend of the experimental mean cross section is properly reproduced and in many cases also their magnitudes there are cases, e.g. in the α-exit channel, where still a large part of cross section is missing.

As the $^{12}C(^{9}Be,\alpha)^{17}O$ reaction corresponds to the 5-nucleon transfer (in forward direction) or to the 8-nucleon transfer (in backward direction) the possibility of a sequential transfer should be taken into account besides a simultaneous transfer[9] (see fig.10). The calculations have been performed using the SATURN-JUPITER code[21]. As this program does not allow for the recoil effects the performed calculations have only the character of an

Fig. 10. *Diagrams of the direct reaction processes included in the calculations.*

estimation. The influence of the recoil effects was simulated through a proper normalisation in the component single stage transfer processes. As can be seen from fig. 9 (curves 3) the general trend of the energy dependence and the magnitude are properly followed by these calculations.

5. Angular distributions

As it follows from the statistical anlysis of the excitation curves only, no clear evidence for the existence of the quasi-molecular resonances can be obtained. The observed maxima in the cross section are not significantly correlated, on contrary they can be quite well explained as fluctuations by the statistical theory. The behaviour of cross sections averaged over these structures could be well reproduced by the Hauser-Feshbach theory with a strong contribution from direct reactions in some reaction channels.

As the deviation function in some reaction channels approaches the 1% significance limit these cases should be especially carefully analysed looking for some additional symptoms of resonances. This was the way followed by some authors investigating the $^9Be + ^{12}C$ system, who have measured the angular distribution at and off maxima in the excitation curves suspected as associated with resonances. The analysing of such angular distributions with a model assuming a resonance amplitude at the angular momentum J_{res} is a generally adopted method for assigning the spin and parity to resonances. Such a method when applied to the system with the channel spin zero and the pure statistical background leads usually to the clear, unambiguous results [22]. However in the system under consideration the situation is quite far from these simple conditions mentioned above. The channel spin is different from zero what can spread the resonance on several L values. What more, the background in the α-particle channel, considered in the analysis, contains only a small contribution from the compound nucleus process. The dominating part comes from the direct reaction which favourises also partial waves corresponding to the surface region. This can be seen in fig. 11 where the $P_7^2 (\cos \theta)$-curve, corresponding to the peripheral L=7 partial wave, gives an equally good reproduction of the experimental results as the DWBA-calculations. This will obscure any contribution from resonances. So e.g. in fig. 8 in ref. [6] no dramatic change in the shape of the angular distribution is visible for on-resonance and off-resonance energies. A similar behaviour was observed also in the ^8Be channel [6].

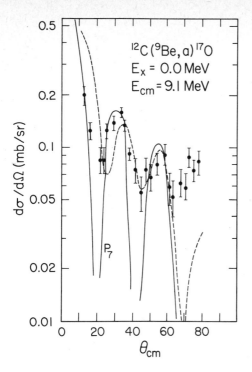

Fig. 11. Angular distribution for $^{12}C(^{9}Be,\alpha)^{17}O$ 5). The dashed line — DWBA analysis from ref. 5, solid line — Legendre polynomial fit.

According to the results of the previous discussion the resonances in $^{9}Be + ^{12}C$ system — even if they exist — can manifest themselves by a small increase of cross section only. Thus there is a little chance that the resonances could be traced by the observation of the angular distribution in the presence of a strong background from the direct processes.

References

1. D.A. Bromley, J.A. Kuehner and E. Almqvist, Phys. Rev. **4** (1960) 365
2. D.L. Hanson, R.G. Stockstad, K.A. Erb, C. Olmer, M.W. Sachs and D.A. Bromley, Phys. Rev. **C9** (1974) 1760
3. N. Cindro and D. Počanić, J. Phys. G Nucl. Phys. **6** (1980) 359
4. J.F. Mateja, A.D. Fowley, A. Roy, J.R. Hurd and N.R. Fletcher, Phys. Rev. **C18** (1978) 2622
5. X. Aslanoglou, G. Vourvopoulos, D. Počanić and E. Holub, Int. Conference on the Resonance Behaviour of Heavy Ion Systems — Aegean Sea — Greece 1980, ed. by Tandem Acc.Lab. Demokritos (Athen, 1980)
6. L.C. Dennis, K.R. Cordell, R.R. Doering, R.L. Parks, S.T. Thornton, J.L.C. Ford,Jr.,J.Gomez del Campo and D. Shapira, Nucl. Phys. **A357** (1981) 521
7. K. Bodek, M. Hugi, J. Lang, R. Müller, E. Ungricht, L. Jarczyk, B. Kamys and A. Strzałkowski, Phys. Lett. **82B** (1979) 369

    M. Hugi, J. Lang, R. Müller, J. Sromicki, E. Ungricht, K. Bodek, L. Jarczyk, B. Kamys, A. Strzałkowski, H. Witała, to be published

8. L. Jarczyk, B. Kamys, A. Magiera, J. Sromicki, A. Strzałkowski, G. Willim, Z. Wróbel, D. Balzer, K. Bodek, M. Hugi, J. Lang, R. Müller and E. Ungricht, Nucl. Phys., in press

    L. Jarczyk, B. Kamys, J. Okołowicz, J. Sromicki, A. Strzałkowski, H. Witała, Z. Wróbel, M. Hugi, J. Lang, R. Müller and E. Ungricht, Nucl. Phys. A325 (1979) 510

9. H. Witała, thesis, Jagellonian University, Cracow 1981

10. P.D. Singh, R.E. Segel, L. Meyer-Schützmeister, S.S. Hanna and R.G. Allas, Nucl. Phys. 65 (1965) 577

11. K.A. Eberhard and A. Richter, in "Statistical Properties of Nuclei" ed. J.B. Garg (Plenum Press, New York - London, 1972), p. 139

12. C. Mahaux, Annual Rev. of Nucl. Sci., 23 (1973) 193

13. D.M. Brink and R.D. Stephen, Phys. Lett. 5 (1963) 77

14. M. Hugi, thesis, Eidg. Technische Hochschule, Zürich 1981

15. R.A. Dayras, R.G. Stockstad, Z.E. Switkowski and R.M. Wieland, Nucl. Phys. A265 (1976) 157

16. A. Van der Woude, Nucl. Phys. 80 (1966) 14

17. P. Braun-Munzinger and J. Barrette, Phys. Rev. Lett. 44 (1980) 719

18. J. Lang, M. Hugi, R. Müller, J. Sromicki, E. Ungricht, H. Witała, L. Jarczyk and A. Strzałkowski, Phys. Lett. 104B (1981) 369

19. T. Tamura and K.S. Low, Computer Phys. Commun. 8 (1974) 349

20. D. Kurath and D.J. Millener, Nucl. Phys. A238 (1975) 269
    D. Kurath, Phys. Rev. C7 (1973) 1390

21. M. Walter, CRC code, University of Munich, unpublished

22. P.T. Debevec, H.J. Körner and J.P. Schiffer, Phys. Rev. Lett. 31 (1973) 171

USE OF THE DEVIATION FUNCTION IN A SEARCH FOR RESONANCES IN THE SYSTEM $^{12}C + ^{16}O$

M. Hugi, J. Lang, R. Müller and J. Sromicki
Laboratorium für Kernphysik, ETH, 8093 Zürich, Switzerland
and
L. Jarczyk, A. Strzałkowski and H. Witała
Institute of Physics, Jagellonian University, 30059 Cracow, Poland
and
K.A. Eberhard
Sektion Physik, Universität München, D-8046 Garching

To decide whether a structure observed in various excitation curves is merely a statistical fluctuation or due to a more interesting phenomenon (e.g. an intermediate resonance) several statistical tests are used. Frequently the energy dependent deviation function is applied in searching correlated structures. In a very recent paper [1] the probability density and confidence limits have been derived for the deviation function. We will use this method for the analysis of the reaction $^{12}C + ^{16}O \to ^{24}Mg^* + \alpha$ and $^{20}Ne + ^{8}Be$ where excitation functions for 10 different final states have been measured for various angles at the University of München [2].

The energy dependent deviation function is defined in the following way

$$D(E) = \frac{1}{N} \sum_{k=1}^{N} (y_k - 1) \; ; \; y_k = \sigma_k(E)/\langle\sigma_k(E)\rangle .$$

It is important to include in this summation only those excitation functions which (for a given exit channel) have been measured outside the relevant coherence angle θ_c. Since for heavy ion reactions θ_c is quite large, in our example $\theta_c \approx 27°$ [3], we have to arrange the measurements into 3 different subsets; in each subset $|\theta_k - \theta_{k'}| \geq \theta_c$ for the same partition.

The probability density of the individual y_k, $p_k(y_k)$, is a function of the direct reaction contributions, $d_k = \sigma_k^d(E)/\langle\sigma_k(E)\rangle$, and the equivalent number n_k of independent channels. n_k is obtained from a Hauser-Feshbach calculation and $d_k = 1 - \sigma_{HF}(E)/\langle\sigma(E)\rangle$. For statistically independent y_k the probability density of the deviation D, p(D), is given by the Fourier transform of the product of the characteristic functions of the original distributions. In Ref.[1] it has been shown, that the upper (f_u) and lower (f_ℓ) fractiles are mainly determined by the variance σ^2 of this distribution and depend only slightly on the individual parameters d_k and n_k as long as $N_{tot} = \sum n_k$ is sufficiently large. The variance given by $\sigma^2 = 1/N^2 \sum \{(1-d_k^2)/n_k\}$ can also be obtained from the experimental autocorrelation coefficients: $\sigma^2 = 1/N^2 \sum R_k (\varepsilon \to 0)$. In the first method suitable correction factors for the averaging procedure have to be calculated; we, therefore, use the second, less ambiguous method, although the two procedures show an excellent agreement in our case.

In fig. 1 the deviation functions, D(E), for the 3 different subsets are given, together with the relevant parameters which characterize the distribution. The dashed lines show the 1% fractiles f_u and f_ℓ. In fig. 2

we compare the theoretical probability density p(D) of subset 2 with the experimental values.

Figs.1 and 2 show that the overall behaviour of the deviation function is in rough agreement with the expectations from the statistical model. At most there is a very weak indication for non-statistical structures at 13.7, 16.6, 20.3 and 22.0 MeV.

References:
[1] J. Lang et al., Phys. Letters 104B (1981) 369
[2] K.G. Bernhardt et al., preprint (1981)
[3] R. Singh, K.A. Eberhard, R.G. Stockstad, Phys.Rev. C22 (1980) 1971

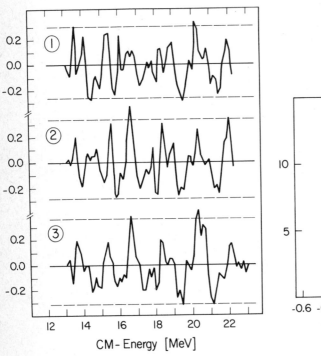

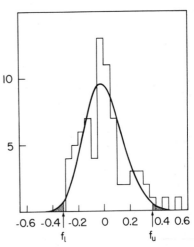

Fig.2. Distribution of the deviation (set 2).

Fig.1. Deviation function, D(E) for the 3 subsets

set	1	2	3
N	13	11	10
N_{tot}	50	48	35
$<d_k>$	0.53	0.35	0.46
σ^2_{exp}	0.0144	0.0180	0.0223

NUCLEON DECAY OF $^{12}C+^{12}C$ AND $^{12}C+^{16}O$: RESONANCES OR STATISTICAL FLUCTUATIONS?

D. Evers
Sektion Physik der Universität München
D-8046 Garching, Germany

I. Introduction

In a molecular resonance we would expect the two reaction partners to form a quasibound state in the continuum thereby keeping for a certain time almost their identy. Consequently the coupling to inelastic or α-exchange channels should be most important in forming a nuclear molecule. This indeed seems to be found experimentally and most of the theoretical treatments go along these lines. How then could a heavy ion resonance be observed in the nucleon decay?

Near the Coulomb barrier observed molecular resonances though embedded in a region of high compound state density can survive because owing to the large structural difference, there is a weak coupling to individual compound states. Therefore the damping of the resonance is small enough that one can observe the statistical decay of a compound nucleus being formed via a molecular doorway state. Final states in the exit channels are then populated according to the available phase space independent of their individual nuclear structure.

At higher incident energies the grazing l values approach the Yrasst line of the compound nucleus and resonances found so far in this regime are mainly concentrated in a region of weak absorption near the grazing angular momenta. If these are close to the Yrast line then by virtue of the fact that a proton or a neutron can carry away only a small amount of angular momentum only high spin states close to the Yrast line in the residual nuclei can be populated. In addition the decreasing life time of molecular states with increasing angular momentum should yield a preferential decay to states with large structural overlap, i.e. large deformation.

Measurements by Cosman et al.[1] published in 1975 showed a very dramatic excitation function for the 9.81,17/2$^+$ Yrast state populated in the $^{12}C(^{12}C,p)^{23}Na$ reaction. This excitation function was interpreted as being evidence for a molecular rotational band formed by two touching ^{12}C nuclei. Our measurements[2] of the neutron decay channel cor-

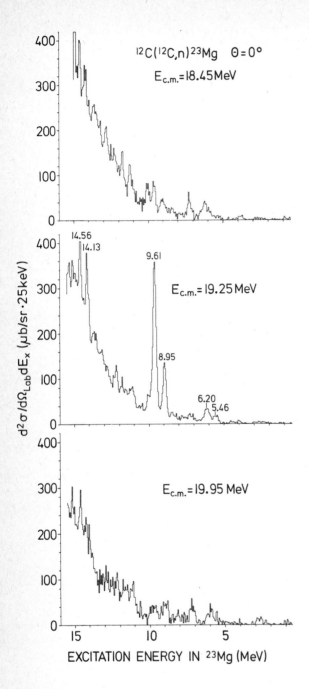

Fig. 1: Neutron time-of-flight spectra converted to double differential cross sections around the 19.3 MeV resonance.

roborated the findings of ref.[1] at E_{cm} = 19.3 MeV. The extreme selectivity found at this energy is demonstrated by our neutron spectra at three different incident energies (s. fig. 1). The resemblence of the proton and neutron spectra can be judged by fig. 2. From the characteristic population as well as the very similar γ-decay through lower lying Yrast states in ^{23}Na and ^{23}Mg, respectively, the 9.81 and 9.61 MeV as well as the 9.04 and 8.95 MeV states could be identified to be mirror states of the A = 23 nuclei. We met a very similar situation in case of the E_{cm} = 19.7 MeV resonance in ^{12}C+^{16}O [3].

From this resemblence one might conclude that isospin is generally conserved in these reactions which however would be a surprise since from other compound nucleus reactions (e.g. ref. 4) a marked reduction of the cross correlation in mirror channels is expected at compound nucleus excitations to be considered here.

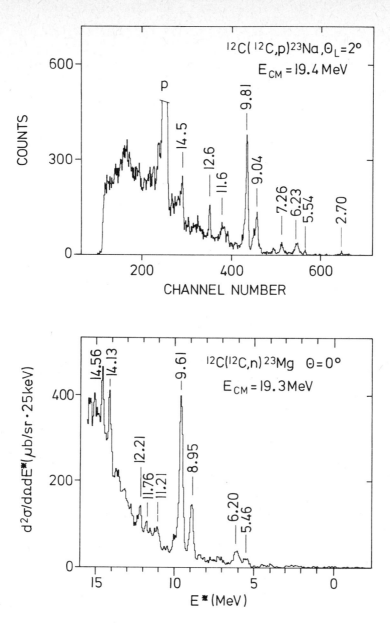

Fig. 2: Comparison of neutron and proton spectra around the E_{cm} = 19.3 MeV resonance.

A study of the proton and neutron decay to mirror states over a longer range of incident energies thus promises interesting results which are as we will see important in detecting heavy ion resonances.

First I will present selected excitation functions for the nucleon decay of $^{12}C+^{12}C$ and $^{12}C+^{16}O$ in the range of E_{cm} = 11-20 MeV, a comparison with Hauser-Feshbach calculations and a statistical fluctuation analysis. Second I will discuss the above mentioned molecular band of $^{12}C+^{12}C$ in view of our new results, third give some new results concerning the E_{cm} = 19.7 MeV resonance in $^{12}C+^{16}O$ and higher incident energies, fourth discuss the origin of the strong selectivity at E_{cm} = 19.3 and 19.7 MeV in the $^{12}C+^{12}C$ and $^{12}C+^{16}O$ system, respectively.

II. Excitation functions of the nucleon decay of $^{12}C+^{12}C$ and $^{12}C+^{16}O$

In our measurements proton spectra were obtained partly at 2° or 0° in a surface barrier detector telescope. The beam was stopped in a gold foil just behind the target. Partly the measurements have been done with an annular surface barrier detector covering a range of $\theta_{lab}=178°-173°$. Particle identification was done is this case by flight time discrimination using a pulsed beam. Neutron time-of-flight measurements were performed using large volume liquid scintillator counters[5].

The excitation functions given in fig. 3 and 4 have been used to check the amount of mirror correlations. Linear cross correlation coefficients

$$r = c(1,2)/(c(1,1) \cdot c(2,2))^{1/2}, \quad \text{where} \quad c(1,2) = \frac{<\sigma_1(E)\sigma_2(E)>}{<\sigma_1(E)><\sigma_2(E)>} - 1,$$

have been calculated for the trend reduced excitation functions and their mean values are compared in table I with those of ref.[4] for ^{26}Al.

Mirror correlations are somewhat weaker in our case than for ^{26}Al which is expected due to the lower excitation energies: $E_x \simeq 26$ MeV for ^{24}Mg, $E_x \simeq 30$ MeV for ^{28}Si and $E_x \simeq 35$ MeV for ^{26}Al. From these values Coulomb mixing matrix elements can be determined which is being done in collab-

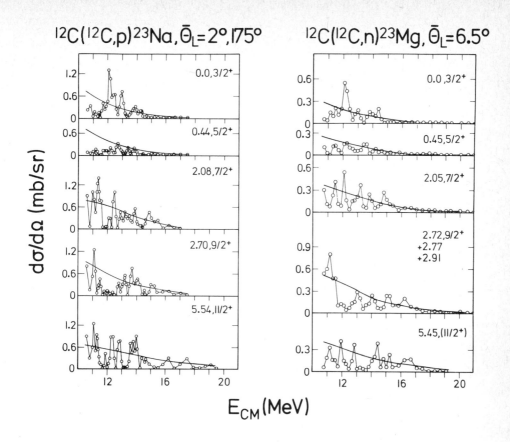

Fig. 3: Excitation functions of $^{12}C(^{12}C,n$ or $p)$ leading to mirror states. The solid curve represents Hauser-Feshbach calculations.

Table I: Linear correlation coefficients (s. text)
Errors are included in brackets.

	^{24}Mg	^{28}Si	^{26}Al
Mirror correlation	0.33(0.05)	0.32(0.06)	0.41(0.06)
Random correlation	-0.13(0.04)	0.11(0.05)	0.10(0.04)

oration with Dr. Harney from MPI Heidelberg. In the present connection I show these values only in order to demonstrate that in our cases mirror correlations are indeed strongly reduced by Coulomb mixing of T = 0 with T = 1 states in the compound nucleus but still are present.

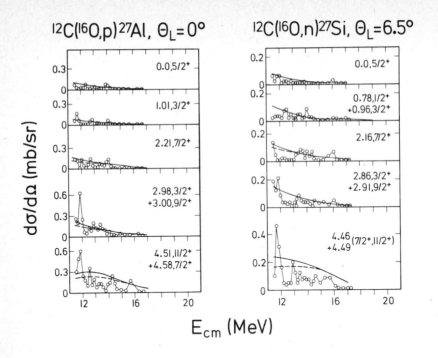

Fig. 4: Excitation functions of $^{12}C(^{16}O,n$ or $p)$ leading to mirror states. Solid and dashed curves are obtained from Hauser-Feshbach calculations with smooth and sharp l cut off in the entrance channel to fit the total experimental fusion cross sections.

We performed fluctuation analyses for states up to E_x = 8.4 MeV in ^{23}Na and ^{23}Mg populated in $^{12}C(^{12}C,p$ or $n)$ and up to E_x = 11 MeV in ^{27}Al and ^{27}Si fed by the $^{12}C(^{16}O,p$ or $n)$ reaction. Results are shown in fig. 5. The deviation function has been calculated with an averaging intervall of δ = 3 MeV. To get the 2 % limit for the deviation function the sum of open channels has been obtained with the code STAT2 [6] and corrected for the smoothing effect of the size of δ [7]. Then the utmost channel number according to the uncertainties induced by the finite sample size [8] has been used. Though the number of maxima M(E) reaches the 1 % limit in some cases the deviation function is consistent with a statistical process. Even when looking only for certain subgroups of states and their mirror states we could not detect a clear enough indication for a nonstatistical process.

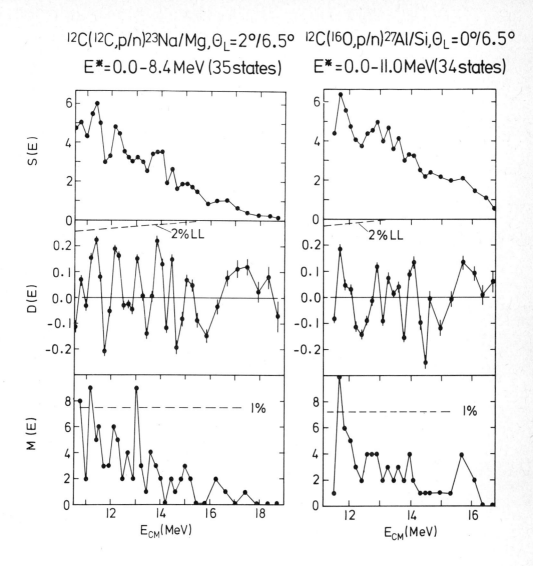

Fig. 5: Fluctuation analyses performed for $^{12}C(^{16}O,n$ or $p)$ and $^{12}C(^{12}C,n$ or $p)$. $S(E)$ summed cross sections in mb/sr, $D(E)$ deviation function, $M(E)$ distribution of maxima.

III. Check of the molecular band in ^{24}Mg proposed by Cosman et al.[1)]

In the last chapter we have excluded the excitation functions of the ^{12}C(^{12}C,n or p) reactions leading to the $J^\pi = 15/2^+$ and $17/2^+$ Yrast states. They will now be treated seperately.

As already said the decay to mirror states should be equal if isospin is conserved in the reactions except for a possible shift of the absolute cross sections due to different neutron and proton transmission coefficients.

For the reaction proceding through a molecular resonance we do expect isospin purity since a strongly deformed T = 0 state will hardly be mixed by the Coulomb interaction with T = 1 compound states. In addition, from fig. 6 it can be seen that for L = 12 $\hbar$ (the proposed reso-

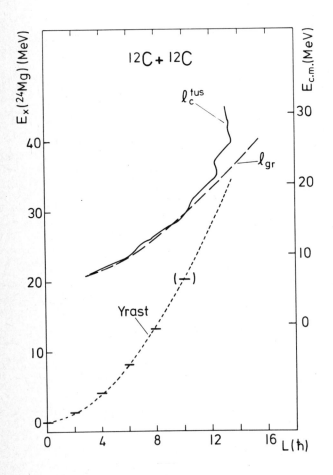

Fig.6:
Plot of some relevant angular momenta versus excitation energies in ^{24}Mg for ^{12}C+^{12}C. l_{gr} marks the grazing 1, l_c^{fus} is the maximum 1 obtained from experimentally determined fusion cross sections[9)] according to the sharp cut off model.

nance spin) at E_{cm} = 19.3 MeV, the excitation energy is some 5 MeV above the Yrast line (extrapolated g.s. band). So it may even be that there are no T = 1 levels available to mix with. (The symmetry term in the semiempirical mass formula gives an isospin splitting of $\Delta E \simeq 4$ MeV). At E_{cm} = 14.3 MeV the proposed L = 10 resonance in ^{24}Mg will be excited about 8 MeV above the Yrast line. So T = 1 states should be available at this energy. However, following the above arguments a molecular resonance should still show up correlated in the mirror channels.

When doing the neutron measurements around E_{cm} = 14.3 MeV we found a strongly reduced correlation especially in the mirror channels of the $17/2^+$ states. Caused by this enigmatical result we remeasured the proton excitation function within an energy range including the peak at E_{cm} = 11.4 MeV and these data are plotted in fig. 7. There are

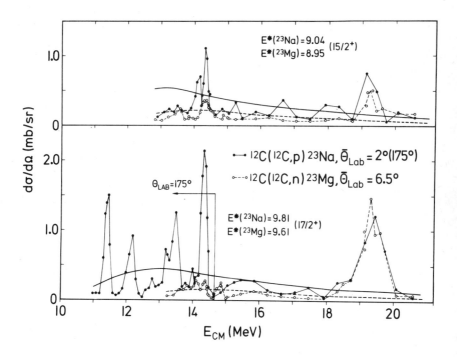

Fig. 7 Excitation functions obtained for the $J^\pi = 15/2^+$ and $17/2^+$ Yrast states in ^{23}Mg and ^{23}Na. Hauser-Feshbach calculations are given by the dashed and solid curves for the neutron and proton exit channels respectively.

significant changes compared to the old data of ref.[1]. Deviation functions obtained from both data sets are given in fig. 8. for the $17/2^+$ states. While the old data of Cosman et al.[1] clearly give a strong hint for a nonstatistical origin of the E_{cm} = 11.4 and 14.3 MeV peaks (close to the 1 ‰ limit) our results are consistent with what one expects from statistical fluctuations. This is also demonstrated by the probability distribution of the fluctuation amplitudes shown in fig. 10. From this we may conclude that there is no molecular resonance at E_{cm} = 14.3 MeV observed in the nucleon decay and probably no at E_{cm} = 11.4 MeV. No check is possible however at this lower incident energy via correlations in the mirror channels since the lower Q-value of the $^{12}C(^{12}C,n)$ reaction (-2.6 MeV) forbids the population of the 8.95 and 9.61 MeV levels in ^{23}Mg.

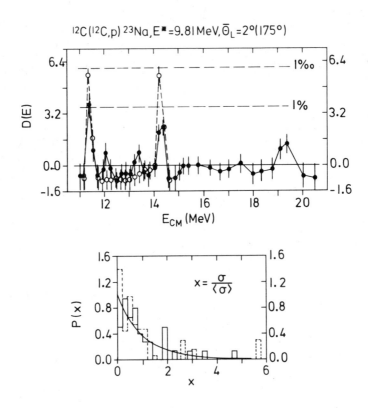

Fig. 8: Deviation function for $^{12}C(^{12}C,p)^{23}Na$ (9.81 MeV) (top) open circles are obtained from the data of ref.[1] Distribution of fluctuation amplitudes (bottom). Experimental values are given by histograms, the broken one according to ref.[1] The solid curve marks the expected distribution of statistical fluctuations.

It should be mentioned that the intensity for the $^{12}C(^{12}C,p)^{23}Na$ (9.81 MeV) reaction at E_{cm} = 11.4 MeV in our data includes the total intensity of a doublet of states at this excitation energy (within $\Delta E_x \approx$ 50 keV) which could not be resolved nor defolded safely. This is concluded from an observed shift of up to 37 keV (more than 3 times the combined statistical and calibrational uncertainties) connected with a proton line width broadening from FWHM = 55 keV up to 74 keV. We guess that the intensity for the 9.81 MeV state will be less than 1 mb/sr around E_{cm} = 11.4 MeV.

The available data from elastic and inelastic, proton and possibly deuteron decay do confirm however a resonance at E_{cm} = 19.3 MeV.[10] No peculiar behaviour is seen in our deviation functions in fig. 8 since the running mean cross section has been obtained from an averaging intervall of δ = 3 MeV which is only about 5 times the resonance width which in turn amounts to 3 times the fluctuation width of Γ = 190 keV obtained from the peak counting method. The width of the other two structures Γ = 200 keV and 150 keV are consistent with them being statistical fluctuations.

For calculating the deviation function around E_{cm} = 19.3 MeV we took the mean cross section from Hauser-Feshbach calculations which are seen to reproduce the mean experimental data "off the resonance" quite well. Then the correlated peak in the deviation function for the 4 states discussed has a probability of clearly less than 1 ‰ to be of statistical origin.

In fig. 9 we compare the measured neutron spectra taken at E_{cm} = 19.3 MeV and 19.6 MeV with a two step evaporation model calculation done with the code GROGI 2 [11]. Evidently not only the resolved high spin states in ^{23}Mg but also unresolved states including those in ^{23}Na above the neutron threshold at E_x = 12.4 MeV get an increased yield at E_{cm} = 19.3 MeV.

So we do confirm the E_{cm} = 19.3 MeV resonance. Due to the missing correlation at the E_{cm} = 14.3 MeV in the population of the mirror states at E_x = 9.81 and 9.61 MeV in ^{23}Na and ^{23}Mg, respectively and because of the significantly corrected excitation function of the former state our experiments clearly show, however, that the pictures of a molecular rotational band cannot be deduced from the $^{12}C(^{12}C,p)^{23}Na$ reaction as done by Cosman et al.[1].

It was argued that the strong reduction of the mirror correlations in the decay to the $J^{\pi} = 17/2^+$ states at $E_{cm}(^{12}C) = 14.3$ MeV might be due to the fact that the neutrons ($E_{cm} = 2.0$ MeV) have much lower transmission coefficients than the protons ($E_{cm} = 6.5$ MeV). However, Hauser-Feshbach calculations limited to $L = 10\,\hbar$ ($T_L = 0.48$) in the entrance channel yielded $d\sigma/d\Omega(0^o) = 0.1$ mb/sr for the neutron and $d\sigma/d\Omega(0^o) = 0.26$ mb/sr for the proton channel. For $E_{cm}(^{12}C) = 19.3$ MeV these calculations give when limited to $L_{in} = 12\,\hbar$ ($T_L = 0.66$), $d\sigma/d\Omega(0^o) = 0.023$ mb/sr in the neutron and $d\sigma/d\Omega(0^o) = 0.10$ mb/sr in the proton channel.

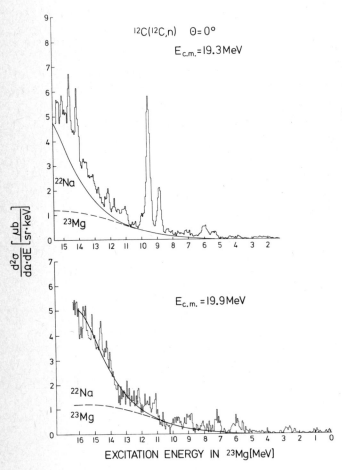

Thus regarding transmission probabilities in the mirror decay channels rather an increase (by a factor of 2) of the neutron to proton decay ratio to the considered mirror states is expected than the observed decrease by a factor of 8 when going from $E_{cm}(^{12}C) = 19.3$ MeV to 14.3 MeV.

This discrepancy persists though reduced by a factor of two when looking at the penetration factors through the Coulomb and centrifugal barriers of a spherical nucleus ($r_o = 1.45$ fm).

Fig.9: Neutron spectra of the $^{12}C(^{12}C,n)$ reaction on and above the $E_{cm} = 19.3$ MeV resonance compared to evaporation model calculations (solid and dashed curves). The dashed curve represents neutrons emitted in the first step, the difference to the solid curve is given by neutrons emitted after proton emission.

IV. ^{12}C+^{16}O at the E_{cm} = 19.7 MeV resonance and higher incident energies

In an earlier paper[3] we confirmed the 19.7 MeV, $J^\pi = 14^+$ resonance[12] found in elastic and inelastic scattering as well as in the proton decay by a correlated appearance in the neutron decay. The resonance could be detected in the neutron channel in that experiment only because of an extremely selective population of three states around E_x = 16 MeV in ^{27}Si. The neutron time of flight spectra in fig. 10 again have been converted to double differential cross sections as a function of excitation energy to be compared with the proton spectra (s. fig. 11) we measured more recently[13] with an annular surface barrier detector at backward angles. Using a pulsed beam, particle identification was done by flight time discrimination. The resolution is worse by a factor of three in the neutron spectra but still the states at E_x = 15.45, 15.77 and 17.13 MeV in ^{27}Si might be identified as the mirror states of those at 15.75, 16.05 and 17.35 MeV in ^{27}Al.

In recent experiments[13] detecting evaporation residues with the Munich rf recoil spectrometer according to their mass, energy and nuclear charge at 0° and evaporated light particles with an annular solid state detector at backward angles we were able to measure the decay branches of particle unstable levels in ^{27}Al. For the E_x = 15.75 MeV level (cf. fig. 11) we found Γ_γ/Γ = 2 %, Γ_α/Γ = 54 %, Γ_n/Γ = 42 %. From the decay branches to the final states and the relevant transmission coefficients obtained from optical model calculations we determined a most probable $J^\pi = 19/2^+$ for this state. This spin is close to the expected window for a decay of a $J^\pi = 14^+$ resonance.

We extended the ^{12}C(^{16}O,p) excitation functions to higher incident energies.[13] With a proton energy resolution of 50-55 keV (Ω = 50 msr) only a few lines stand out from the continuum of unresolved states. Their excitation functions are shown in fig. 12.

I want to draw your attention to three points. First the E_{cm} = 19.7 MeV resonance shows up very strongly in the excitation function of the E_x = 15.75 MeV state. Our cross sections agree perfectly with the corresponding ones of Shapira et al.[12] close to the resonance. Our data, however, cover a wider range. The fine structure of the resonance has been known already and shows up clearly in our measurements. Note also correlated structures at other excitation energies.

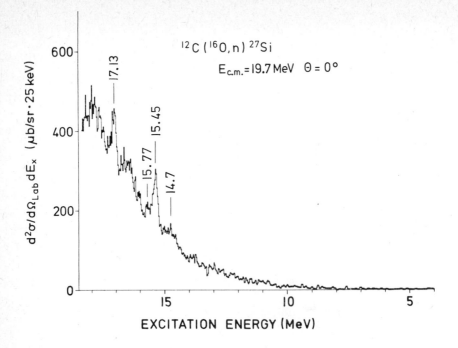

Fig. 10: Neutron spectra obtained at E_{cm} = 19.7 MeV in the $^{12}C(^{16}O,n)$ reaction.

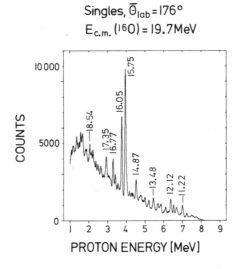

Fig. 11: Proton spectra of the $^{12}C(^{16}O,p)$ reaction at E_{cm} = 19.7 MeV.

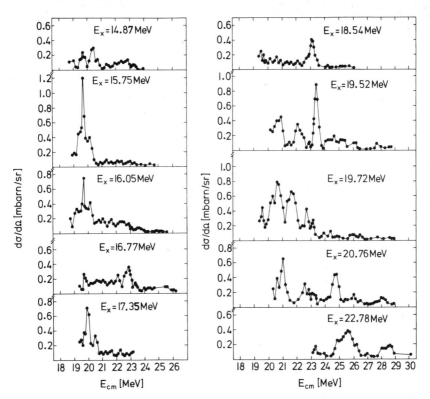

Fig. 12: Excitation function of the $^{12}C(^{16}O,p)$ reaction above E_{cm}= 18.7 MeV. Our excitation energies differ from those given by Cosman et al.[12]. For our values we get an overall uncertainty of $\Delta E_x = \pm$ 10 keV.

Second the selectivity is much stronger than one could have foreseen from a grazing collision picture. E.g. the state at E_x= 19.52 MeV is excited more than five times stronger than any other state up to $E_x \sim$ 23 MeV at E_{cm} = 23.47 MeV. One would be tempted to scent a resonance but up to now we found no other correlated phenomena.

Third the mean cross sections are surprisingly high for most of the shown excitation functions. Hauser-Feshbach calculations reproduce quite well the cross sections for the E_x= 15.75 MeV (J^π = 19/2$^+$) state above E_{cm}= 20.6 MeV. But keeping L_{max} in the entrance channel at 15 $\hbar$ as determined from fusion experiments[9] the maximum Hauser-Feshbach cross section at E_{cm} = 24 MeV never exceeds 35 µb/sr assuming spins up to 25/2 ℏ for the states shown in fig.12 . At E_{cm}= 19.7 MeV this calculated maximum cross section is 70 µb/sr for L_{max} = 14 ℏ.

The observed order of magnitude deviation from these predictions is highly suggestive of other than compound nucleus processes. It might be interesting to look for forward-backward asymmetries for shedding light on this question. Forward-backward asymmetries as a function of incident energy we obtained for the $^{12}C(^{13}C,n)$ reaction[16] indicate that direct processes of the type suggested by Noble[14] might be important.

V. Possible origin of the extraordinary selectivity at E_{cm} = 19.3 and 19.7 MeV in the $^{12}C+^{12}C$ and $^{12}C+^{16}O$ system

In our earlier paper[2] we reported on sort of a backbend observed in the g.s. band of ^{23}Mg. From this we suggested that there might be an increase of deformation at the $J^\pi = 15/2^+$ and $17/2^+$ states which favors in addition to l matching the population of these states from a deformed molecular state.

Coming back to this question it is of course important to check the Coulomb energy shifts of the mirror states. An increase of deformation would lower the excitation energy in the proton richer of the two mirror nuclei.

In table II the Coulomb energy shift differences for the Yrast states in ^{23}Na and ^{23}Mg are listed. An increasing shift is indeed observed starting at $J^\pi = 15/2^+$.

Table II: Yrast states in ^{23}Mg and ^{23}Na

J^π	^{23}Mg	^{23}Na	$\Delta E_x(^{23}Mg-^{23}Na)$
$3/2^+$	0	0	0
$5/2^+$	0.4507	0.4399	+0,0108
$7/2^+$	2.051	2.0764	-0.025
$9/2^+$	2.715	2.7037	+0.011
$11/2^+$	5.455	5.536	-0.081
$13/2^+$	6.200	6.236	-0.036
$15/2^+$	8.945	9.042	-0.097
$17/2^+$	9.610	9.807	-0.197

all energies are given in MeV.

According to M.H. Mcfarlane [15)] the Coulomb energy shifts are connected to deformation by means of the formula

$$\Delta E_c(\delta) \simeq (1 - \frac{4}{45} \delta^2) \Delta E_c(0)$$

or for the difference of the excitation energies ΔE_x given in table II:

$$\Delta E_x \simeq \frac{4}{45} \Delta E_c(0) (\delta^2_{g.s.} - \delta^2_{exc.})$$

where $\Delta E_c(\delta)$ and $\Delta E_c(0)$ are the Coulomb energy shifts for a deformed and a spherical nucleus, respectively, and δ is the deformation parameter ($\delta_{g.s.}$ and δ_{exc} is for the ground and excited state, respectively). Since $(1-\frac{4}{45} \delta^2) \ll 1$ $\Delta E_c(0)$ can be calculated from the ground state binding energy differences corrected for the proton-neutron mass difference.

Taking $\delta_{g.s.} = 0.4$ we obtain for the $J^\pi = 17/2^+$ states $\delta_{exc} = 0.79 \pm 0.03$ which is close to the value $\delta = 0.84$ expected for a spheroid with an axis ratio 2:1 ! Much less is known about the Yrast states for the mirror nuclei ^{27}Si-^{27}Al. Taking $\delta_{g.s.} = 0.3$ we get for the excitation energy difference of the $E_x = 15.45$ and 15.75 MeV $J^\pi = 19/2^+$ states in ^{27}Si and ^{27}Al, respectively, $\delta_{exc} = 0.83 \pm 0.04$, a value which again is very close to that for an axis ratio 2:1.

These results give a strong hint that the nucleon decay of the $E_{cm} = 19.3$ and $E_{cm} = 19.7$ MeV resonances in ^{12}C+^{12}C and ^{12}C+^{16}O indeed occurs mainly to states with a deformation very similar to that expected for nuclear molecules.

VI. Summary

From our measurements of the nucleon decay of the ^{12}C+^{12}C system we found that the picture of a molecular ^{12}C+^{12}C rotational band deduced from the ^{12}C(^{12}C,p) reaction cannot be supported by the data for the neutron decay. We pointed out that by means of missing correlations in strongly populated mirror decay channels the anomaly at $E_{cm} = 14.3$ MeV is nothing but the result of statistical fluctuations. Our new excitation function of the proton decay to the $E_x = 9.81$ MeV state in ^{23}Na in the range of $E_{cm} = 11$ to 15 MeV could well be explained by statistical fluctuations while that of ref. [1)] gave erroneously strong indications of nonstatistical phenomena at $E_{cm} = 11.4$ and 14.3 MeV. Also from the nucleon decay to other states we measured, no nonstatistical processes could be detected.

Our measured nucleon decay data revealed no nonstatistical processes also for the $^{12}C+^{16}O$ system. In the range of E_{cm}= 11-20 MeV for both the $^{12}C+^{12}C$ and the $^{12}C+^{16}O$ system there were only two but marked exceptions: the E_{cm}= 19.3 MeV resonance in $^{12}C+^{12}C$ and the E_{cm}= 19.7 MeV resonance in $^{12}C+^{16}O$. The very strong resonantly enhanced and correlated nucleon decay to very probably considerably deformed high spin states (close to an axis ratio of 2:1) can be taken as an important piece of evidence that nuclear molecules have been formed in these two cases.

Remarkable selectivity and rather high mean cross sections for strongly excited states have been found in the $^{12}C(^{16}O,p)$ reaction in the range E_{cm}= 20-30 MeV. The data seem not to be explainable by compound nucleus processes in the mean.

I like to thank my colleagues Dr. G. Denhöfer, Dr. P. Konrad, Prof. K.E.G. Löbner, Dr. K. Rudolph, Prof. S. Skorka, Dr. P. Sperr, Mrs. I. Weidl which took part in the experiments I talked about and especially Dr. W. Assmann whose thesis work covered a large part of my talk.

References:

1) E.R. Cosman et al., Phys. Rev. Lett. 35 (1975) 265

2) P. Sperr et al., Phys. Lett. 49 B (1974) 345
   D. Evers et al., Z. Phys. A 280 (1977) 287

3) P. Sperr et al., Phys. Lett. 57 B (1975) 438

4) J.J. Simpson et al., Phys. Rev. Lett. 40 (1978) 154

5) D. Evers et al., Nucl. Instr. and Meth. 124 (1975) 23

6) R.G. Stokstad, STAT2, Yale University, NSL-Report 52, 1972

7) D. Shapira et al., Phys. Rev. C 10 (1974) 1063

8) P.J. Dallimore et al., Nucl. Phys. 88 (1966) 193

9) P. Sperr et al., Phys. Rev. Lett. 37 (1976) 321
   P. Sperr et al., Phys. Rev. Lett. 36 (1976) 405

10) K. van Bibber et al., Phys. Rev. Lett. 32 (1974) 687
    T.M. Cormier et al., Phys. Rev. Lett. 38 (1977) 940
    T.M. Cormier et al., Phys. Rev. Lett. 40 (1978) 924
    B.R. Fulton et al., Phys. Rev. C 21 (1980) 198

11) J. Gilat, Brookhaven National Laboratory, BNL-Report 50246 (T-580), 1980

12) R.E. Malmin et al., Phys.Rev.Lett. 28 (1972) 1590;
    E.R. Cosman et al., Phys.Rev.Lett. 29 (1972) 1341;
    D. Shapira et al., Phys. Rev. C 12 (1975) 1907

13) D. Evers et al., Annual Report, Beschleunigerlaboratorium d. Universität und Technischen Universität München, 1980, p. 20

14) J.V. Noble, Phys. Rev. Lett. 28 (1972) 111

15) M.H. Macfarlane, ANL, Inf. Rep. PHY-1966B, 1966

16) W. Assmann et al., Annual Report, Beschleunigerlaboratorium d. Universität und Technischen Universität München, 1977, p. 25

Resonant Structures in the $^{16}O+^{16}O$-System near the Coulomb Barrier*)

G. Gaul, W. Bickel, W. Lahmer, R. Santo
(Institut für Kernphysik, Universität Münster, W.-Germany)

In order to look for resonances of the $^{16}O+^{16}O$ system close to the Coulomb barrier the reaction $^{16}O(^{16}O,\alpha)^{28}Si$ has been measured in the energy range 9.5 MeV $\leq E_{CM} \leq$ 12.5 MeV in steps of ΔE_{CM} = 100 keV [1]. The data reveal that at certain energies the angular distributions show a regular P_L^2 behaviour (Fig. 1), suggesting the existence of resonances (L=2 at E=9.7 MeV, L=4 at E=10.7 MeV, L=6 at E=11.3 MeV, L=8 at E=12.0 MeV). These resonances would form a band with a slope of 25 keV/L(L+1). Recent microscopic calculations [2] taking into account dynamical and quantuum mechanical corrections give a band with similar slope.

In order to establish that the structures observed in the α-channel correspond to resonances of the $^{16}O+^{16}O$ system a high precision measurement (statistical error $\leq$ 0.5 %) of the elastic cross section has been performed between E_{CM} = 9.5 MeV and 12.0 MeV in steps of ΔE_{CM} = 25 keV using the high current ($\approx$ 30 µA $^{16}O^{4+}$) of the Dynamitron Tandem at Bochum and the unique properties of a windowless jet gas target.

In the elastic excitation function measured at Θ_{CM} = 90° (Fig. 2) one can clearly observe three pronounced resonancelike structures at E_{CM} = 9.8 MeV, E_{CM} = 10.1 MeV and E_{CM} = 11.0 MeV. These structures are strongly correlated with the channel averaged cross section (Fig. 3) of the reaction $^{16}O(^{16}O,\alpha)^{28}Si$ at Θ_{LAB} = 7°.

Thus an interpretation of the observed structures in the elastic channel as genuine resonances of the $^{16}O+^{16}O$ system in the subcoulomb region is strongly favoured.

References:

[1] Arbeitsberichte 1976, 1977, 1978; Institut für Kernphysik, Universität Münster
Int. Conference on Resonant Behaviour of Heavy Ion Systems, Athen 1980

[2] J. Urbano, K. Goeke, P. Reinhard, Nucl. Phys. A370 (1981) 329

*) Supported by the Bundesministerium für Forschung und Technologie

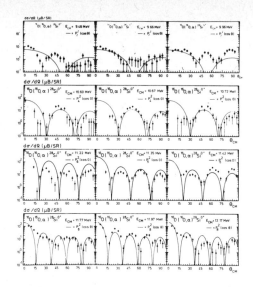

Fig. 1:

Angular distributions of the reaction $^{16}O(^{16}O,\alpha)^{28}Si_{g.s.}$ at some selected energies together with P_L^2-curves. The central column shows those angular distributions which are best described by the square of a Legendre polynomial.

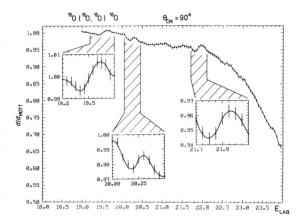

Fig. 2:

Excitation function of elastic $^{16}O+^{16}O$ scattering measured at $\Theta_{CM} = 90°$ in steps of $\Delta E_{CM} = 25$ keV (statistical error $\approx 0.5\%$).
The insets show the cross section at the indicated energy regions on a different scale.

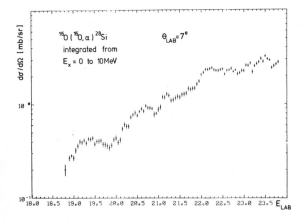

Fig. 3:

Excitation function of the channel averaged cross section[+] of the reaction $^{16}O(^{16}O,\alpha)^{28}Si$ at $\Theta_{LAB} = 7°$.
([+] Summed cross section of all excited states in ^{28}Si from $E_x = 0$ to 10 MeV).

PARTIAL COHERENCE IN HEAVY-ION REACTIONS

K.M.Hartmann, Hahn-Meitner-Institut für Kernforschung, 1000 Berlin 39

W.Dünnweber, Sektion Physik, Universität München, 8046 Garching

W.E.Frahn, Physics Department, University of Cape Town, Rondebosch 7700

First- and second-order interferences are investigated for fluctuating cross sections with average angular distributions not necessarily symmetric about 90°. This is expected to be relevant for reactions between light complex nuclei where discrete final states are populated through an intermediate deep-inelastic or pre-compound process.

In order to discuss the wave-optical properties of such reactions we introduce the angular coherence function $<f(\theta)f^*(\theta')>$ ($f(\theta)$ is the reaction amplitude) in terms of which the energy-averaged cross section is given by

$$<\sigma(\theta)> = <f(\theta)f^*(\theta)> \qquad (1)$$

and the angular cross-correlation function by[1]

$$C(\theta,\theta') = \frac{|<f(\theta)f^*(\theta')>|^2}{<\sigma(\theta)><\sigma(\theta')>} . \qquad (2)$$

Confining ourselves to spin zero nuclei in the entrance and exit channels and using the partial wave expansion for $f(\theta)$, we obtain the angular coherence function

$$<f(\theta)f^*(\theta')> = \frac{1}{4k^2} \sum_{\ell,\ell'} (2\ell+1)(2\ell'+1) P_\ell(\cos\theta) P_{\ell'}(\cos\theta') <S_\ell S_{\ell'}^*> \qquad (3)$$

in terms of the angular momentum coherence function $<S_\ell S_{\ell'}^*>$. In ref.2 we motivate the following parametric form for $<S_\ell S_{\ell'}^*>$:

$$<S_\ell S_{\ell'}^*> = S_o^2 \exp\left[-\left(\frac{\ell-L}{\Delta}\right)^2 - \left(\frac{\ell'-L}{\Delta}\right)^2\right] \exp\left[-\left(\frac{\ell-\ell'}{\sqrt{2}\delta}\right)^2\right] \exp\left[i(\ell-\ell')\Theta\right] \qquad (4)$$

where the first exponential on the RHS of eq(4) describes the strong localization in ℓ-space of the partial wave transition probability $<|S_\ell S_\ell^*|>$[3] and the other exponentials account for the fact that deep-inelastic or pre-compound reactions display angular focussing which implies a mean scattering angle, $\Theta(L)$, and a finite correlation length, δ, between neighbouring partial waves. Eq(4) reduces to the usual limits for direct ($\delta \to \infty$) and compound ($\delta \to 0$) reactions.

The energy-averaged cross section is obtained as a sum of fully coherent and fully incoherent contributions. In the forward angle region,

$$\sin\theta <\sigma(\theta)> \simeq g(\theta)\left[\sigma^{(-)}(\theta) + \sigma^{(+)}(\theta) + 2\sqrt{\sigma^{(-)}(\theta)\sigma^{(+)}(\theta)} \sin(2L\theta)\right]$$
$$+ [1-g(\theta)]\left[\sigma^{(-)}(\theta) + \sigma^{(+)}(\theta)\right] \qquad (5)$$

where the contributions from either side of the interaction region are

$$\sigma^{(\pm)}(\theta) = \exp\left[-\frac{1}{2}\frac{\Delta^2\delta^2}{\Delta^2+\delta^2}(\theta\pm\Theta)^2\right] \qquad (6)$$

and the degree of coherence is measured by

$$g(\theta) = \exp\left[-\frac{1}{2}\frac{\Delta^4}{\Delta^2+\delta^2}\theta^2\right].\tag{7}$$

Thus $\langle\sigma(\theta)\rangle$ remains coherent (i.e. exhibits interferences) within an angle $\theta_c \approx \sqrt{\Delta^2+\delta^2}/\Delta^2$ of $0°$. Note that $\theta_c \to \infty$ for direct reactions and $\theta_c \to 1/\Delta$ for compound reactions.

The angular cross-correlation function exhibits damped oscillations due to the interference between the intensities $\sigma^{(-)}$ and $\sigma^{(+)}$,

$$C(\theta,\theta') = \exp\left[-\left(\frac{\Delta\phi}{2}\right)^2\right]\frac{\sigma^{(-)}(\bar{\theta})^2 + \sigma^{(+)}(\bar{\theta})^2 + 2\sigma^{(-)}(\bar{\theta})\sigma^{(+)}(\bar{\theta})\cos(2L\phi)}{[\sigma^{(-)}(\theta)+\sigma^{(+)}(\theta)][\sigma^{(-)}(\theta')+\sigma^{(+)}(\theta')]}\tag{8}$$

where $\phi = \theta-\theta'$ and $\bar{\theta} = \frac{1}{2}(\theta+\theta')$. A characteristic feature of eq(8) is that the minima in $C(\theta,\theta')$ do not reach zero if $\delta>0$. If a compound nucleus is the intermediate phase in the reaction ($\delta=0$) then

$$C(\theta,\theta') = \exp\left[-\left(\frac{\Delta\phi}{2}\right)^2\right]\cos^2(L\phi),\tag{9}$$

with which a coherence angle $\phi_c = 1.7/\Delta$ may be associated (cf.refs. 1,3).

The dots in the figures give the experimentally determined values of $C(\theta,\theta')$ and $\langle\sigma(\theta)\rangle$ for $\alpha+^{24}$Mg elastic scattering near $E_{lab} = 17$ MeV[4]. The thin solid line gives $C(\theta,\theta')$ assuming the compound nucleus as intermediate phase: L=7.2, Δ=2.0, δ=0 (see eq (9)) while the thick solid line for $C(\theta,\theta')$ and $\langle\sigma(\theta)\rangle$ results from a pre-compound intermediate phase: L=7.2, Δ=2.0, δ=1.0, $|\Theta|$=60°.

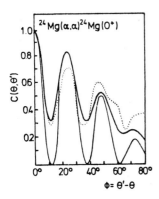
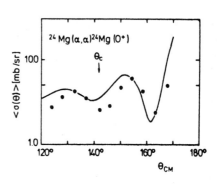

1) D.M.Brink, R.O.Stephen, N.W.Tanner, Nucl.Phys. underline{54} (1964) 577
2) K.M.Hartmann, W.Dünnweber, W.E.Frahn, to be published
3) P.Braun-Munzinger, J.Barrette, Phys.Rev.Lett. underline{44} (1980) 719
4) K.A.Eberhard, C.Mayer-Böricke, Nucl.Phys. A142 (1970) 113

III. RESONANCE STUDIES IN PARTICULAR REACTIONS -

CARBON-OXYGEN MASS REGION AND LIGHTER

Search for resonances in light heavy ion systems

H. Fröhlich, P. Dück, W. Treu, and H. Voit
Physikalisches Institut der Universität
Erlangen-Nürnberg, D8520 Erlangen, W.-Germany

There are different reasons to look for resonances in heavy ion reactions. In systems like $^{12}C+^{12}C$ were the existence of resonances is well established it is necessary to find every single resonance in order to test existing models. For other systems one just wants to explore - as a first step - if there exist resonances at all in order to learn more about the conditions for the existence of resonances and finally - via this detour - about the nature of the molecular resonances itself.

In this contribution we want to report results of a search for resonances being performed with the two aspects given above. We have tried to pin down all resonances in the $^{12}C+^{12}C$ and $^{16}O+^{12}C$ systems in the Coulomb barrier region and to find out if resonances exist in the systems $^{12}C+^{11}B$ and $^{12}C+^{24}Mg$.

If one wants to establish the existence of a resonance one is immediately faced with a serious problem. This problem concerns the relevant criteria for a true resonance and has been discussed to some extent in the literature (1,2). The resonance criterion we have chosen is that a true resonance has to show up in a carefully measured excitation function for the total reaction cross section σ_R.

Two different types of measurements have been used in our investigations to determine σ_R: (i) γ-yield measurements and (ii) precise measurements of the elastic scattering and subsequent application of the optical theorem. We prefer the latter method because it allows to deduce absolute values for σ_R and to extract angular momenta for resonances via phase shift analyses. Method (i) was applied to the systems $^{12}C+^{11}B$ and $^{12}C+^{24}Mg$, method (ii) to $^{16}O+^{12}C$ and $^{12}C+^{12}C$.

$^{16}O+^{12}C$

The total reaction cross section for $^{16}O+^{12}C$ was determined between E(cm)=8.5-15MeV from elastic data and subsequently used in order to pin down resonances in this system. The possibility to deduce σ_R for charged particles from elastic data was first explored by Holdeman and Thaler (3). Starting from a slightly modified optical theorem they arrived at an expression which was shown by Wojciechowski et al. (4) to reduce further to

(1) $$\sigma_R = 2\pi \int_{\theta_0}^{\pi} (\sigma_{Cb}(\theta) - \sigma_{el}(\theta)) \sin\theta \, d\theta$$

in case of heavy ion reactions. In eq.(1) $\sigma_{Cb}(\theta)$ is the differential Coulomb cross section, $\sigma_{el}(\theta)$ is the measured elastic cross section and θ_o an angle which should be smaller than θ_{Cb}, the angle for which $\sigma_{el}(\theta)$ starts to deviate from $\sigma_{Cb}(\theta)$. The original expression given in ref.(3) contains two additional terms which depend on the nuclear scattering amplitude at zero degree $f_N(0)$. Since the absolute value of $f_N(0)$ is small for low energy heavy ion reactions with large values of the Sommerfeld parameter these terms can be dropped. It must be noted that eq.(1) represents therefore an approximation which, however, yields accurate cross sections for reactions fulfilling the above condition.

It is obvious from eq.(1) that measurements of elastic angular distributions must be extended to angles $\theta_o \leq \theta_{Cb}$. This means that the application of eq.(1) for evaluation of σ_R is limited to relatively small energies (θ_{Cb} decreases with increasing energy). In cases where $\sigma_{el}(\theta)$ starts to oscillate about $\sigma_{Cb}(\theta)$ for $\theta < \theta_{Cb}$ eq.(1) can also be used if θ_o is chosen to be equal or smaller than θ_{Cb}. This has been shown by Oeschler et al.(5).

We have measured approximately 100 angular distributions between E(cm)=8.5 and 15MeV in steps of 85keV. Measurements of such large numbers of angular distributions can only be done in a reasonable time if complete angular distributions are measured simultaneously. We therefore have used a multidetector array (see fig.1) in which 32 very thin Si-detectors were positioned on both sides of the beam axis both above and below the horizontal plane. The angular range (lab) covered was $+6° \leq \theta \leq \pm 50°$. The 4 detectors placed at the most forward angles ($6° \leq \theta \leq 10.5°$; see fig.1) were used to correct the data for the effect of small changes in the beam spot position. For the absolute normalization of each angular distribution the cross sections of the most forward angles were used which are pure Coulomb cross sections in the energy range considered.

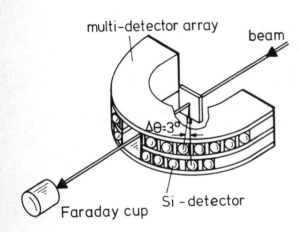

Fig.1: Multidetector array used to measure elastic angular distributions.

Besides the scattered ^{16}O ions also ^{12}C recoils had to be detected in order to get backward angle data. Unfortunately the ^{12}C recoils have roughly the same energy as the ^{16}O ions scattered from the oxygen contamination in the target. They could be separated, however, due to different energy losses in a $290\mu g \cdot cm^{-2}$ Mylar foil placed in front of the detectors (see fig.2). The ^{12}C targets used had a thickness between 10 and $15\mu g \cdot cm^{-2}$.

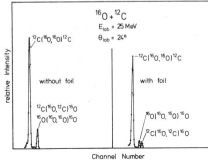

Fig.2: Elastic spectrum for $^{16}O + ^{12}C$ obtained with (right side) and without a $290\mu g \cdot cm^{-2}$ Mylar foil in front of the detectors.

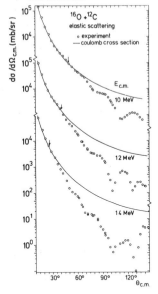

Fig.3: Angular distributions of the elastic scattering $^{16}O + ^{12}C$. Solid lines represent the Coulomb cross section. The arrows lable the cut off angles θ_o.

Fig.3 shows three out of the hundred measured angular distributions. The experimental data seem to follow the Coulomb cross section (solid line) at forward angles. Fig.4 shows however, that $\sigma_{el}(\theta)$ oscillates about $\sigma_{Cb}(\theta)$. The arrow in fig.4 marks the cut off angle θ_o used in eq.(1) to determine σ_R. θ_o was chosen to be θ_{Cb} throughout the present investigation. The total reaction cross section obtained in this way is displayed in fig.5 together with fusion cross

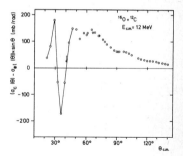

Fig.4: Difference of measured elastic cross section and Coulomb cross section as function of the angle- The arrow marks θ_o.

sections and reaction cross sections obtained from γ-yield measurements. Agreement with the latter data is rather satisfactory as far as absolute cross sections are concerned. As expected, the fusion cross sections fall below σ_R for higher energies.

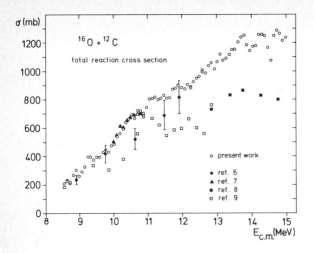

Fig.5: Total reaction cross section for $^{16}O+^{12}C$ as obtained in the present work together with previously reported data.

We have investigated the dependence of the absolute value of σ_R on the choice of the cut off angle θ_o using different values for θ_o. It turned out that the cross section obtained differ at most by 20% as long as θ_o is chosen to be smaller or equal to θ_{Cb}. For $\theta_o > \theta_{Cb}$ unreasonable results were obtained. The limited range of our angular distributions ($\theta_{max}=140°$) has also some influence on σ_R. In fact if one sums up the differences between σ_{Cb} and σ_{el} only between θ_o and θ_{max} one obtains a value for σ_R which is too small. The missing cross section is largest ($=\Delta\sigma_R^{max}$) if $\sigma_{el}(\theta)=0$ for all $\theta > \theta_{max}$ and is given by

$$\Delta\sigma_R^{max} = 2\pi \int_{\theta_{max}}^{\pi} \sigma_{Cb}(\theta) \sin\theta\, d\theta .$$

$\Delta\sigma_R^{max}$ is plotted as a function of energy in fig.6. The figure shows that the maximum missing cross section poses a serious limit on the smallest energies for which this method can be used. Fortunately the actual missing cross section σ_R has a value between 0 and $\Delta\sigma_R^{max}$. In fact in case of the 15MeV angular distribution for which data up to $\theta=170°$ were available (10) $\Delta\sigma_R$ could be determined to be one third of $\Delta\sigma_R^{max}$. This $\Delta\sigma_R$ was accounted for using $\theta_{max}=180°$ and assuming $\sigma_{el}(\theta)=\sigma_{el}(140°)$ for $140 < \theta \leq 160$ and $\sigma_{el}(\theta)=0$ for $\theta > 160°$ in our analysis. In this way the error could be kept negligible small compared with the uncertainty introduced by θ_o. From these investigations we conclude that our total reaction cross section is accurate within ±20%.

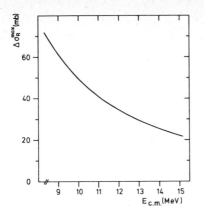

Fig.6: The maximum missing cross section for $\theta_{max}=140°$ as a function of energy.

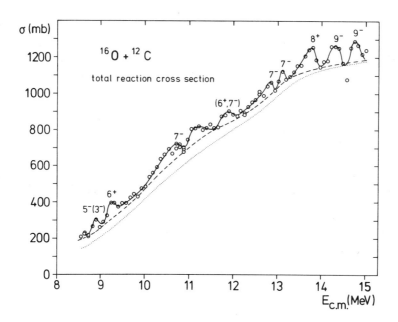

Fig.7: Total reaction cross section for $^{16}O+^{12}C$. The solid line is just to guide the eye. The dashed and dotted lines represent assumed background contributions to σ_R. The dashed line was used for the actual calculations.

Fig.7 shows that the total reaction cross section contains pronounced structures superimposed on a smooth background. It should be noted that these structures are to a large extend independent of the choice of the cut off angle θ_o as long as $\theta_o \leq \theta_{Cb}$. We observe deviations of the order of 10% to 40% for σ_R at the resonance energies from an average cross section $\langle\sigma_R\rangle$ (averaged over 1MeV). The probabilities that these deviations are statistical fluctuations have been calculated using N_{eff} values (number of effective channels) obtained from Hauser-Feshbach calculations (parameters of ref.11). The calculated probabilities are smaller than 10^{-4} in all cases. We therefore conclude that these structures are of nonstatistical origin and true resonances of the A=28 system. In fact all resonances found in the total reaction cross section have been observed already in other investigations as anomalies. Their occurence in the total reaction cross section is an ultimate proof for them being true resonances. Table 1 lists the resonances observed in this investigation.

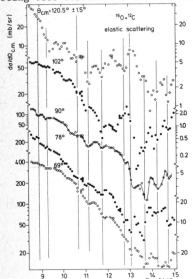

Fig.8: Excitation functions of the elastic scattering $^{16}O+^{12}C$ measured at different angles. The vertical lines mark the position of molecular resonances.

It is interesting to note that the fingerprints of these resonances can be found already in the 90°-elastic excitation function. Fig.8 shows that pronounced minima exist at all resonance energies (solid vertical lines). The same observation has been made for $^{12}C+^{12}C$ resonances (see fig.9).

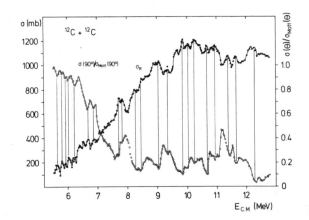

Fig.9: 90°-excitation function of the elastic scattering $^{12}C+^{12}C$ together with the $^{12}C+^{12}C$ total reaction cross section.

The observation of resonances in the total reaction cross section offers a unique possibility to extract elastic partial widths Γ_{el} for quasi-molecular resonances. We notice that the composite A=28 system is populated at excitation energies where many levels overlap strongly. The widths Γ_{cn} of these states (11,12) are comparable with resonance widths observed in this work.

Since the experimental energy resolution δE has to be smaller than Γ_{cn} and Γ in order to observe resonances one will also find strong interference effects between "normal" compound nuclear states and resonance states in the excitation functions of individual exit channels. A description of measured cross sections using an energy averaged scattering matrix which decouples resonances from compound nuclear states in cases where $\Gamma_{cn} < \delta E < \Gamma$ is impossible. Thus, the extraction of partial widths Γ_{el} is almost impossible from the data of individual channels. It will be shown, however, that the total reaction cross section does not contain interference effects if certain conditions are fulfilled.

The off-diagonal elements of the scattering matrix for a transition from an initial state a with total angular momentum J to a final state c can be written (using standard notation) as follows:

$$(2) \quad S_{ac}^{J} = -i e^{i(\delta_a^J + \delta_c^J)} \left\{ \sum_{\lambda \neq \rho} \frac{g_{\lambda a}^J g_{\lambda c}^J}{E - E_\lambda + \frac{1}{2}i\Gamma_\lambda^J} + \frac{g_{\rho a}^J g_{\rho c}^J}{E - E_\rho + \frac{1}{2}i\Gamma_\rho^J} \right\}$$

where the index ρ labels a "true" resonance state embedded in a continuum of "normal" compound nuclear states labeled λ. We assume that resonance states do not overlap. Eq.(2) does not include an energy independent direct term since we know from statistical analyses (14,15) that direct contributions to the reaction cross section are negligible at energies in the vicinity of the Coulomb barrier.

With eq.(2) the total reaction cross section σ_R can be written as follows:

$$(3) \quad \sigma_R = \pi \lambdabar^2 \sum_{c \neq a} \sum_{\ell = J, \ell', s'} (2\ell+1) \, |S_{ac}^J|^2 =$$

$$= \pi \lambdabar^2 \sum_{\ell = J, \ell', s'} (2\ell+1) \left\{ \sum_\lambda \frac{\Gamma_{\lambda a}^J \sum_c \Gamma_{\lambda c}^J}{(E-E_\lambda)^2 + (\frac{1}{2}\Gamma_\lambda^J)^2} + \frac{\Gamma_{\rho a}^J \sum_c \Gamma_{\rho c}^J}{(E-E_\rho)^2 + (\frac{1}{2}\Gamma_\rho^J)^2} + \right.$$

$$+ 2 \operatorname{Re} \sum_{\lambda \neq \lambda'} \frac{g_{\lambda a}^J \cdot g_{\lambda' a}^{J*} \sum_c g_{\lambda c}^J \cdot g_{\lambda' c}^{J*}}{(E-E_\lambda + \frac{1}{2}i\Gamma_\lambda^J)(E-E_{\lambda'} - \frac{1}{2}i\Gamma_{\lambda'}^J)} +$$

$$\left. + 2 \operatorname{Re} \sum_\lambda \frac{g_{\rho a}^J \cdot g_{\lambda a}^{J*} \sum_c g_{\rho c}^J \cdot g_{\lambda c}^{J*}}{(E-E_\rho + \frac{1}{2}i\Gamma_\rho^J)(E-E_\lambda - \frac{1}{2}i\Gamma_\lambda^J)} \right\}$$

The quantities $g^J_{\lambda c}$, $g^J_{\lambda' c}$ are normally distributed (with respect to λ) with mean value zero. They are, however, not a priori statistically independent in case of $\Gamma_{CN}/D_{CN} \gg 1$ due to the unitarity of the scattering matrix. The mutual dependence decreases, however, with increasing number N_c of channels c in which the states λ, λ' decay. If this number is large as in the present case, $g^J_{\lambda c}$ and $g^J_{\lambda' c}$ can be considered to be independent with respect to c. This means that also products like $g^J_{\lambda c} \cdot g^J_{\lambda' c}$ are independent variables with mean value 0. Then $\sum_c g^J_{\lambda c} \cdot g^{J*}_{\lambda' c}$ is normally distributed with mean value zero if N_c is a large number. The same holds for $\sum_c g^J_{\rho c} \cdot g^{J*}_{\lambda c}$. Thus, eq.(3) reduces to

$$(4) \quad \sigma_R = \pi \lambdabar^2 \left[\sum_{J,s',\ell'} (2\ell+1) \sum_{\lambda \neq \rho} \frac{\Gamma^J_{\lambda a} \Gamma^J_{\lambda reac}}{(E-E_\lambda)^2 + (\frac{1}{2}\Gamma^J_\lambda)^2} + (2J+1) \frac{\Gamma^J_{\rho a} \Gamma^J_{\rho reac}}{(E-E_\rho)^2 + (\frac{1}{2}\Gamma^J_\rho)^2} \right]$$

where $\Gamma_{\lambda reac} = \sum_{c \neq a} \Gamma_{\lambda c}$ and $\Gamma_{\rho reac} = \sum_{c \neq a} \Gamma_{\rho c}$. It can be assumed that the first term in eq.(4) has a smooth energy dependence if the number of states λ laying within dE is large. This is the case both for the $^{16}O+^{12}C$ and the $^{12}C+^{12}C$ reaction in the energy range studied (16). The important consequence of these considerations is that the total reaction cross section σ_R decomposes into a background term and a resonance term σ_R^{res} and that therefore the determination of $\Gamma^J_{\rho a} = \Gamma_{el}$ can be readily achieved using eq.(5)

$$(5) \quad \sigma_R^{res} = 4\pi \lambdabar^2 (2J+1) \frac{\Gamma_{el}(\Gamma_{el}-\Gamma)}{\Gamma^2}$$

where $\Gamma = \Gamma^J_\rho$.

We have determined Γ_{el}-values for the most prominent $^{16}O+^{12}C$ resonances with known angular momentum J. The background contribution which was substracted from σ_R in order to get σ_R^{res} is shown as dashed line in fig.7 (it represents a hand drawn line). From the two solutions of eq.(5) the smaller one was used since it is felt that it represents the physical solution. As a result of an estimate using different absolute values for σ_R (within the estimated limits) and different background contributions (for instance dotted line in fig.7) the Γ_{el}-values are accurate to 20-30%. Table 1 gives the Γ_{el}-values obtained. From these values reduced elastic widths γ^2_{el} have been calculated (using r_0=1.5fm). They represent a large fraction of the single particle widths γ^2_w obtained in the Wigner limit. This means that the A=28 system populates strongly the $^{16}O+^{12}C$ channel at the resonance energies and that the configuration of those resonances most probably is that of a $^{16}O+^{12}C$ nuclear molecule. The large values found for γ_{el}^2 can be viewed at as a direct experimental proof for the fact that $^{16}O+^{12}C$ dinuclear molecules are formed during the $^{16}O+^{12}C$ collision in the vicinity of the Coulomb barrier.

These conclusions are supported by the observation that the Γ_{el}/Γ-values exceed $(\Gamma_{el}/\Gamma)_{HF}$ values obtained from Hauser Feshbach calculations (18) by far (see table 1).

$^{12}C+^{12}C$

The $^{12}C+^{12}C$ total reaction cross section has been determined from elastic data along the lines described above. Due to the identity of the particles in the entrance channel a few dissimilarities are to be noted: (i) there are no problems with the missing cross section and (ii) eq.(1) reads in case of identical particles

$$\sigma_R = 2\pi \int_{\theta_0}^{\pi/2} (\sigma_{Cb}(\theta) - \sigma_{el}(\theta)) \sin\theta d\theta.$$

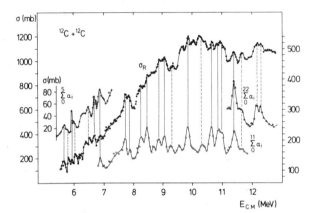

Fig.10: Total reaction cross section σ_R for the reaction $^{12}C+^{12}C$ together with angle integrated cross sections of the α-particle exit channel.

Fig.10 shows the total reaction cross section together with previous result for the α-particle exit channel (1,22,23). There is a clear correspondence between resonances already known from the α-particle channel and resonances in the total reaction cross section. The latter contains additional resonances which are not very pronounced in the α-particle channel. Fig.11 shows again σ_R together with the background contribution (dashed line) used for the determination of the elastic partial widths.

Partial widths and ratios $(\Gamma_{el}/\Gamma)/(\Gamma_{el}/\Gamma)_{HF}$ obtained from Hauser-Feshbach calculations are given in table 2. It is obvious from this table that resonances in the close vicinity of the Coulomb barrier are good candidates for a $^{12}C+^{12}C$ dinuclear molecule.

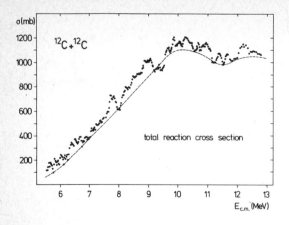

Fig.11: Total reaction cross section for $^{12}C+^{12}C$. The dashed line is the assumed background.

$^{12}C+^{11}B$

We have performed γ-yield measurements in order to obtain quasi reaction cross sections for the most prominent reaction channels of the reaction $^{12}C+^{11}B$. The experimental technique is standard. A schematic of the experimental setup is shown in fig.12.

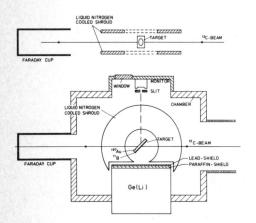

Fig.12: Experimental setup for the γ-yield measurements.

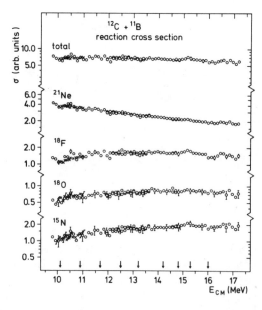

Fig.13: Quasi reaction cross sections for the most prominent reaction channels of $^{12}C+^{11}B$. The arrows mark the positions of resonance structures reported in ref.24.

The reaction cross sections are shown in fig.13. We observe no resonance structures in contrast to Frawley et al. (24) who have reported 9 resonances in the energy range studied in this work.

$^{12}C + ^{24}Mg$

Fig.14 shows the results of γ-yield measurements for the reaction $^{12}C + ^{24}Mg$. The data are normalized to total reaction cross sections obtained from optical model calculations performed with parameters which fit elastic angular distributions measured at three different energies. The excitation functions, in particular the $\alpha - ^{32}S$ exit channel, do not exhibit resonance structures in contrast to results of Cindro et al. (25).

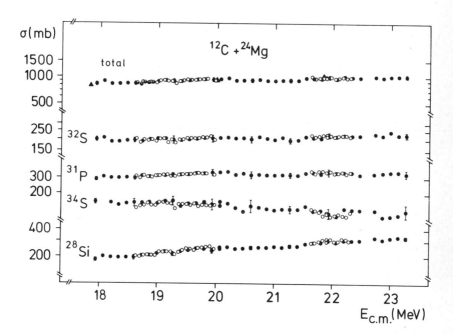

Fig.14: Results of γ-yield measurements for the reaction $^{12}C + ^{24}Mg$. The data are normalized to total reaction cross sections (triangles in the uppermost curve) obtained from optical model calculations performed with parameters which fit elastic angular distributions measured at E(lab)=27, 30 and 33 MeV.

Conclusion

It has been found that excitation functions of total reaction cross sections are a powerful tool in order to pin down resonances in heavy ion reactions. We think we have identified all resonances in the systems $^{16}O+^{12}C$ and $^{12}C+^{12}C$ in the energy range studied. We find no indication for the existence of resonances in the systems $^{12}C+^{11}B$ and $^{12}C+^{24}Mg$ for which resonances have been reported previously. It is shown that the total reaction cross section offers a unique possibility to determine elastic partial widths for resonances in case of many overlapping compound nuclear resonances having widths comparable to the resonance widths.

Elastic widths for $^{16}O+^{12}C$ and some $^{12}C+^{12}C$ resonances represent a large fraction of the total widths. This can be taken as a direct experimental proof that these resonances have a dinuclear structure.

The authors thank H. Hofmann for several very valuable discussions.

This work was supported by the Deutsche Forschungsgemeinschaft, Bonn, W.-Germany.

References

1   H. Voit, W. Galster, W. Treu, H. Fröhlich, and P. Dück,
    Phys.Lett. 67B, 399 (1977)

2   P. Taras, in Clustering Aspects of Nuclear Structure and Nuclear Reactions,
    ed. by W.T.H. Van Oers (AIP, New York, 1978) p. 234

3   J.T. Holdeman and R.M. Thaler, Phys.Rev.Lett. 14, 81 (1965)

4   H. Wojciechowski, D.E. Gustafson, L.R. Medsker, and R.H. Davis,
    Phys.Lett. 63B, 413 (1976)

5   H. Oeschler, H.L. Harney, D.L. Hills, and K.S. Kim,
    Nucl. Phys. A325, 463 (1979)

6   P.R. Christensen, Z.E. Switkowski, and R.A. Dayras,
    Nucl. Phys. A280, 189 (1976)

7   B.N. Nagorcka, G.D. Symons, P.B. Treacy, and I.C. Maclean,
    Aust. J. Phys. 30, 149 (1977)

8   P. Sperr, S. Vigdor, Y. Eisen, W. Henning, D.G. Kovar, T.R. Ophel, and
    B. Zeidmann, Phys.Rev.Lett. 36, 405 (1976)

9   H. Fröhlich, P. Dück, W. Galster, W. Treu, H. Voit, H. Witt, W. Kühn,
    and S.M. Lee, Phys.Lett. 64B, 408 (1976)

10 W. von Oertzen, H.H. Gutbrod, M. Müller, U. Voos, and R. Bock,
    Phys.Lett. 26B, 291 (1968)

11 L.R. Greenwood, K. Katori, R.E. Malmin, T.H. Braid, J.C. Stoltzfus, and
    R.H. Siemssen, P.R. C6, 2112 (1972)

12 M.L. Halbert, F.E. Durham, and A. van der Woude, Phys.Rev. 162, 899 (1967)

13 L.R. Greenwood, R.E. Segel, K. Raghunathan, M.A. Lee, H.T. Fortune,
    J.R. Erskine, Phys.Rev. C12, 156 (1975)

14 W. Galster, P. Dück, H. Fröhlich, W. Treu, H. Voit, and S.M.B. Lee,
    Phys.Rev. C22, 515 (1980)

15 G. Hartmann, Diploma thesis, University of Erlangen-Nürnberg, (1972),
    unpublished

16 The number of states within δE has been estimated using the level density
    formula of ref.17 to be 40 and 700 for the $^{12}C+^{12}C$ and the $^{16}O+^{12}C$ reaction,
    respectively, for the lowest energies studied

17 A. Gilbert and A.G.W. Cameron, Can.J.Phys. $\underline{43}$, 1446 (1965)

18 We used the relation $(\Gamma_{el}/\Gamma)_{HF} = T_1/G^J$ where T_1 is the entrance channel transmission coefficient and G^J is the Hauser-Feshbach denominator. The parameters used for the calculation are from refs. 11 and 13

19 W. Treu, W. Galster, H. Fröhlich, H. Voit, and P. Dück, Phys.Lett. $\underline{72B}$, 315 (1977)

20 F. Soga, J. Shimizu, H. Kamitsubo, N. Takahashi, K. Takimoto, R. Wada, T. Fujisawa, and T. Wada, Phys.Rev. $\underline{C18}$, 2457 (1978)

21 J.R. Hurd, N.R. Fletcher, A.D. Frawley, and J.F. Mateja, Phys.Rev. $\underline{C22}$, 528 (1980)

22 W. Galster, W. Treu, P. Dück, H. Fröhlich, and H. Voit, Phys.Rev. $\underline{C15}$, 950 (1977)

23 W. Treu, H. Fröhlich, W. Galster, P. Dück, and H. Voit, Phys.Rev. $\underline{C13}$, 2148 (1978)

24 A.D. Frawley, J.F. Mateja, A. Roy, and N.R. Fletcher, Phys.Rev. $\underline{C19}$, 2215 (1979)

25 N. Cindro, J.D. Moses, N. Stein, M. Cates, D.M. Drake, D.L. Hanson, and J.W. Sunier, Phys.Lett. $\underline{84B}$, 55 (1979)

Table 1

$^{16}O + ^{12}C$ quasimolecular resonances observed in the total reaction cross section together with resonance parameters.

E_{res} (MeV)	J^π	σ_R^{res} (mb)	Γ (keV)	Γ_{el} (keV)	γ_{el}^2/γ_W^2 (%)	$\left(\frac{\Gamma_{el}}{\Gamma}\right)/\left(\frac{\Gamma_{el}}{\Gamma}\right)_{HF}$
8.86	$5^-(3^-)$ [a]	73	131	25	15	324
9.26	6^+ [a]	91	241	52	32	39
10.71	7^- [b]	72	217			
11.11		89	322			
11.85	$6^+, 7^-$ [c]	66	292			
12.79	7^- [b]	77	171	37	4	14
13.07	7^- [c]	82	95	20	2	20
13.70	8^+ [b,c]	116	259	109	11	42
14.27	9^- [b,c]	115	253	87	10	73
14.74	9^- [b,c]	114	238	86	9	84

a) from ref. 19   b) from ref. 20   c) from ref. 21

Table 2
Resonance parameters for resonance in the $^{12}C+^{12}C$ reaction.

E_{res} (MeV)	J	Γ (keV)	Γ_{el} (keV)	γ^2_{el}/γ^2_w %	$\dfrac{\Gamma_{el}/\Gamma}{(\Gamma_{el}/\Gamma)_{HF}}$
5.65	2	147	22.4	25	32
5.93	4	74	4.6	9.4	57
6.04	4	107	8.9	15	61
6.39	2	128	27.3	25	14
6.68	4	107	10.1	6.1	17
6.87	4	95	4.0	1.9	5
7.72	4	77	14.5	3.1	7
7.82	4	156	25.3	4.9	6
8.87	6	136	19.0	4.3	9
9.05	6	168	21.3	4.3	7
9.66	8	115	5.0	2.1	6
9.85	8	89	7.1	2.6	9
10.03	6	128	8.3	1.0	2
10.25	8	176	13.1	3.6	6
10.42	6	102	5.9	0.6	3
10.66	8	80	4.5	1.0	4
10.96	8	118	10.1	1.8	5
11.39	8	146	11.1	1.6	4
12.31	8	100	3.7	0.4	1

^{12}C + ^{12}C Resonances Studied in the

Elastic, Inelastic, and Transfer Channels[*]

T. M. Cormier[†]

Nuclear Structure Research Laboratory
Department of Physics and Astronomy
University of Rochester
Rochester, New York, U.S.A.

ABSTRACT

The occurrence of resonances in ^{12}C + ^{12}C elastic, inelastic and transfer reactions is reviewed at energies from ∼2 to 6 times the Coulomb barrier. The resonances are characterized by large partial widths for decay into ^{12}C + ^{12}C, ^{12}C + ^{12}C(2^+) and ^{12}C(2^+) + ^{12}C(2^+). Assuming that ^{12}C + ^{12}C inelastic scattering data defines the gross features of the ^{24}Mg molecular spectrum, the possibility of directly exciting these states in the ^{12}C(^{16}O,α) reaction at high energy is considered.

[*] Supported in part by the National Science Foundation under grant PHY-79-23307

[†] Alfred P. Sloan Foundation Fellow

The ^{24}Mg molecular spectrum is now well studied at energies above the barrier and several clear systematic features have emerged. Prompted by the observation of broad resonacnes in the ^{12}C + ^{12}C total fusion cross section by Sperr et al.[1] several years ago, we undertook the measurement of ^{12}C + ^{12}C total inelastic scattering cross sections. Our first results,[2] obtained with a single 10" × 10" NaI detector are shown in Figure 1. Broad structures similar in width to those seen in the fusion data are also seen here as well as a hint of finer structure although the energy step size in these first data is marginally too large to see all the details. In subsequent measurements a second 10" × 10" NaI detector was added and the total mutual inelastic cross section was measured via coincident observation of two 4.44 MeV γ-rays. Very similar structure was observed in the mutual inelastic channel.

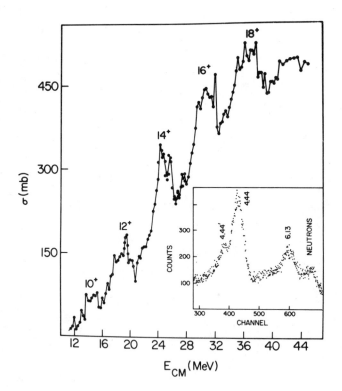

Figure 1. Excitation function of the yield of 4.44 MeV γ-rays from ^{12}C + ^{12}C Reactions.

The measurements have since been refined[3] through the direct observation of inelastically scattered particles over the entire angular range and the results are summarized in Fig. 2. Several other channels including the 0^+_2, 3^-_1 and 4^+_1 have been investigated[4] in less detail and these are also shown in Figure 2.

Our earliest speculation regarding these data was that we were observing a strength function phenomenon in which some simple characteristic (eg. Γ_c) is spread over the more numerous and presumably more complex states of the system. The reason for this suggestion is most evident in the single 2^+_1 inelastic excitation function below $E_{cm} \cong 25$ MeV. The cluster of resonances near $E_{cm} \equiv 19$ MeV are mostly $J^\pi = 12^+$ while those near $E_{cm} \cong 14$ MeV are largely 10^+. Presumably, then, the higher lying broad structures represent progressively higher spins. We have noted[2] that their spacing corresponds quite closely to the $^{12}C + ^{12}C$ grazing partial wave sequence.

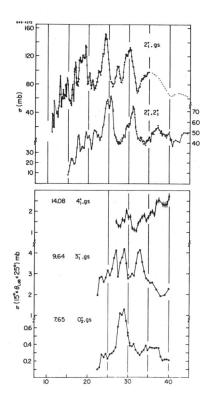

Figure 2. Summary of available $^{12}C + ^{12}C$ total inelastic scattering cross sections.

It is generally accepted that the individual narrow fragments do not correspond to single statistical ^{24}Mg compound states. For example, a conventional Fermi-gas level density predicts ~ 50 levels per MeV for $J^\pi = 12^+$ at $E_x(^{24}Mg) \cong 33$ MeV whereas the observed resonances have a density ~ 2 per MeV. It may not be appropriate, however, to compare the number of observed resonances with the total level density in the compound nucleus. Rather, if we assume, as in fission, that the broad molecular state cannot couple directly to compound nucleus states with deformations typical of the ground state band, then we should consider the density of ^{24}Mg states with deformations close to the saddle point.

The density of states with this deformation is expected to be considerably smaller.

Extensive elastic scattering data is also available across much of this energy range, and there have been a few attempts to extract molecular partial widths for some of these resonances. Figure 3 shows a typical case,[3] the well studied E_{cm} = 19.3 MeV resonance. In this example the carbon partial width is deduced from Breit-Wigner plus optical model fits to elastic excitation functions at 16 angles. The data strongly support the $J^\pi = 12^+$ assignment for this resonance and yield $\Gamma_c \cong 75$ keV. The uncertainty in this number resulting from the optical model parameterization of the background is substantial, probably ∼±25 keV, but it is actually in good agreement with an independent determination based on various total cross section measurements. Taken together these measurements imply:

$$\Gamma(^{12}C + ^{12}C) \cong 75 \text{ keV}$$
$$\Gamma(^{12}C + ^{12}C^*) \cong 125 \text{ keV}$$
$$\Gamma(^{12}C^* + ^{12}C^*) \lesssim 20 \text{ keV}$$
$$\Gamma_R \cong 180 \text{ keV}.$$

Γ_R is the partial width associated with all reaction channels other than the two inelastic channels listed explicitly and corresponds almost entirely to light particle evaporative decay. The two molecular-like channels taken together comprise ∼50% of the total width of this resonance. This number is comparable to or greater than those observed near the Coulomb barrier. Similar results have recently been obtained by Cosman et al.[5] across a very broad energy range indicating that large partial widths for decay into fission-like

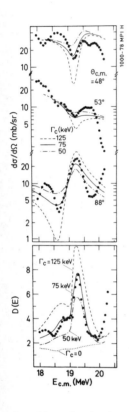

Figure 3. Fits obtained to $^{12}C + ^{12}C$ elastic scattering excitation functions with an optical model plus an isolated $J^\pi = 12^+$ Breit-Wigner resonance. $\Gamma_c \cong 75$ keV is deduced.

channels is a common feature of the narrow resonances.

It is interesting to compare the narrow structures observed in the elastic and inelastic scattering. This is done in Figure 4. Many of the pronounced features of both curves below $E_{cm} \lesssim 25$ MeV are correlated and indeed, for the most part, correspond to resonances previously identified in various reaction channels.

The elastic deviation function has been previously the subject of statistical analysis. The conclusion of such an analysis[6] is that all of the observed structure is consistent with statistical fluctuations although resonance behavior can not be ruled out.

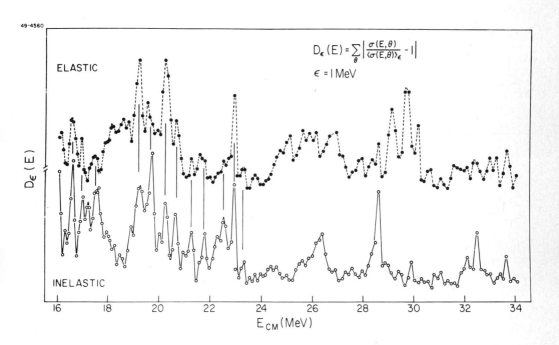

Figure 4. Deviation functions of $^{12}C + ^{12}C$ elastic and 2_1^+ inelastic scattering.

The additional evidence seen here, however, and the correlations with the now very extensive reaction data[5] suggests that statistical fluctuations play a far less important role in $^{12}C + ^{12}C$ elastic scattering than had previously been thought. At energies $\gtrsim 25$ MeV the deviation functions damp down and the correlations are less obvious. This is at least partially due to the difficulty of separating the narrow and gross structures as the widths become more and more similar.

To complete the picture of the two body decays of these resonances we have surveyed the two body channels ^{9}Be + ^{15}O, ^{10}B + ^{14}N, ^{11}B + ^{13}N, ^{10}C + ^{14}C, ^{13}C + ^{11}C across the energy range E_{cm} = 20 to 40 MeV. Throughout this range, <u>only</u> the ^{10}B + ^{14}N channel exhibits any significant resonant behavior. Figure 5 shows an example of these data.

The total cross sections in the ^{10}B + ^{14}N channels are small compared to the inelastic channels but the difference is readily accounted for by the poor penetrability of the ^{10}B + ^{14}N channel. This is intriguing for it suggests, and detailed analysis[7] confirms, that the reduced partial widths for ^{10}B + ^{14}N decay are comparable to those for ^{12}C + ^{12}C decay. The absence of resonances in the other two body channels is not consistent with penetrability arguments alone. In particular, given the observation of strong resonances in the ^{10}B* + ^{14}N channel we would, on the basis of penetrabilities alone, expect resonances in all of the other channels surveyed with intensities up to 5 to 10 times stronger than ^{10}B* + ^{14}N. The experimental observation, however, is that no significant resonances in any other channels are observed.

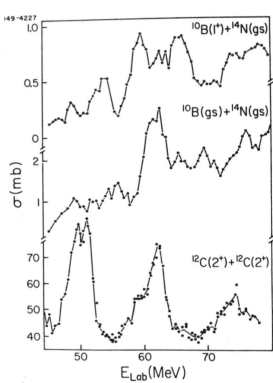

Figure 5. Total cross section excitation functions for two ^{10}B + ^{14}N exit channels compared to the ^{12}C(2^+) + ^{12}C(2^+) channel.

This situation argues strongly against simple statistical decay and for a non-trivial structural effect in the decay of the narrow resonances. The nature of the structural connection is not clear at present, though it seems that the shell structure of the separated fragments can not account for the observed and unobserved decays.

Attempts to understand the ^{24}Mg molecular spectrum working in a simple two-body (i.e. molecular) basis have had marginal success. The calculations of Kondo et al.[8] emphasize the importance of band crossing regions where various zeroth order molecular configurations become degenerate. A band crossing diagram for ^{12}C + ^{12}C is shown in Figure 6. In the energy range of the $J^\pi = 10^+$ and 12^+ resonances the model predicts strong mixing of the ^{12}C + ^{12}C, ^{12}C + ^{12}C(2^+), and ^{12}C(2^+) + ^{12}C(2^+) configurations with higher lying configurations becoming important at higher excitation energies.

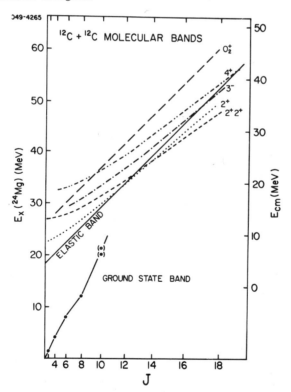

Figure 6. Band crossing diagram for aligned ^{12}C + ^{12}C inelastic bands.

The results of a full coupled channels calculation are shown in Figure 7 and compared with $^{12}C + ^{12}C$ inelastic excitation functions. The model does a reasonable job of describing the average behavior of the inelastic cross sections over the entire energy range but the details of the spectrum of resonances, their widths and the total number of narrow resonance is not even remotely reproduced. Further problems with the model are clearly its inability to account for the observed resonances in mismatched channels such as $^{12}C + ^{12}C(0_2^+)$ or $^{10}B + ^{14}N$ for which band crossing does not occur and the apparent impossibility within a simple two-body basis to account for the selectivity of $^{12}C + ^{12}C$ and $^{10}B + ^{14}N$ channels over all other fission-like channels.

It thus seems that a major theoretical breakthrough is still needed in this field. The fact that of all possible two body channels only $^{12}C + ^{12}C$ and $^{10}B + ^{14}N$ have large reduced widths suggests that the single particle structure of the intermediate state may be of essential importance. It may be more appropriate to begin to consider various microscopic sources of fission-isomeric or other shape-isomeric structures in ^{24}Mg. Some preliminary steps in this direction have already been reported by Chandra and Models[9] and Ragnarsson et al.[9]

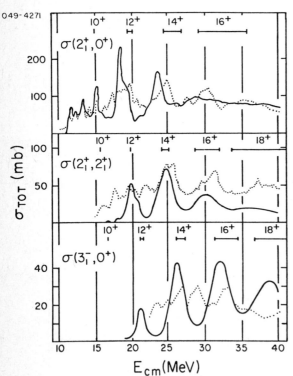

Figure 7. Band crossing model fits to $^{12}C + ^{12}C$ inelastic scattering excitation functions.

An obvious direction for future study would seem to be the direct excitation of the molecular spectrum as the final states in high energy two-body reactions.

Nagatani et al.[11] recently reported the observation of ^{24}Mg molecular states excited directly as final states in the ^{12}C(^{16}O,α)^{24}Mg reaction at E(^{16}O) = 145 MeV. Figure 8 shows some of their α spectra obtained at θ_{lab} = 7, 15, and 40°. An enormous continuum of α particles is observed extending all the way to the ^{24}Mg ground state. On top of this continuum is a series of broad peaks which are seen more clearly after background subtraction. The correspondance of these peaks with the 10^+ through 18^+ gross structure resonances is suggested by the brackets over the peaks. If this speculation were correct it would suggest a number of new exciting ways to probe the molecular structure of nuclei.

In an attempt to verify the hypothesis of Nagatani et al., Rae et al.[12] have recently published the results of a search for the ^{12}C + ^{12}C decay of the states seen in the α spectrum. Their result was negative, no evidence for a ^{12}C + ^{12}C final state interaction was found. Unfortunately, the result is not definative since it is possible that the enormous background yield seen in the α-spectrum could obscure the final state correlation due to the much weaker peaks.

Actually, there was an earlier claim by Lazzarini et al.[13] that the (^{16}O,α) might excite ^{24}Mg molecular states. In these experiments at somewhat lower energies it was suggested that one could

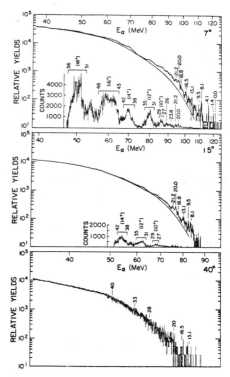

Figure 8. α spectra from the ^{12}C(^{16}O,α)^{24}Mg reaction at E(^{16}O) = 145 MeV.

observe with high resolution the individual intermediate structure fragments. Quite interestingly, when Lazzarini's data are averaged over excitation energy, gross oscillations reminiscent of the ^{24}Mg gross structure resonances are observed.

It was subsequently demonstrated by Branford et al.[14], however, that the structures observed by Lazzarini et al. could not correspond to ^{24}Mg molecular states. This was shown by measuring the total widths of the peaks in the alpha spectrum with very high resolution as summarized in Table 1.

Table 1. Total widths of ^{24}Mg states seen in the ^{12}C(^{16}O,α) reaction

E_x (MeV)	Γ (keV)
22.93	62±13
23.23	35±13
44.37	21±7
25.18	163±6
26.05	< 13
26.45	115±20

The known ^{24}Mg molecular states in this range of excitation energy all have Γ > 200 keV. Thus at least in Lazzarini's experiment, direct excitation of molecular states seems unlikely. But what then of the gross structure which is observed when the α-spectrum is energy averaged?

We have investigated this question at Rochester in collaboration with the University of São Paulo by looking in detail at the properties of energy average α spectra from the ^{12}C(^{16}O,α)^{24}Mg reaction. Before looking at energy averaged α spectra we should remember what a spectrum looks like with good resolution. Figure 9, taken from Greenwood et al.[15] is typical of what one sees at lower bombarding energies. The strong peaks seen here are understood[15] as high spin states populated selectively as a result of strict angular momentum matching constraints in the decay of the ^{28}Si compound nucleus.

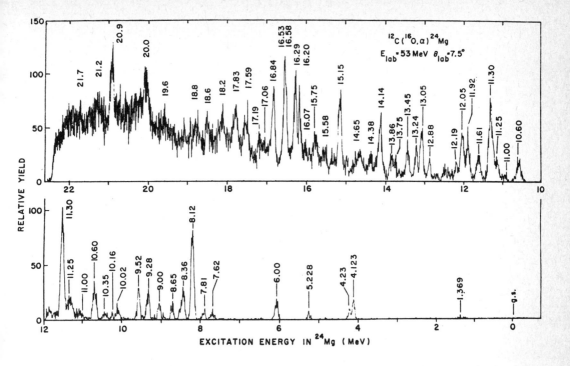

Figure 9. An α spectrum from the $^{12}C(^{16}O,\alpha)^{24}Mg$ reaction at low energy with good resolution.

Now Figure 10 shows energy averaged (ΔE_x = 1 MeV) spectra obtained for nearly the same beam energy as Fig. 9. In order to remove Ericson fluctuations the spectra are integrated over angle and also averaged over the three beam energies shown. After subtracting the smooth background illustrated by the dashed line, a quite striking spectrum consisting of three gross structure features remains.

Figure 11 shows a summary of the available energy averaged α spectra extending from quite low energy to the $E(^{16}O)$ = 145 MeV data of Nagatani et al. A clear pattern of gross structure peaks emerges extending from $E_x(^{24}Mg) \sim$ 12 to 55 MeV with no change whatever in the qualitative characteristics of the peaks.

It seems clear from Branford's work that the broad peaks in the lower half of this energy range are not related to ^{24}Mg molecular structure. We can speculate then that the higher energy data of Nagatani et al. is also not related to these

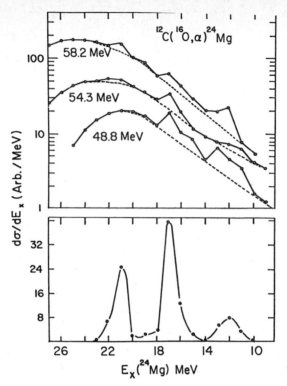

Figure 10. Top: Angle integrated and excitation energy averaged α spectrum.
Bottom: The result of beam energy averaging and background subtraction.

states in spite of the qualitative similarities in the energies of the broad structures seen here and those of the ^{24}Mg gross structure resonances. In any case the broad structures seen in these energy averaged α spectra demand an explanation.

We have devised a simple model based on the high spin selectivity of α particle decay of the ^{28}Si compound nucleus. As a result of the high spin selectivity of this reaction, the cross section for exciting a state of spin J falls very quickly with increasing excitation energy. Thus on the average states closest to the yrast line are most strongly populated. What is remarkable, is that the cross section for a single 8^+ state, say, falls by a factor of ∼10 from $E_x(^{24}\text{Mg}) = 12$ to 18 MeV (for $E(^{16}\text{O}) = 50$ MeV) and thus relatively narrow features might survive

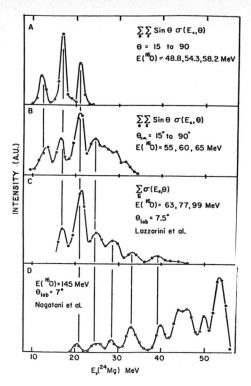

Figure 11. Summary of available energy averaged α spectra.

energy averaging and would correspond to the yrast and near yrast states of spin J. Figure 12 shows an E_x vs $J(J + 1)$ plot for the gross structures assuming a $\Delta J = 1$ sequence compared to an extrapolation of the known ^{24}Mg yrast band. The sequence fits quite well.

We have calculated[16] the energy averaged and background subtracted α spectra expected for compound nucleus decay by devising a level density which explicitly includes the yrast states. Thus we modify the usual Fermi gas level density ρ_{FG} by taking:

$$\rho(E_x, J) = \rho_{FG}(U, J) + \rho_y(E_x, J)$$

where $\quad U = E_x - \Delta$

$$\rho_y = \begin{cases} 2 \text{ levels/MeV} & E_x \geq E_y(J) \\ 0 & E_x < E_y(J) \end{cases}$$

with $\quad E_y(J) = \dfrac{\hbar^2}{2\mathcal{J}} J(J+1).$

Δ is the usual pairing gap and $\mathcal{J}$ is the moment of inertia implied by Fig. 12. The choice of 2 levels/MeV for the density of states near the yrast line is arbitrary but consistent with available data near $J = 8$ to 10.

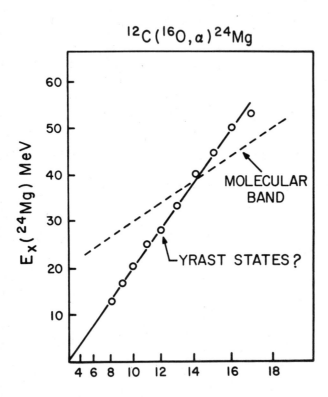

Figure 12. Comparison of the energies of the gross structure peaks in the α spectra with an extrapolation of the ^{24}Mg yrast sequence.

The alpha spectra calculated in the Hauser-Feshbach formalism using this levels density are shown in Figure 13. Remarkably, the simple model discussed here gives a good qualitative account of the observed α spectra over the entire $E(^{16}O)$ bombarding energy range. We conclude, therefore, that it may be possible to understand the new high energy $(^{16}O,\alpha)$ data in the same framework as the older low energy data although it is still not possible to completely rule out some direct excitation of molecular states.

If it turns out that molecular states are not excited in $(^{16}O,\alpha)$ reaction we may still have (as a consolation prize) new insights into the ^{24}Mg yrast line. For example, Figure 12 seems to suggest that the ^{24}Mg yrast line is rigid rotor-like all the way to 17^+, actually it shows that the molecular states become yrast above J = 14. It must be emphasized, of course, that Figure 12 is not unique - no spin assignments have been made.

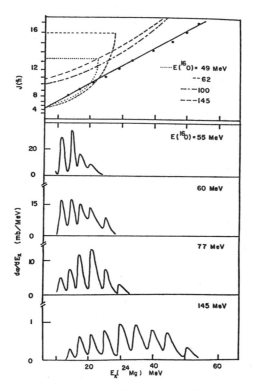

Figure 13. Energy averaged α spectra calculated in the Hauser-Feshbach Formalism for the level density discussed in the text.

In summary, the properties of resonances at energies well above the barrier in the ^{12}C + ^{12}C system are becoming rather well defined. The most significant property of these states is their large reduced widths for decay into various ^{12}C + ^{12}C and ^{10}B + ^{14}N fission-like channels and the absence of such decays in all other heavy two-body channels. This feature is not immediately forthcoming from any of the existing models of ^{24}Mg molecular structure and suggests the need for a microscopic approach.

The possibility of direct excitation of the ^{24}Mg spectrum in the conventional two-body reaction ^{12}C(^{16}O,α) has been examined. It appears that, given certain assumptions concerning the density of states in ^{24}Mg near the yrast line, broad structures may appear in the α spectrum of the ^{12}C(^{16}O,α) reaction which have a trivial origin in the high spin selectivity of compound nucleus α-decay. More thorough study of these high energy α spectra will be necessary to substantiate if direct excitation of molecular states has already been observed.

References

1. P. Sperr, S.E. Vigdor, Y. Eisen, W. Henning, D.G. Kovar, T.R. Ophel and B. Zeidman, Phys. Rev. Lett. 36, 405 (1976).

2. T.M. Cormier, J. Applegate, G.M. Berkowitz, P. Braun-Munzinger, P.M. Cormier, J.W. Harris, C.M. Jachcinski, L.L. Lee, J. Barrette and H.E. Wegner, Phys. Rev. Lett. 38, 940 (1977).

3. T.M. Cormier, C.M. Jachcinski, G.M. Berkowitz, P. Braun-Munzinger, P.M. Cormier, M. Gai, J.W. Harris, J. Barrette and H.E. Wegner, Phys. Rev. Lett. 40, 924 (1978).

4. B.R. Fulton, T.M. Cormier and B.J. Herman, Phys. Rev. C21, 198 (1980).

5. E.R. Cosman, R. Ledoux and A.J. Lazzarini, Phys. Rev. C21, 2111 (1980).

6. D. Shapira, R.G. Stokstad and D.A. Bromley, Phys. Rev. C10, 1063 (1974).

7. T.M. Cormier and B.R. Fulton, Phys. Rev. C22, 565 (1980).

8. Y. Kondo, Y. Abe and T. Matsuse, Phys. Rev. C19, 1356 (1979). See also Ref. 4.

9. H. Chandra and U. Mosel, Nucl. Phys. A298, 151 (1978).

10. I. Ragnarsson, S. Åberg and R.K. Sheline, Nobel Symposium 50, Nuclei at Very High Spin - Sven Gösta Nilsson in Memorium, Örenäs, Sweden, June 23-27, 1980.

11. K. Nagatani, T. Shimoda, D. Tanner, R. Tribble and T. Yamaya, Phys. Rev. Lett. 43 1480 (1979).

12. W.D. Rae, R.G. Stokstad, B.G. Harvey, A. Dacal, R. Legrain, J. Mahoney, M.J. Murphey and T.J.M. Symons, Phys. Rev. Lett. 45, 884 (1980).

13. A.J. Lazzarini, E.R. Cosman, A. Sperduto, S.G. Steadman, W. Thoms and G.R. Young, Phys. Rev. Lett. 40, 1426 (1978).

14. D. Branford, M.J. Levine, J. Barrette and S. Kubono, Phys. Rev. C23, 549 (1981).

15. L.R. Greenwood, K. Katori, R.E. Malmin, T.H. Braid, J.C. Stoltzfus and R.H. Siemssen, Phys. Rev. C6, 2112 (1972).

16. A. Szanto de Toledo, M.M. Coimbra, N. Carlin-Filho, T.M. Cormier and P.M. Stwertka, Phys. Rev. Lett. 47, 632 (1981).

THE SPINS AND SPECTROSCOPY OF $^{12}C + ^{12}C$ INTERMEDIATE STRUCTURE RESONANCES

E.R. Cosman, R.J. Ledoux, M.J. Bechara,[*] C.E. Ordonez, and H.A. Al-Juwair

Department of Physics and Laboratory for Nuclear Science
Massachusetts Institute of Technology
Cambridge MA 02139, USA

In this paper, the systematics of $^{12}C + ^{12}C$ resonances will be discussed. A review of known resonance positions and spins will be given, and new data from the MIT group on the $^{12}C + ^{12}C$ elastic and $^{12}C(^{12}C,\alpha)^{20}Ne$ reactions will be reported which add some important spectroscopic information on the nature of these resonances. It is suggested that all the intermediate structures from the Coulomb barrier to much higher energies have a common origin, and that a plausible explanation of them is as shape-isomeric states in ^{24}Mg derived from the deformed shell model.

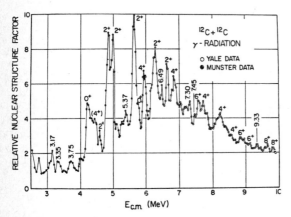

Fig. 1: $^{12}C + ^{12}C$ nuclear structure factor near the Coulomb barrier.

Let us begin with the data near the Coulomb barrier shown in Fig. 1 [1]. The $^{12}C + ^{12}C$ resonances were first discovered in this region by Bromley and co-workers at Chalk River [2]. It is clear that these anomalies are real resonances because they show strong cross correlations among the few open p, n, and α reaction channels. It is a remarkable feature of these states that they have such large ^{12}C widths, and that there are so few of them in view of the high density of the $J^{\pi}=0^{+}$, 2^{+}, and 4^{+} levels in ^{24}Mg that should exist at these excitation energies. We also emphasize a relatively unnoticed feature of these data that there appears to be a threshold at $E(CM) \simeq 4.3$ MeV or $E_x(^{24}Mg) \simeq 18.2$ MeV where the strong 0^{+} and 2^{+} states begin. The nuclear structure factor in Fig. 1 is derived by dividing the cross section by a sum of weighted penetrabilities, and thus is dependent on the radius parameter used [3]. However, the strong 0^{+} and 2^{+} states from 4 to 6 MeV have peak-to-background ratios of 2 or 3 to 1 compared to the peaks below 4 MeV which are typically 1 to 1. Thus, a variation of the smooth penetration factor curve should not alter the appearance of an onset of strength at 4.2 MeV. We will make the argument later in this paper that this threshold is a manifestation of a rotational band head and that these $^{12}C + ^{12}C$ resonances at the Coulomb barrier are simply the lowest spin members of a sequence of resonances that persist to energies well above the barrier, all of them having a common nuclear structure origin.

Fig. 2 shows the $^{12}C(^{12}C,p)^{23}Na$ reaction data from our group at MIT [4,5] which showed for the first time that prominent, relatively narrow and isolated resonances with rather high spins exist in the $^{12}C + ^{12}C$ system at energies well above the Coulomb barrier. Here, many more channels are open. The resonances were identified by following excitation functions of unusual high spin states in ^{23}Na and by demonstrating cross correlations among them and with other d and α reaction channels. There was evidence that the states near E(CM)=11.3, 14.3, 19.3, and 25 MeV have spins of $J^\pi = 8^+$, 10^+, 12^+, and 14^+, respectively, and it was argued [5] that they form a rotational band with projected band head at E(CM)≃18 MeV. Subsequently, great amounts of p, n, α, ^8Be, elastic, inelastic, and total reaction data have been measured in this region, and more candidates for resonances have been found. We will summarize these results here and attempt to locate the positions of cross channel correlations.

Fig. 3 shows several $^{12}C + ^{12}C$ reaction channels: the 90° CM elastic from Shapira, et al [6]; the summed 7.5° Lab proton transitions to $11/2^+$, $13/2^+$, $15/2^+$, and $17/2^+$ states in ^{23}Na (see Fig. 2) from Cosman, et al [5]; the angle-integrated alpha yields

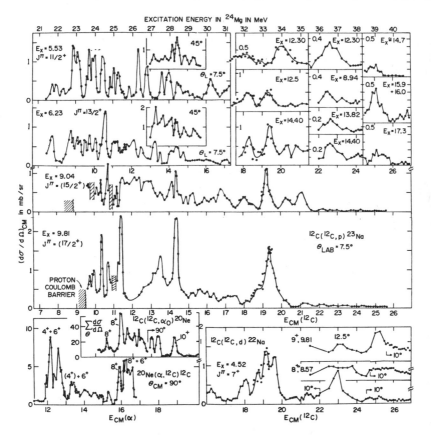

Fig. 2: Selected excitation functions for $^{12}C + ^{12}C$ p, d, and α reactions.

to ^{20}Ne from Voit, et al [7] (on the left) and Kolata, et al [15] (on the right); and angle-integrated ^{8}Be yields to ^{16}O(g.s.) from James, et al [8]. The J^π values above the alpha curve are from ^{12}C(^{12}C,α_o)^{20}Ne(g.s.) angular distributions of Borggreen, et al [9] and Voit, et al [7], and those above the ^{8}Be curve are from ^{12}C(^{12}C,^{8}Be)^{16}O (g.s.) angular distributions. These and other data are shown in Fig. 4, including: (on the left) the sum of 3.75° Lab alphas to the excited states in ^{20}Ne at E_x = 6.72 MeV(0^+), 7.20 MeV(0^+), 7.42 MeV(2^+), 7.83 MeV(2^+), and 9.03 MeV(4^+) from Middleton, et al [10]; 90° CM elastic from Emling, et al [11]; angle-integrated ($0^+,2^+$) inelastic from [11]; (on the right) 5° Lab summed alphas to ^{20}Ne from Greenwood, et al [12]; and further ($0^+,2^+$) inelastic data from [11]. The dashed lines in Fig. 4 indicate positions where a visual inspection of the graphs shows evidence for correlated maxima. Including some lower states from Fig. 3, these correspond to the following values of E(CM), in MeV, and J^π (estimated from Fig. 2): 7.8(4^+), 8.2(4^+), 8.9(6^+), 9.1(6^+), 9.8(8^+), 9.9(8^+), 10.4(8^+ + 6^+), 10.9(8^+), 11.4(8^+), 12.1(8^+), 13.1(8^+), 13.5(10^+), 14.3(10^+), (14.8(?)), (15.2(10^+)), 16.1-16.3((10^+)), (17.0(?)), (17.4(12^+)), 18.3((12^+)), 19.3(12^+), 20.3((14^+)), 21.0(?). Below

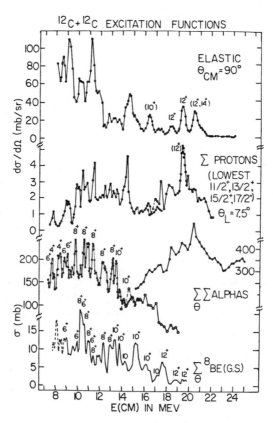

Fig. 3: Comparison of ^{12}C + ^{12}C elastic and reaction data. Spin assignments discussed in text.

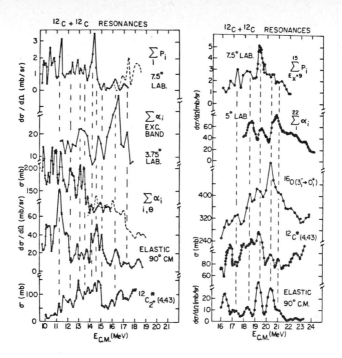

Fig. 4: Comparison of $^{12}C + ^{12}C$ resonances in various reaction channels. Dashed lines indicate apparent correlations.

E(CM)=15 MeV, these correspond almost identically to positions claimed by Treu, et al [7] for correlations in their alpha spectra. From E(CM)≃16 to 22 MeV, excellent correlations occur in most channels.

Clearly, in the region from E(CM)≃8 to 18 MeV, there appear to be resonances almost everywhere. To prove this, one requires better data in most channels than are shown in Figs. 3 and 4, but in cases where very complete angle-integrated data have been taken, such as for $^{12}C(^{12}C,\alpha)^{20}Ne$ between E(CM)=7 and 15 MeV by Voit, et al [7], this is borne out. It is also indicated in the elastic, inelastic, and alpha data of Fig. 4 that additional finer structures may exist within the groups cited above. Thus, as in the barrier region, the reaction cross sections in this region show a wealth of real resonances that do not appear dominated by fluctuations.

We have done cross correlation analysis of the data in Figs. 3 and 4, and these are shown in Fig. 5. The proton-alpha case shows strong positive correlations near most of the positions listed above. The p-^{8}Be and α-^{8}Be cases though are not as clear in the regions of the claimed resonances. There are frequently large excursions, either positive or negative. This could easily be the result of background interference which can give a resonance a fore-aft asymmetry and thus shift its maximum. This is especially true of the elastic-alpha correlation between E(CM)=8 to 15 MeV

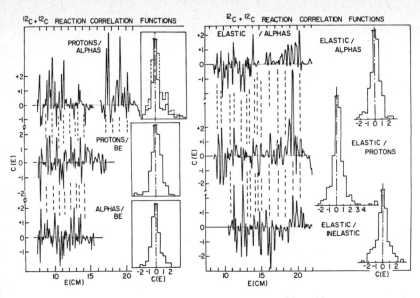

Fig. 5: Correlation functions between $^{12}C + ^{12}C$ reaction channels. Averaging width was 1 MeV.

where it appears more negative near the resonances. Of course, the Coulomb background is very large compared to the resonances. Such "negative correlations" are in fact consistent with the analysis of Frohlich, et al [13] in this conference. Note, however, that at higher energies, E(CM)=16 to 21 MeV, the situation changes markedly, and the elastic-alpha correlation becomes quite positive. This reflects the larger peak-to-background ratio there. No comparison to predictions of the statistical model and Hauser-Feshbach calculations has been made for these cross correlations.

We have continued the search to higher energies, and in a previous paper [14] have shown the correlation among intermediate structures in the elastic, inelastic, summed alpha, and total reaction channels. These resonances are many, as in the lower region, but they are sufficiently narrow and sparse in density as to be individually identifiable. New $^{12}C(^{12}C,\alpha)^{20}Ne$ data shown in Fig. 6 demonstrate this, but these data also show some surprising final state selectivity which we will discuss below. As usual, such single-angle data for a given transition do not easily show the structure; however, when summed over all the states, it does. Fig. 7 compares the sum of data in Fig. 6 to the total alpha yields ($^{20}Ne-\gamma$ and $^{16}O(3^- \rightarrow 0^+)$) of Kolata, et al [15] and the 90° CM elastic data of Shapira, et al [6]. First, the single- and summed-angle data agree, proving that the former faithfully represent the resonances. Then, there is a clear, visual correlation of positions where both the alpha and 90° CM elastic data show prominent intermediate structures especially near E(CM)≃20, 25, and 30 MeV. Correlation functions of the elastic with Kolata's data, Fig. 8, and the MIT

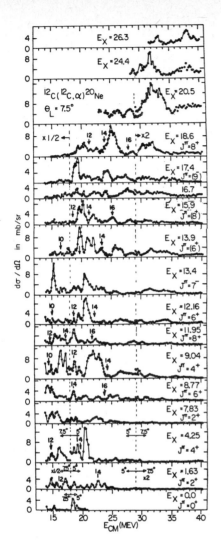

Fig. 6: Selected excitation functions for the $^{12}C(^{12}C,\alpha)^{20}Ne$ reaction. Numbered arrows indicate locations of the ℓ-grazing condition.

data, Fig. 9, show strongly positive correlations as reported in [14].

In view of these results, we attempted in ref. [14] to reproduce the 90° CM elastic yield with an optical model background plus Breit-Wigner resonances of assumed spins. The result in Fig. 10 is encouraging. However, the optical potential used was derived from fitting the average elastic data, and that may be incorrect for use when the resonances are explicitly separated out. Thus, we decided to measure complete angular distribution excitation functions from E(CM)=14 to 40 MeV and fit them with a phase shift analysis which might yield more accurate background and resonance spins, widths, and phases. Part of the data in the range E(CM)$\approx$20 MeV is shown in Fig. 11. The phase shift fitting program consists of a gradient search on the phase shifts η_ℓ and δ_ℓ to minimize the χ^2 fit to the angular distributions and the total reaction cross section as measured by Kolata, et al [15]. To arrive at an acceptable solution, we also required that only one ℓ-wave be resonant at each anomaly, that the resonance wave describes a good Argand phase diagram, and that other ℓ-waves be smoothly varying over the anomaly. One further must assume some reasonable starting conditions on η_ℓ and δ_ℓ. Full details of this procedure will be published elsewhere [16]. Despite these constraints, the general difficulty with such a procedure is the uniqueness of the extracted solution in view of the large number of free parameters. As we will show in the region of the E(CM)=20.3 MeV resonance, it is possible to achieve fits for $J^\pi=12^+$ or 14^+ for this state with associated differences in background phase shifts. On the other hand, in the case of the E(CM)=19.3 MeV resonance, our confidence in the $J^\pi=12^+$ fit is much greater, since one cannot achieve a good fit with $J^\pi=14^+$ without forcing the neighboring ℓ-values to vary rapidly. In neither case is a fit with $J^\pi \neq 12^+$ or 14^+ possible.

The results of two solutions which were found will be given here. In the first

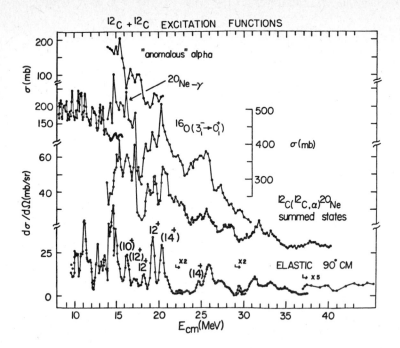

Fig. 7: Comparison of various alpha and the 90° CM elastic yields. J^{π} values discussed in text.

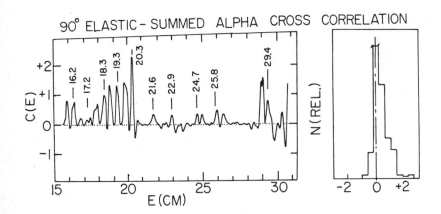

Fig. 8: Correlation function of the 90° CM elastic and summed-alpha yields.

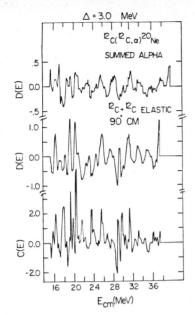

Fig. 9: Deviation and correlation functions for the elastic (ref. [6]) and summed alpha (ref. [16]) reaction yields.

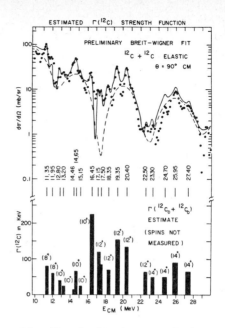

Fig. 10: Breit-Wigner analysis of the 90° CM elastic data.

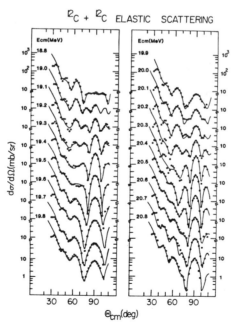

Fig. 11: $^{12}C + ^{12}C$ elastic angular distributions with phase shift analysis.

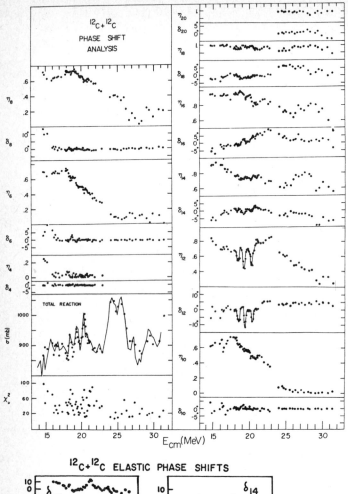

Fig. 12: One set of phase shift parameters obtained from an analysis of elastic scattering. Also shown is fit to the total reaction cross section (solid curve).

NOTE: $S_\ell = \eta_\ell e^{2i\delta_\ell}$ = nuclear elastic S-matrix element.

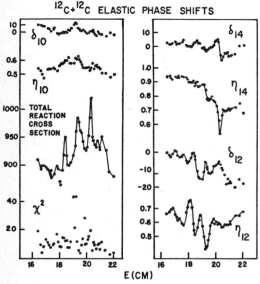

Fig. 13: Alternate set of phase shift parameters obtained by allowing large phases in the $\ell=6$ and 8 partial waves.

case, we started by having η_ℓ for $\ell=4$ to 18 vary according to the sharp cut-off model ($\eta_\ell=0$ for $\ell=0$ and 2, and were not varied), and with $\delta_\ell=0$ for all waves. This resulted in the solution shown in Fig. 12 and the solid line fits of Fig. 11. As can be seen, the reflection coefficient η_{12} has dips at the E(CM)=18.5, 19.3, and 20.3 MeV resonances with proper resonant form for δ_{12} also. There is also a hint of a resonance in η_{10} at 16.5 MeV, and both the η_{14} and η_{16} contribute near the doublet at 25 MeV. In this case, the phases δ_ℓ for $\ell=6$ and 8 remain near zero during the χ^2 search.

As a second solution, we started with phase shifts which were found by a similar analysis by Emling, et al [11] at lower energies (they were not trying to fit resonances, but merely the average background). In this solution, the initial conditions for the phase shift parameters are similar to those that would be obtained with a strongly absorbing potential. The phases are allowed to start at, or attain large values. The values of δ_ℓ for $\ell=6$ and 8 were in the range $50° \le \delta_\ell \le 110°$ for E(CM) > 14 MeV. The resulting best fit parameters for $\ell \ge 10$ are shown in Fig. 13 for the region near E(CM)=20 MeV. Again, the E(CM)=18.5 and 19.3 MeV states emerge with $J^\pi=12^+$, but the 20.3 MeV state is fit with a 14^+. The χ^2 also is about a factor of three better in the second case. Independent decay selection rule information would favor a 12^+ and 14^+ assignment at E(CM)=19.3 and 20.3 MeV, respectively. The 19.3 MeV case is discussed in ref. [5]. For the 20.3 MeV case, we note that it has no decay to the 9.81 MeV($17/2^+$) state in ^{23}Na, no decay to the 4.45 MeV(7^+) state in ^{22}Na, and decays strongly to the 4.25 MeV(4^+), 12.16 MeV(6^+), and other $J^\pi \ge 4^+$ states in ^{20}Ne, but only weakly to the 1.63 MeV(2^+) and 7.83 MeV(2^+) states in ^{20}Ne, which on the basis of penetration factors would favor its spin being $J^\pi \ge 14^+$. From the above considerations, we have indicated the J^π values which are most likely in Fig. 7. We are in the process of taking more elastic and alpha data with the objective of further illuminating these questions of spin.

We now turn to the implications of the above results to some model predictions. Fig. 14 shows a schematic diagram of possible processes that could explain the existence of the resonances. To be sure, these models may have some common features and use different languages to describe the same physics. That is, to some degree, they may be consistent with each other. On the other hand, there may be aspects of the data which can discriminate between the models and make it more natural to favor one of these pictures over the others. To date, most attention has been paid to "entrance and exit channel models" to explain the $^{12}C + ^{12}C$ gross and intermediate structures. In particular, the potential shape resonance models of Arima, et al [17], the band-crossing or double resonance models of Nogami and Imanishi [18], Abe, Kondo, and Matsuse [19], and Fink, Scheid, and Greiner [20], the two-center shell model calculations of Chandra and Mosel [21] reproduce certain aspects of the data, but not all. The origin of the narrow (∼ 100 to 500 keV) intermediate structures is still a puzzle. Feshbach [22] proposed the doorway state model as a framework for understanding it, but the specific physics is still not clear. We should like to propose here

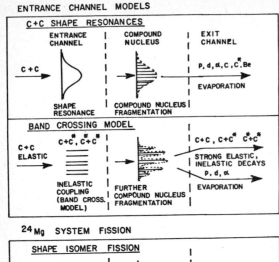
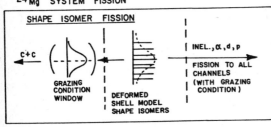
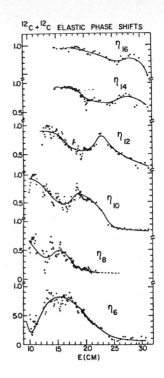

Fig. 14: Schematic representation of possible $^{12}C + ^{12}C$ reaction mechanisms.

Fig. 15: Reflection coefficients obtained from phase shift analysis of the $^{12}C + ^{12}C$ elastic scattering.

that the phenomena in $^{12}C + ^{12}C$ and other resonant heavy ion systems may best be understood in terms of the deformed shell model for ^{24}Mg, and that the intermediate structures and decays can be understood as fission from shape-isomeric states in a secondary minimum of the ^{24}Mg potential energy surface.

We shall now run through a series of observations about the data which bear on these model predictions. In Fig. 15, we show the $^{12}C + ^{12}C$ elastic reflection coefficients from our work (above $E(CM) \simeq 16$ to 18 MeV) and continue to lower energies from Emling's phase shift analysis [11]. Notwithstanding the ambiguities discussed above, we are optimistic that the average trends of η_ℓ will be an invariant. If so, we demonstrate from Fig. 15 that each ℓ-wave has a 2 to 4 MeV-wide region of enhanced absorption. Such a gross structure is common to all the above models, but it shows that real resonant effects are occurring over a localized range. Of course, our quest is to show the unique separation of each η_ℓ gross structure into individual intermediate structures with 100 to 500 keV widths.

As we have seen, this is more difficult. An interesting aspect of Fig. 15 is that we may also find the background phases which may lead to insight into the actual

$^{12}C+^{12}C$ interaction potential. That may be crucial to our understanding of the overall problem. We note in passing that each η_ℓ in Fig. 15 shows a far less steep fall toward absorption than the standard optical potentials for $^{12}C+^{12}C$ predict, which suggests that previous fits to elastic scattering may not be physical.

The assignment of $J^\pi=14^+$ versus 12^+ for the strong E(CM)=20.3 MeV resonance is a crucial one. If our favored value of 14^+ proves to be true, one must reconsider the simple picture of a strength function grouping of states of the same J^π. It would also be outside the band-crossing model predictions. If we plot our best estimates of resonance energies versus $J(J+1)$, as in Fig. 16, we see a clear trend from the states at the Coulomb barrier up to higher energies. This seems to us most suggestive that the intermediate structures have the same origin throughout. The predicted positions of $^{12}C+^{12}C$ potential shape resonances [23] and tri-axial ^{24}Mg shape isomers [24] are also shown in Fig. 16. They both are near grazing condition for $^{12}C+^{12}C$, which is not surprising since both involve resonances near the barrier top for each ℓ-wave. It is notable, however, that the deformed shell model does predict such states per se, and as Chandra and Mosel [21] have proven, this axially asymmetric solution should have strong

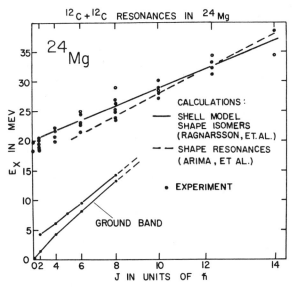

Fig. 16: Spectrum of known $^{12}C+^{12}C$ intermediate structures as compared to shape resonance and shape isomer predictions.

decays to $^{12}C+^{12}C$, $^{12}C+^{12}C^*(2^+)$, and $^{12}C^*(2^+)+^{12}C^*(2^+)$, consistent with the inelastic data of Cormier, et al [25] and the elastic width estimates of Cosman, et al [14].

Our new information from the $^{12}C(^{12}C,\alpha)^{20}Ne$ reaction may play a central role in discriminating between models. Notice in Fig. 6 that the preponderance of alpha yields from the resonances go to excited rotational band levels in ^{20}Ne, once the latter are above threshold. These decays are highly non-statistical, since if they followed penetration factors, they should strongly favor decays to lower excited states of a given J^π (see ref. [26]). Indeed, comparing the reduced alpha widths for the states of the excited bands to those of the ground band for the 19.3 and 20.3 MeV resonances, the former are from 10 to 100 times stronger than the latter [16]. This would indicate a structural connection between the resonances and the excited rotational band of ^{20}Ne. Moreover, the ratios of the reduced widths to the Wigner limit for the non-statistical alpha decays are comparable to those for the $^{12}C + ^{12}C$ elastic and inelastic channels, indicating that the former are as significant as the latter in the full description of the ^{24}Mg states. The positions of the grazing condition ℓ-values are indicated in Fig. 6, and these suggest that a necessary condition for an alpha transition to be prominent over the resonances is that the grazing condition is approximated. However, there are examples in the data of transitions which are not prominent on resonance (not shown in Fig.6) for which the grazing condition is met. Thus, it is not a sufficient condition for prominence, suggesting that other structural selectivity is present.

A possible explanation for the non-statistical alpha decays can be found in the deformed shell model calculations of Leander and Larsson [27]. Their calculated potential energy surfaces for ^{24}Mg and ^{20}Ne are shown in Fig. 17. The axially asymmetric

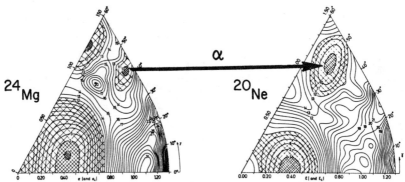

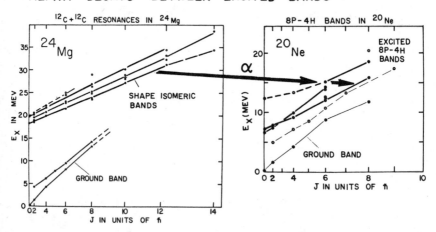

Fig. 17: Results of the deformed shell model potential energy surface calculations [27]. Arrows indicate possible selective alpha decays from shape-isomeric states in ^{24}Mg to those in ^{20}Ne.

secondary minima are shown with the alpha decay arrow between them. States in this ^{24}Mg minimum are the ones which should best be populated by the ^{12}C +^{12}C entrance channel [21]. States in this minimum in ^{20}Ne are predicted [27] to include the 8p-4h excited bands with band heads at E_x=6.72 and 7.20 MeV. These bands include the prominent states in Fig. 6. Further, these ^{24}Mg and ^{20}Ne states differ by a (2p,2n) configuration [27] which should make the alpha transition between them an enhanced one. Thus, the model of alpha particle decays between shape-isomeric bands provides a plausible explanation of the data.

On a related point, Hindi, et al [28], have shown that some of the states in ^{20}Ne* in Fig. 1 which we find have the most dramatic resonances, viz., E_x=15.9, 18.6, 20.5, and 24.4 MeV, also decay strongly to α +^{16}O*($3^-,0^+$) and ^{12}C +^{8}Be. Thus, there appears to be a family of enhanced decays related by alpha-like clusters as illustrated in Fig. 18. This suggests that an alpha-cluster expansion of these states, as proposed by Michaud and Vogt [29], and Arima, et al [17], may be possible.

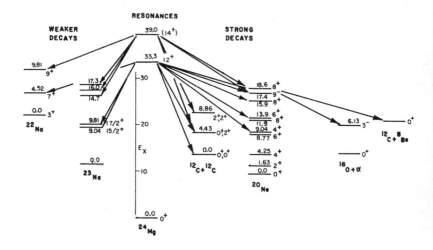

Fig. 18: Decay scheme of ^{12}C +^{12}C intermediate structures and subsequent decays to alpha-cluster states.

Demonstration of underlying patterns in the ^{12}C +^{12}C resonance spectra would be a strong discriminator between models. We have previously proposed a scheme of rotational band multiplets [30] and shall mention this again here. Fig. 19 shows the positions of the strongest resonances at the Coulomb barrier. Allowing for some

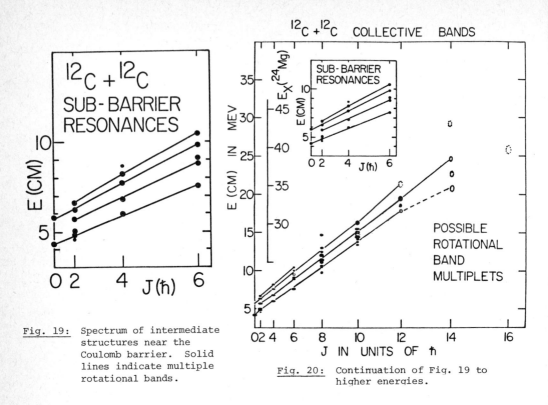

Fig. 19: Spectrum of intermediate structures near the Coulomb barrier. Solid lines indicate multiple rotational bands.

Fig. 20: Continuation of Fig. 19 to higher energies.

fine structure on top of the intermediate structures, a pattern of multiple rotational bands fits the data remarkably well. Extending this scheme to higher energies in Fig. 20, the general trend of groups of states continues, but a more dramatically increased number of $J^\pi = 8^+$ and 10^+ states makes a clear connection difficult. Definitive location of 12^+ and 14^+ resonances would be most important in proving such a scheme.

The pattern at the Coulomb barrier appears to be consistent with a rotational-vibrational spectrum, the 0^+ state at E(CM)=4.2 MeV being the head of a $K^\pi = 0^+$ band, and the 2^+ state at 4.9 MeV being the head of a $K^\pi = 2^+$ γ-vibrational band. The second 0^+ state near 5.7 MeV may be a β-vibrational band head. It would be most interesting to study the deformed shell model predictions further to determine what vibrations or particle excitations of the intrinsic collective structures are possible, and whether the widths observed for them are reasonable.

We would like to propose another mechanism by which J^π multiplets may arise -- namely, that they are the natural spectrum of a rotor with equilibrium axial asymmetry [31]. As noted above, one of the ^{24}Mg shape-isomers which should be seen in ^{12}C + ^{12}C reactions is the axial asymmetric solution, so that we must expect to see J^π multiplets. In this case, the number of states for a given J^π increases with J, but since K is limited in the entrance channel to values near K = 0, not all states should be

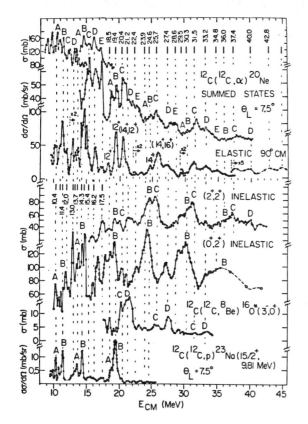

Fig. 21: Several reaction excitation functions at high excitation energies.

seen. This may account for all or part of the multiplet structure seen here, and is an alternative view to that of simple fragmentation which appears to increase with increasing J^{π}. It is interesting to speculate that if tri-axial shape isomers are a more universal phenomenon, then they may be expected to be seen as heavy ion resonances even in the region of medium and heavy nuclei. J^{π} multiplets may also be a natural feature of such resonance spectra and could explain the large numbers of resonances seen in $^{12}C + ^{16}O$, $^{16}O + ^{16}O$, $^{16}O + ^{28}Si$, $^{12}C + ^{32}S$, $^{28}Si + ^{28}Si$, etc.

Finally, the $^{12}C + ^{12}C$ reaction spectra at higher energies show strong selectivity and should provide clues to the underlying structures. To illustrate this, we show in Fig. 21 a collection of reaction channel data: alphas from Voit, et al [7], Kolata, et al [15], and the present work; elastic data from Emling, et al [11], and Shapira, et al [6]; inelastic data from Cormier, et al [25], and Emling, et al [11]; $^{8}Be + ^{16}O^{*}$ data from Weidinger, et al [32]; and proton data from Cosman, et al [5]. There are hints throughout the data of groupings of levels, repeating patterns within groups, and differential selectivity as a function of channel. We have attempted to note some of these by the lettering of the peaks. Note, as an example, that the elastic, $^{12}C(0^+) + ^{12}C^*(2^+)$, and $p + ^{23}Na^*$ structures are almost anti-correlated with those of $^{8}Be + ^{16}O^*(3^-,0^+)$. These may signal some intrinsic difference among ^{24}Mg states such as symmetry or shape and that decays may be sensitive to the configurational similarities to final channels.

Work supported in part by the U.S. Department of Energy, Contract No. DE-AC02-76-ER03069.

*Permanent address - Institute de Fisica, Universidade de São Paulo, C.P. 20516, Brazil. Partially supported by FAPESP-Brazil.

REFERENCES

[1] Taken from a preprint of K. Erb and D.A. Bromley, and a contribution by these authors to the conference.
[2] D.A. Bromley, et al, Phys. Rev. Lett $\underline{4}$, 365 (1960).
[3] K. Erb, private communication.
[4] E.R. Cosman, et al, in Proccedings of the International Conference on Nuclear Physics, Munich, 1973, edited by S. de Boerand and H.J. Mang (North Holland, Amsterdam), p. 542; and, K. Van Bibber, et al, Phys. Rev. Lett. $\underline{32}$, 687 (1974).
[5] E.R. Cosman, et al, Phys. Rev. Lett. $\underline{35}$, 265 (1975).
[6] D. Shapira, et al, Phys. Rev. $\underline{C10}$, 1063 (1974).
[7] H. Voit, et al, Phys. Lett. $\underline{67B}$, 399 (1977); W. Treu, et al, Phys. Rev. $\underline{C18}$, 2148 (1978); W. Galster, et al, Phys. Rev. $\underline{C22}$, 515 (1980).
[8] D.R. James and N.R. Fletcher, Phys. Rev. $\underline{C12}$, 2248 (1978).
[9] J. Borggreen, et al, K. Dan. Viden. Selsk, Mat. Fys. Medd. $\underline{34}$, No. 9 (1964).
[10] R. Middleton, et al, Journal de Physique $\underline{C6}$, 6 (1971).
[11] H. Emling, et al, Nucl. Phys. $\underline{A211}$, 600 (1973); H. Emling, et al, Nucl. Phys. $\underline{A239}$, 172 (1975).
[12] L.R. Greenwood, et al, Phys. Rev. $\underline{C17}$, 156 (1975).
[13] H. Frohlich, et al, invited paper to this conference.
[14] E.R. Cosman, et al, Phys. Rev. $\underline{C21}$, 2111 (1980).
[15] J.J. Kolata, et al, Phys. Rev. $\underline{C21}$, 579 (1980).
[16] R.J. Ledoux, E.R. Cosman, M.J. Bechara, C.E. Ordoñez, H.A. Al-Juwair, and R. Valicenti, submitted to Phys. Rev.
[17] A. Arima, et al, Phys. Rev. Lett. $\underline{25}$, 1043 (1970).
[18] M. Nogami, unpublished; and B. Imanishi, Nucl. Phys. $\underline{A125}$, 33 (1969).
[19] Y. Abe, et al, Prog. Theor. Phys. $\underline{59}$, 1393 (1978).
[20] H.J. Fink, et al, Nucl. Phys. $\underline{A188}$, 259 (1972).
[21] H. Chandra and U. Mosel, Nucl. Phys. $\underline{A298}$, 151 (1978).
[22] H. Feshbach, J. Phys. Colloq. $\underline{37C5}$, 177 (1976).
[23] A. Arima, et al, Phys. Rev. Lett. $\underline{40B}$, 7 (1972).
[24] I. Ragnarsson, et al, Contribution to Nobel Symposium 50, Nuclei at Very High Spin, LUND-MPH-80/19, 1980.
[25] T.M. Cormier, et al, Phys. Rev. Lett $\underline{40}$, 924 (1978).
[26] L.R. Greenwood, et al, Phys. Rev. $\underline{C12}$, 156 (1975).
[27] G. Leander and S.E. Larsson, Nucl. Phys. $\underline{A239}$, 93 (1975).
[28] M.M. Hindi, et al, preprint, Yale-3074-579 (1980).
[29] G.J. Michaud and E.W. Vogt, Phys. Rev. $\underline{C5}$, 350 (1972).
[30] E.R. Cosman, et al, MIT preprint, unpublished; and, BAPS (Baltimore Meeting) Vol $\underline{26}$, No. 4, 610 (1981).
[31] A. Bohr and B. Mottelson, Nuclear Structure, Vol II, (W.A. Benjamin, Reading), pp. 175-198 (1975).
[32] A. Weidinger, et al, Nucl. Phys. $\underline{A257}$, 144 (1976).

DIRECT OBSERVATION OF ^{12}C-^{12}C CONFIGURATION STATES IN THEIR ^{12}C DECAY

K.KATORI
Laboratory of Nuclear Studies, Osaka University, Toyonaka, Osaka, 560
and
K.FURUNO, J.SCHIMIZU, Y.NAGASHIMA, S.HANASHIMA and M.SATO
Institute of Physics and Tandem Accelerator Center, University of Tsukuba, Ibaraki, 305

Natural C target of 112 μg/cm^2 thickness was bombarded with ^{16}O^{6+} ions at $E_{lab}(^{16}O)$=70 MeV using the 12UD Pelletron Tandem Accelerator at University of Tsukuba. As for the α-detector in the ^{12}C(^{16}O,α)^{24}Mg reaction, a conventional counter telescope was used and placed at θ_{lab}=4.5±2.0° with a solid angle of 4.4 msr. ^{12}C particles coincident with the α-particles were detected by using the magnetic spectrograph with a solid angle of 3.0 msr for better identification of low energy ^{12}C. ^{12}C^{6+} particles were focused on the gas-proportional focal plane detector. The spectrograph was placed at θ_{lab}=6°,7°,8° and 9°.

Results were drawn in Dalitz-plot representation between energies of α and those of ^{12}C ions. Two loci are clearly observed. According to three-body kinematics, the outer locus corresponds to the ^{12}C(gs)-^{12}C(gs)-α process in the three-body final states with Q=-7.16 MeV, while the inner locus corresponds to that of ^{12}C(2+)-^{12}C(gs)-α with Q=-11.59 MeV. The spectra projected on the α-energy axis are shown in Fig.1 for each process. Little continuous background was observed in both spectra.

In order to confirm that such peaks are really related to the states in ^{24}Mg, angular dependence for the peak energy position was studied. The largest peak in the ^{12}C(gs)-^{12}C(gs)-α process shifted to lower α-energy as the detector angle moved backward. Thus, this peak does not correspond to a state in ^{24}Mg, but identified as coming from the ^{12}C(gs)-^{16}O(4+,10.35 MeV) process decaying to the same three-body final state. On the otherhand, since (1) two small peaks stand sharply,(2) the energy position does not shift as the angle changes and (3) relevant excited states of ^{16}O emitting α-particles to the ground state of ^{12}C are rare in the present limitted energy range, those two peaks can be identified as states of ^{24}Mg, whose excitation energies are E_x=28.7±0.25 and 29.3±0.25 MeV. The total width was extracted to be Γ=400 keV in the center-of-mass system as shown in Fig.1 by dotted lines. This width included experimental spread of 310 keV (FWHM).

For the ^{12}C(2+)-^{12}C(gs)-α process, we summed the yield over the angles measured. Although peaks were not statistically significant, three peaks could be identified and their excitation energies could be obtained at E_x=30.1±0.3, 31.1±

0.3 and 32.1±0.3 MeV. The total widths of those peaks could be extracted to be 700keV.

The summed spectrum was compared with that obtained by subtracting reasonably smooth continuous background from the single α-spectrum measured at the same angle of θ=4.5°. Peak positions obtained from the ^{12}C-coincident α-spectrum correspond fairly well to those obtained from the single α-spectrum as shown in Fig.2.

The angular correlation measurement gave us information on spin of the states. The Legendre polynomial squared for L less than 8 did not fit to the data. The most probable spin was suggested to be $10\hbar$ for two states, but $12\hbar$ could not be ruled out. Branching ratio of $\Gamma_{12C(gs)-12C(gs)}/\Gamma$ could be extracted for the 28.7 and 29.3 MeV states assuming the relative orbital angular momentum of $L=10\hbar$. The values extracted are 0.43 and 0.38 for the 28.7 and 29.3 MeV states, respectively. Since the Hauser-Feshbach theory calculation predicts the branching ratio of 10^{-2} for $L=10\hbar$ states, there is an obvious enhancement by a factor of 10-30. The state-parameters obtained for the 28.7 and 29.3 MeV states in the present correlation experiment were compared with those extracted in the excitation function for the elastic scattering of ^{12}C on ^{12}C measured at $\theta_{cm}=90°$[1] and good correspondence was obtained: (1) excitation energies at $E_x(^{24}Mg)=28.58$ and 29.08 MeV and total widths of 200-400 keV extracted in the excitation function of the elastic scattering correspond to those at $E_x=28.7±0.25$ and $29.3±0.25$ MeV with total widths less than 400 keV, (2) the spin assignment is consistent with that (10+) from the elastic scattering experiment, (3) the branching ratios of 0.43 and 0.38 for the 28.7 and 29.08 MeV states are comparable to the partial widths of 60 and 20 keV for the 28.58 and 29.08 MeV states with the total widths of 200-400 keV.

In summary, two states of ^{24}Mg at $E_x=28.7±0.25$ and $29.3±0.25$ MeV, which have a large fraction of ^{12}C(gs)-^{12}C(gs) configuration, have been first confirmed in their ^{12}C decay by applying the angular correlation technique. These states correspond to intermediate-structure resonances at $E_{cm}=14.65$ and 15.15 MeV observed in the elastic scattering of ^{12}C on ^{12}C.[1]

1) E.R.Cosman et al. Phys.Rev. C21 (1980) 2111.

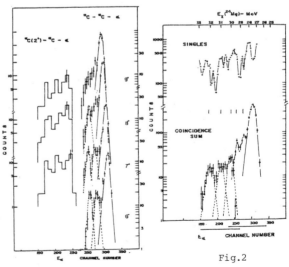

Fig.1

Fig.2

ELASTIC AND INELASTIC SCATTERING OF $^{14}C+^{14}C$ AND $^{12}C+^{14}C$ §

D.Konnerth, K.G.Bernhardt, K.A.Eberhard, R.Singh(a), A.Strzalkowski(b)
W.Trautmann, and W.Trombik

Sektion Physik, Universität München, D-8046 Garching, FRG

The elastic scattering of some light heavy-ion systems, e.g. $^{12}C+^{12}C$, $^{12}C+^{16}O$, $^{16}O+^{16}O$ has revealed pronounced gross structures in the excitation function which in some of these systems are considerably fragmented. Until now the origin of the narrower intermediate structure is not really understood but most of the proposed reaction mechanisms[1] proceed from a coupling of the elastic entrance channel to more complex degrees of freedom of the system, in particular to the inelastic excitation of one or both nuclei. Here we report a study concerned with the role of the inelastic channels for the occurence of intermediate structure in $^{14}C+^{14}C$ and $^{12}C+^{14}C$.

The 90° (c.m.) elastic scattering excitation function of $^{14}C+^{14}C$ measured between 6 and 35 MeV (c.m.) is dominated by a sequence of regular pronounced gross structures of 2-3 MeV width[2,3]. This is interpreted as a consequence of the extended surface transparency of the system[2,3]. Intermediate structures in the elastic scattering are weak and are seen only above ≅ 25 MeV (c.m.). They are accompanied by a general decrease of the elastic cross section and by a series of pronounced structures in the single and mutual inelastic scattering to the 3⁻ (6.73 MeV) state in ^{14}C and in transfer channels[2,4]. Structures at 26 MeV, 28.5 MeV, and 31 MeV seem to be correlated in several channels.

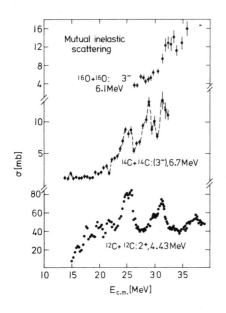

Fig. 1

A general feature of the inelastic and transfer channels in $^{14}C+^{14}C$ are the relatively small cross sections of several mb. The absolute cross sections for inelastic scattering are by a factor 5-10 smaller than for $^{12}C+^{12}C$ (Fig.1). Therefore it was concluded[2] that in $^{12}C+^{12}C$ the strong coupling to the collective 2^+ (4.43 MeV) state is responsible for the irregularities in the elastic scattering excitation function whereas in ^{14}C+^{14}C (and ^{16}O+^{16}O) the coupling is weak, and the regular gross structures are not strongly affected.

In view of this systematic behaviour, ^{12}C+^{14}C represents a particularly interesting system. Here the elastic excitation function exhibits strong fragmentation of the gross structure and irregularities similar to ^{12}C+^{12}C. This is expected on the basis of the possibility of strong coupling to the 2^+ (4.43 MeV) state in ^{12}C. However, the inspection of inelastic channels reveals that scattering to the 2^+ (4.43 MeV) state is weak whereas scattering to the 3^- (6.73 MeV) state in ^{14}C is stronger by about one order of magnitude (fig.2). It also shows pronounced structures in the excitation function as already observed in γ-ray experiments[5]. A possible solution to this puzzling situation may come from the observation that the cross sections of transfer channels (^{13}C+^{13}C) are in the same order of magnitude as that of the inelastic 3^- (6.73 MeV) excitation which would emphasize the role of the 2n-transfer in the elastic and inelastic channels of ^{12}C+^{14}C.

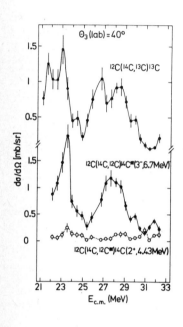

Fig.2

§ supported by the BMFT
(a) North Eastern Hill University, Shillong, India
(b) Jagellonian University, Cracow, Poland

1) see e.g. K.A.Erb et al., Comun.Nucl.Part.Phys. 8 (1978) 11
2) D.Konnerth et al., Phys.Rev.Lett. 45 (1980) 1154
3) D.M.Drake et al., Phys.Lett. 98B (1981) 36
4) R.M.Freeman et al., Phys.Rev. (to be published)
5) R.M.Freeman et al., Phys.Lett. 90B (1980) 229

Microscopic investigation of the $^{14}C + ^{14}C$ interaction

D. Baye and P.-H. Heenen

Physique Théorique et Mathématique - CP 229, Campus de la Plaine U.L.B., Bd du Triomphe
B 1050 Brussels, Belgium.

The generator coordinate method (GCM) has made possible microscopic studies of heavy ion reactions. Interesting results have been obtained about quasimolecular resonances for light systems. However, the calculations are limited to the elastic channel or to a small number of open channels. The comparison with experiment is therefore always indirect. In this communication, we want to show that a phenomenological imaginary part added to a nucleus-nucleus interaction calculated microscopically enables one to obtain quantitative agreement with experiment. The $^{14}C + ^{14}C$ collision provides a good example since it has been predicted theoretically to be a good candidate for the occurrence of quasimolecular resonances [1]. Recent experimental data confirm this prediction [2,3].

The nucleus-nucleus interactions calculated with the GCM are non local. The energy curves $V_L(R)$ (defined for a given value of the generator coordinate R as the matrix element of the microscopic Hamiltonian between projected Slater determinants) provide a convenient local approximation of the real part of an optical potential. The energy curves obtained with the interaction B1 of Brink and Boeker, a zero-range two-body spin-orbit force with a strength parameter equal to 85 MeV fm^5 and a harmonic oscillator parameter b = 1.70 fm are shown in fig. 1. The phase shifts obtained with these curves as a real optical potential (including Coulomb and centrifugal terms) are compared in fig. 2 with the phase shifts calculated by the GCM (dashed lines). The agreement between both calculations is satisfactory for each partial wave in the energy range where it contributes to the elastic scattering. An imaginary part has been determined by fitting the excitation functions at three angles [2]. The results obtained with the Woods-Saxon parameters :

$$W = 1.5 + 0.3 \, E_{cM} \qquad r_I = 1.27 \text{ fm} \qquad a_I = 0.21 \text{ fm}$$

are shown in fig. 3. The overall agreement with experiment is very satisfactory. The same quality of fit is obtained for the differential cross sections at 15.5, 19.5, 24. and 28. MeV. The dominant partial waves at the peaks of the 90° excitation function ($|S_L|^2 = \frac{1}{2}$ at 15., 18.8, 23.3 and 28.3 MeV for L = 12, 14, 16, and 18 respectively) agree with the assignment of Konnerth et al [2]. This is an improvement compared to the experimental optical potentials [2,3] for which no single partial wave dominates.

The present microscopic calculation gives a good approximation of the real part of the optical potential, in a collision with a transparent imaginary part. We have checked that the properties of this real part (depth, range, mass parameter) can only be varied in a limited range. This result gives confidence in the microscopic calculations of nucleus-nucleus interaction limited to the elastic channel.

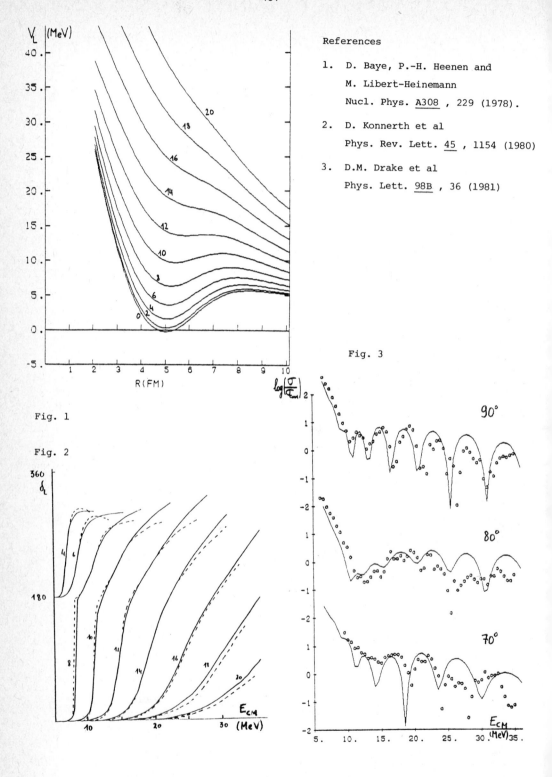

References

1. D. Baye, P.-H. Heenen and
   M. Libert-Heinemann
   Nucl. Phys. A308 , 229 (1978).

2. D. Konnerth et al
   Phys. Rev. Lett. 45 , 1154 (1980)

3. D.M. Drake et al
   Phys. Lett. 98B , 36 (1981)

Fig. 1

Fig. 2

Fig. 3

^{10}B + ^{14}N AND ^{12}C + ^{12}C REACTION DATA NEAR MOLECULAR RESONANCES *

W. Hoppe, E. Klauß, D. Sprengel, J. Drevermann, R. Isenbügel,
H. v. Buttlar, N. Marquardt

(Institut f. Exp.-Physik III, Ruhr-Universität Bochum, Germany)

We are performing a detailed investigation of ^{24}Mg resonances observed previously at 33.2, 38.9 and 44.4 MeV excitation and of resonances far below the Coulomb barrier around 17 MeV $\leq E_x(^{24}\text{Mg}) \leq$ 21 MeV. Whereas the resonances at high excitation energy, which are believed to have spins 12^+, 14^+ and 16^+, respectively, are populated by the two possible heavy-ion entrance channels ^{12}C + ^{12}C and ^{10}B + ^{14}N, the resonances of spins $\leq 6\,\hbar$ at extreme sub-Coulomb energies can only be investigated by ^{12}C + ^{12}C.

It is our intention to compare fine-step excitation functions and angular distributions obtained with the two different entrance channels

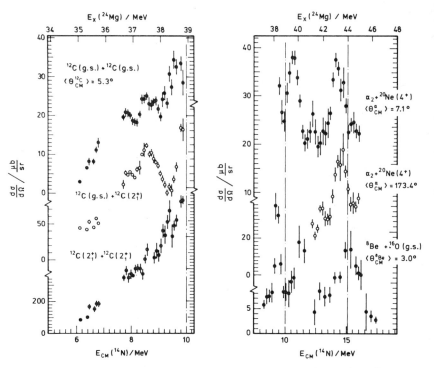

Fig. 1: Selected excitation functions close to 0° and 180° of the reactions ^{10}B(^{14}N,^{12}C)^{12}C (on the left) and ^{10}B(^{14}N,α)^{20}Ne and ^{10}B(^{14}N,^{8}Be)^{16}O (on the right) near ^{24}Mg resonances.

---

* Supported by the BMFT, Bonn, Germany

but leading to the same final channels $\alpha + {}^{20}Ne, {}^{8}Be + {}^{16}O$ and ${}^{12}C + {}^{12}C$. Such detailed work and the extraction of reduced partial width near the high-spin resonances should help to clarify the internal structure and to determine the spins of these resonant states.

The measurements are done at angles close to $0°$ and $180°$, where cross sections of high-spin resonances are expected to be largest. Whereas solid-state detectors are used for measuring α and ${}^{8}Be$ channels, the low-energy ${}^{12}C$ particles are identified by a new kind of annular $\Delta E-E$ telescope. A large s.f.b. annular detector is mounted in the gas volume of a cylindrical ionization chamber with an inner anode tube for the accelerator beam to pass.

Some selected results of the ${}^{10}B + {}^{14}N$ reactions are shown in fig. 1. Whereas the ${}^{12}C(g.s.) + {}^{12}C(g.s.)$ transition exhibits a rather smooth energy behaviour near the 38.9 MeV resonance in agreement with the ${}^{12}C({}^{12}C,{}^{10}B){}^{14}N$ results of Clover et al. (Phys. Rev.Lett. **43** (1979) 256), there appears to be structure of intermediate width in the ${}^{10}B({}^{14}N,{}^{12}C(2^+)){}^{12}C$ channel not accessible to the reverse reaction. There is also evidence for correlated structure in selected α and ${}^{8}Be$ channels near the 44.4 MeV resonance.

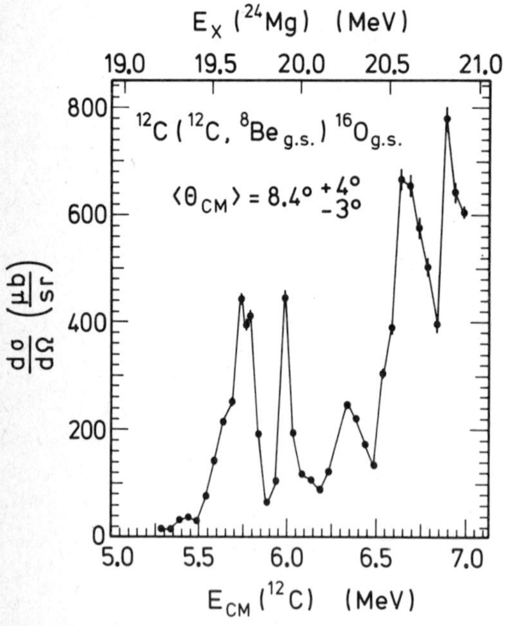

Evidence has recently been presented (R. Isenbügel et al., Contrib. Intern. Conf., Berkeley (1980)) for the existence of a 4^+ resonance at a ${}^{24}Mg$ excitation energy as low as 17.07 MeV. We report here on an extension of this low-energy ${}^{12}C({}^{12}C,\alpha)$ measurements by investigating the reaction ${}^{12}C({}^{12}C,{}^{8}Be_{g.s.}){}^{16}O_{g.s.}$. A new technique of measuring ${}^{8}Be$ excitation functions with large efficiency at angles close to $0°$ has been applied which consists in measuring coincidences between the signals of a large-area annular detector splitted into isolated halves. First results are shown in Fig. 2.

Fig. 2: Excitation function of the ${}^{12}C({}^{12}C,{}^{8}Be_{g.s.}){}^{16}O_{g.s.}$ reaction measured with a split annular detector.

HIGH-RESOLUTION EXCITATION FUNCTIONS OF ^{14}N + ^{14}N REACTIONS NEAR RESONANCES IN ^{28}SI *

M. Treichel, R. Isenbügel, H. v. Buttlar, N. Marquardt

(Institut f. Exp.-Physik III, Ruhr-Universität Bochum, Germany)

Preliminary results on high-resolution measurements of excitation functions of the ^{14}N(^{14}N,α)^{24}Mg and ^{14}N(^{14}N,^{8}Be)^{20}Ne reactions near $\Theta_{CM} = 0°$, $90°$ and $180°$ are reported. A three-stage differentially pumped jet gas target has been constructed which allows measurements of low-energy charged particles in the whole angular range. The special feature of measuring close to the beam direction is of importance for investigations of reaction mechanisms and for the search for high-spin resonances. It is realized with the present set-up by insertig on opposite sides of the windowless gas jet two annular-detector telescopes of either typs, ΔE-E- s.f.b. detectors or gas-ionization chamber with solid-state E detector. For this purpose a new kind of ionization chamber of coaxial type has been constructed for the first time. This annular ionization chamber is provided with a small central tube passing through its active gas volume.

The high-current beam of the Bochum 4 MV-Dynamitron Tandem was carefully collimated and focussed through a series of tubes belonging to the gas-jet target and the annular-detector telescopes. An energy range of 8 MeV < $E_{CM}(^{14}$N$)$ < 10 MeV has so far been investigated in steps of 100 keV (c.m.) with a target thickness of about 20 μg/cm^2. Whereas α particles have been measured with single annular s.f.b. detectors at mean c.m. reaction angles $\langle\Theta_{CM}\rangle = 14°$ and $173°$, ^{8}Be nuclei have been identified with the ionization-chamber telescope at $\langle\Theta_{CM}\rangle \approx 7°$. Single detectors close to $90°$(c.m.) for the detection of α particles and for monitoring purposes have also been used.

The idea of the experiment is to search with the ^{14}N + ^{14}N entrance channel for the strong 14^+ resonance observed in ^{12}C + ^{16}O reactions by many research groups at $E_x(^{28}$Si$)$ = 36.5 MeV. Recently a 10^+ resonance has also been discovered nearby at $E_x(^{28}$Si$)$ = 35.6 MeV by the reaction ^{12}C(^{16}O,^{8}Be$_{g.s.}$)^{20}Ne (K.A. Eberhard et al., Phys. Rev. Lett. 42 (1979) 432). Therefore, an investigation of the energy range between 35 < $E_x(^{28}$Si$)$ < 37 MeV by the rather strongly mismatched reactions ^{14}N(^{14}N,α) and ^{14}N(^{14}N,^{8}Be) appears to be interesting.

Preliminary data concerning ^{8}Be excitation functions (not shown here) exhibit a smooth energy behaviour. However, there appear to be

---

* Supported by the BMFT, Bonn, Germany

indications of correlated structure in some α channels (Fig. 1) at the energies of interest. Moreover, the energy variation of cross sections is different from that expected from the influence of the Coulomb barrier. Definitely, more data and measurements of other reaction channels are necessary before final conclusions can be drawn.

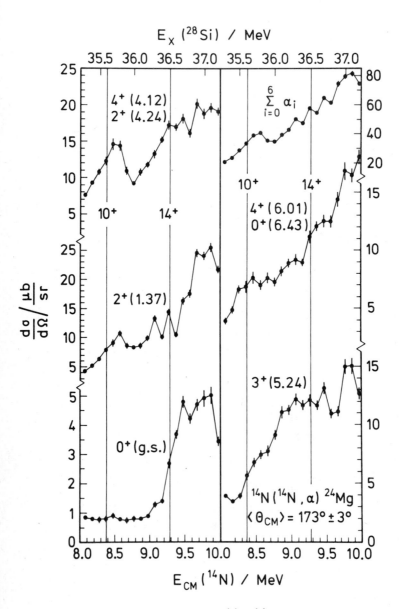

Fig. 1: Excitation functions of the $^{14}N(^{14}N,\alpha)$ reaction at 173°.

INTERMEDIATE AND FINE STRUCTURE STUDIES IN THE SYSTEM $^{16}O+^{12}C$*

P. Braun-Munzinger and H. W. Wilschut

Department of Physics
State University of New York
Stony Brook, New York 11794

ABSTRACT

New high resolution elastic and inelastic scattering data near the $E_{cm}=19.7$ MeV structure in the $^{16}O+^{12}C$ system are discussed. Attention is devoted to the isolation of 'nonstatistical' effects and to the extraction of (nearly) model independent S-matrix elements.

INTRODUCTION

Resonance-like phenomena are observed[1] in a large number of heavy ion reactions ranging from light systems like $^{12}C+^{12}C$ up to systems like $^{28}Si+^{28}Si$. Despite the wealth of data, however, no satisfactory model with predictive power has emerged from these studies. This is certainly in part due to the fact that, for essentially all systems, the observed structures bear some statistical features. For example, the distributions of the measured cross sections around their mean, whenever determined, is approximately gaussian and very few of the structures appear correlated over many channels. On the other hand, there are also experimental results, especially for the systems $^{12}C+^{12}C$ and $^{12}C+^{16}O$, which do not fit easily into the framework of the standard statistical model. A prime example[2] of such a situation is the region near $E_{cm}=20$ MeV in the system $^{12}C+^{16}O$. There, prominent structures of width $\Gamma \approx 400$ keV have been found in elastic and inelastic scattering. These structures appear to be much wider than the coherence width of the underlying compound nuclear levels ($\Gamma_{coh} \approx 100\text{-}150$ keV) and show up in several exit channels and over an angular range which is much larger than expected from the statistical model. Despite these facts, however, a 'model independent' proof that these structures correspond to isolated resonances, does not exist up to now.

In the present manuscript, we would like to discuss our recent measurements of elastic and inelastic angular distributions and excitation functions in that energy region. These experiments were undertaken to fill just the gap mentioned above. In Chapter 2 we will briefly review the evidence for the existence of non-statistical effects in the system $^{12}C+^{16}O$. In order to do so, the dominant exit channels, namely, elastic and inelastic scattering, light particle exit channels and the fusion cross section will be discussed and the data compared to qualitative predictions of the statistical model.

Chapters 3 and 4 will deal more specifically with angular distributions and excitation functions for elastic and inelastic scattering and their analysis and possible interpretation in terms of resonances. A summary and conclusion is given in the last chapter.

STATISTICAL CONSIDERATIONS

The strongest evidence for the 19.7 MeV structure to be non-statistical comes from $^{12}C+^{16}O$ elastic and inelastic scattering data. In Fig. 1, we show near $E_{cm}=20$ MeV, the elastic deviation function D(E) taken from Ref. 2 together with the angular integrated ($80^\circ \leq \theta_{cm} \leq 150^\circ$) elastic and inelastic cross sections. In all channels except the one leaving the ^{12}C in its (2^+,4.43 MeV) state a clear structure is observed near $E_{cm}=19.7$ MeV. More quantitatively, the dashed line in the deviation function panel shows the expected variance of D(E), deduced from the cross section distributions and taking into account that the coherence angle $\theta_c \approx 20°$, so that some data points included in D(E) are not statistically independent. The structure near 19.7 MeV then consists of a least 6 points more than one standard deviation outside of zero; the chance that this happens as a pure fluctuation is $<(0.32)^6 \approx 10^{-3}$. If one were to include the inelastic channels, the significance of this 'deviation' would be still increased as is obvious from Fig. 1. Surprisingly, however, the situation gets more complicated if one includes light particle exit channels[3-5] (p,d,α). Because of the high density of peaks in these excitation functions, the actually observed number of peaks in these channels near 19.7 MeV barely exceeds the value expected for chance correlations. The correlation might improve substantially if one chooses to include only high spin states in ^{27}Al and ^{27}Si, (see Refs. 4 and 5).

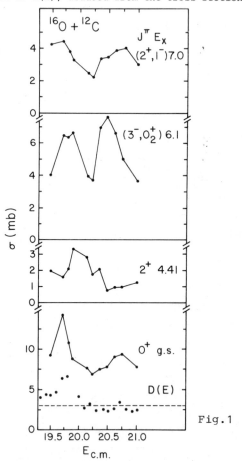

Fig. 1

It is also significant and puzzling to note that in the elastic deviation function there is little (if any) evidence for further non-statistical structures. This is in marked contrast to the recent finding[6] of strong intermediate structure in the total reaction cross section near $E_{cm}=20$ MeV (but not at 19.7 MeV). This conclusion is mainly based on the fact that the effective number of independent channels, deduced from the cross section distributions is relatively small ($N_{eff}^{exp} \approx 5000$) compared to the numbers quoted in Ref. 6, on the basis of statistical considerations. However, it turns out that, because of unitarity, the statistical model also predicts a fairly small number of N_{eff}. In fact, N_{eff} for the total reaction cross section can be expressed entirely in terms of the fluctuating and optical model part of the elastic S-matrix (see Appendix) and, in the absence of direct reactions, is predicted to be between 500 and 1000. The above experimental number is, therefore, entirely compatible with the predictions of the statistical model, assuming some direct reaction contribution to the total reaction cross sections. We note in passing that, for the same reason, at least the major part of the fine structure in the fusion cross section at lower energies[7] is then also in agreement with statistical model predictions. From these date alone one cannot draw conclusions about the existence of intermediate structure!

ANGULAR DISTRIBUTIONS

Figure 2 shows the results of eleven elastic angular distribution measurements in the energy region near the 19.7 MeV structure. The curves are results of a quasi-phase shift analysis to be discussed below. These data were taken with a 20 μg/cm^2

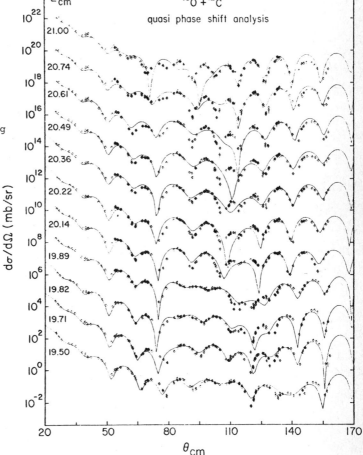

Fig. 2 Elastic scattering angular distribuitons near the 19.7 MeV structure.

thick ^{12}C target corresponding to an energy averaging of approximately $\delta E_{cm} \approx 50$ keV. The most striking feature of these data is the strong energy dependence observed in the angular range $60 \leq \theta_{cm} \leq 140°$ for energies $19.5 \leq E_{cm} \leq 20.2$ MeV in marked contrast to the relatively smooth behavior at the higher energies. A further noteworthy feature of these data is that, for the lower energies, the last 3 minima in the elastic angular distribution closely coincide with the zeros of $P_{13}(\cos\theta)$, while the minima at the higher energies seems to be closer to the zeros of $P_{12}(\cos\theta)$. The first result had already been noticed by Malmin et al.[8] and it was only the fact that the 19.7 MeV does show up in the 90° excitation function (where $P_{13}(\cos\theta)=0$) which lead then to a tentative spin assignment of $J^{\pi}=14^{+}$. The tentative nature of this assignment can be more appreciated by inspecting Fig. 3, where we have plotted the elastic cross sections as a function of energy at angles corresponding to zeros of various Legendre polynomials. The structure near 19.7 MeV is indeed visible in the 90° excitation function but, at the same time, excitation functions at angles corresponding to zeros of $P_{14}(\cos\theta)$ are by no means structureless, as would have been expected for the isolated resonance with $J^{\pi}=14^{+}$.

We would also like to note in passing that, the observation of slower frequencies (corresponding to $P_{12}(\cos\theta)$) at the higher energies, i.e., above 20.2 MeV, is consistent with the previously obtained result[9] that near $E_{cm}=20.5$ MeV, the particle-γ ray angular correlation for the $^{12}C+^{16}O(3^{-},6.1$ MeV) channel can be well reproduced by assuming a $J^{\pi}=12^{+}$ assignment for this structure.

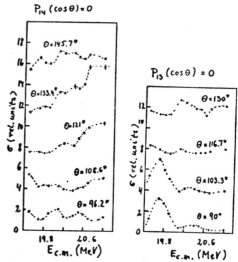

Fig. 3 Elastic scattering excitation functions at angles corresponding to zeros of P_{13} and P_{14}.

In order to analyze the elastic angular distributions more quantitatively and to isolate possible resonant amplitudes, a quasi-phase shift analysis of the data has been performed. The qualifier 'quasi' indicates that not all scattering amplitudes are allowed to vary independently in the analysis. Rather, the elastic nuclear S-matrix elements were expressed as

$$S_{\ell} = S_{\ell}^{0} + A_{\ell} e^{i\phi_{\ell}},$$

where S_ℓ^o are suitably chosen 'background' S-matrix elements. The additional amplitudes A_ℓ and phases ϕ_ℓ were determined from a χ^2-fit to the data. In order to keep the parameter space limited, A_ℓ was set equals to zero for $0 \leq \ell \leq 11$ and $16 \leq \ell$ so that only the four most important partial wave amplitudes were allowed to vary during the search.

The results of this procedure are the values of A_ℓ and ϕ_ℓ and the corresponding errors (from the error matrix) for each energy. Of course, if there is an isolated resonance in one partial wave, then the corresponding amplitude, plotted in an Argand diagram, should follow a circle with radius Γ_{el}/Γ in a counter clockwise motion as a function of energy. Such Argand diagrams are shown in Fig. 4 for some of the partial waves for which amplitudes were determined. The numbers label the different energies and are explained in Table I. The most interesting aspect of

Table I

List of energies for which angular distributions were measured.

Distribution Number	E_{lab}	$E_{cm}^{corrected}$
1	45.50	19.48
2	46.00	19.69
3	46.25	19.80
4	46.40	19.87
5	47.00	20.12
6	47.20	20.21
7	47.50	20.34
8	47.80	20.47
9	48.10	20.51
10	48.40	20.72
11	49.00	20.98

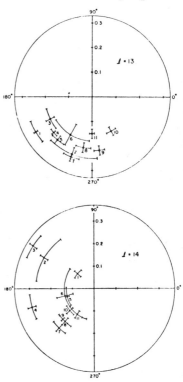

Fig. 4 Argand diagrams.

Fig. 4 is seen in the plot for partial wave L=14, where the first 6 S-matrix elements indeed quite closely follow a circle in a counter clockwise way, while the remaining points seem to scatter essentially randomly. This result is in striking contrast to the behavior of the other partial wave amplitudes which show very little systematic variations. The behavior of the L=14 partial wave indeed is a very strong indication that near $E_{cm}=19.7$ MeV, there is an isolated resonance in the $^{12}C+^{16}O$ system with quantum numbers $J^\pi=14^+$. No other resonances are immediately identifiable from these data.

The first question which comes to mind in assessing the relevance of these results is if the solutions of the χ^2 procedure displayed in Fig. 4 are unique. In performing the χ^2 fits a compromise had to be struck between keeping the number of free parameters (and, thereby, the number of solutions) small, while at the same time not introducing too much model dependence via the background S-matrix elements. In the present case, the background parameters were determined using various optical potentials. Specifically, the results of Fig. 4 have been obtained using a folding potential[10] but other potentials gave similar results. In particular, the strong and regular energy dependence observed in the L=14 partial wave amplitude seems to be independent of the choice of the background parameters[11]. It should also be pointed out that the prize one pays to restrict the number of free parameters is, that the final fits to the data are not perfect. We felt, however, that the ability to identify a continuous set of solutions as a function of energy is more important than a 'perfect fit' at each given energy.

With these caveats in mind, the results presented in Fig. 4 very strongly indicate that the "19.7 MeV structure" is, indeed, an isolated resonance. Moreover, from the S-matrix elements we can determine the resonance parameters for this state. A compilation of these numbers is found in Table II. Note that these parameters do not completely agree with the results quoted in Ref. 2 from a analysis of only three angular distributions.

Table II

Comparison of resonance parameters deduced from the phase shift analysis with the values taken from Ref. 2.

	Phase shift Analysis	Ref. 2
E_R (MeV)	19.8±0.1	19.7
Γ (keV)	300±100	380
Γ_{el}/Γ	0.12±0.03	0.25±.05
ϕ_R	0-10°	0-30°
J_R^π	14^+	14^+

It may, perhaps, be surprising that only one resonance can be identified from the phase shift analysis of the elastic angular distributions. From inspection of the inelastic data of Ref. 2, one would have expected another resonance at $E_{cm} \approx 20.5$ MeV, possibly in the $\ell=12$ partial wave.[9] The absence of such a resonance in the present S-matrix elements could be explained if the partial elastic width is sufficiently small ($\Gamma_{el}/\Gamma \lesssim 0.05$) so that the expected circle in the Argand diagram has a radius comparable to the error bars and, therefore, cannot be extracted from the data.

Because this 'resonance' shows up in the inelastic excitation function with similar strength as the 19.7 MeV state, this would imply a much larger inelastic width at 20.5 MeV. Note that, because $\Gamma_{inel.}/\Gamma \lesssim 1$, the strength of the resonance in the inelastic channel is bounded by $4\Gamma_{el}\cdot\Gamma_{in}/\Gamma^2 \lesssim 4\cdot\Gamma_{el}/\Gamma$ which is equal to the strength in the elastic channel. However, the observed resonant cross section at 20.5 MeV is approximately[12] 6.5 mb corresponding to $4\cdot\Gamma_{el}\cdot\Gamma_{in}/\Gamma^2 \approx 0.05$, assuming a spin J=12 for this structure. This value is still safely below the limit obtained by assuming $\Gamma_{el}/\Gamma \lesssim 0.05$. The increased inelastic width compared with the inelastic width at 19.8 MeV might be explainable in terms of the increased transmission through the angular momentum barrier (if $J^\pi=12^+$). Further, even more complete, data would be necessary to resolve this question.

In Fig. 5, we present as a function of energy angular distributions leading to the ($^{16}O, 3^-, 6.1$ MeV) channel. Of course, because the data were taken using ionization chambers and solid state detectors, this cross section includes the transition to the nearby 6.06, 0^+ state. However, the latter transition is strongly mismatched in angular momentum and is estimated[12] to comprise less than 10% of the measured cross section.

There are several noteworthy features in these inelastic angular distributions. Firstly, despite the large channel spin, pronounced oscillations are visible at all energies. This shows that interference among several partial waves in the entrance and exit channels is present even near 19.8 MeV, where we know from the phase shift analysis that the $J^\pi=14^+$ resonance dominates the cross section. Looking more closely, it is apparent that the amplitude of the oscillations gets largest at energies corresponding to maxima in the cross section, i.e., near 19.8 and 20.5 MeV (see also Fig. 1). At present, because of the many parameters required, a phase shift analysis of these data seems rather unfeasible. We feel, however, that these and similar data for other strongly populated exit channels, which we cannot discuss here for reasons of space limitations, place very stringent tests on theories like the band crossing model[13,14].

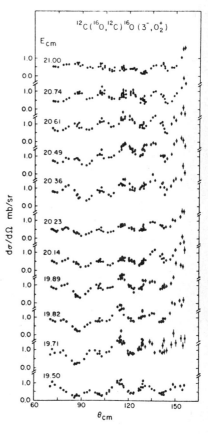

Fig. 5

EXCITATION FUNCTIONS

While the angular distribution data discussed in the previous section indeed establish the presence of a $J^\pi=14^+$ resonance near 19.8 MeV, there are still many puzzling features in the data. We refer, e.g., to Fig. 1, which shows that the peaks in the angle integrated elastic and inelastic cross sections differ considerably in their position and width. Furthermore, inspection of the elastic deviation function and energies below 19.5 MeV (see, e.g., Ref. 2) reveals the presence of a large number of narrow ($\Gamma \lesssim 150$ keV) structures which do not seem to be correlated over many angles and exit channels and are generally interpreted as due to statistical fluctuations[2]. The question then arises why such fluctuations are apparently not observed near 20 MeV. In order to try to shed more light on these question, we have undertaken a search for possible fine structure near the 19.8 MeV resonance in the $^{12}C+^{16}O$ system[11]. The experiments used targets of approximately 5 μg/cm^2 thickness, equivalent to an energy averaging of about 12 keV in the c.m. system. Data were taken in steps of δE_{cm}=11 keV at the two angles, $\theta_{cm} \approx 128°$ and $\theta_{cm} \approx 104°$ with an angular aperture of about $\Delta\theta_{cm} \approx 15°$, i.e., of the order of the coherence angle expected from statistical model calculations.[14]

The results of these fine structure excitation functions for the ($^{16}O,3^-$, 6.1 MeV) channel are presented in Fig. 6. Because of some overall systematic differences in the data, repeated excitation functions have not been averaged. The most conspicous feature, visible in all excitation functions at both angles, is the presence of two structures centered near E_{cm}=19.72 and 19.85 MeV, respectively, which modulate the 400 keV wide structure previously identified as the "19.7 MeV resonance". In addition there seems to be some evidence for more structure, albeit with somewhat reduced amplitude. Similar features also exist in the excitation functions leading to other exit channels. They are compared to each other in Fig. 7. Especially prominent is the structure in the ($^{16}O,2^+1^-$,7 MeV) channel. In contrast, the elastic scattering excitation functions at both angles shows a peak with width of $\Gamma \approx 150$ keV centered around 19.65 MeV, but certainly no prominent substructure.

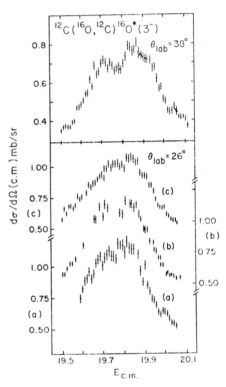

Fig. 6 Fine step excitation functions.

In the following, we would like to discuss the implications of these results and present a (semi) quantitative analysis of the measured excitation functions.

The observed substructure could be caused by two different mechanisms. The first possibility is that the structures correspond to individual states in the compound system ^{28}Si. While this possibility cannot be completely ruled out on the basis of the present data, it seems fairly unlikely in view of the absence of any visible correlation between the elastic and (6.1 MeV,3^-) channel; there is a possible correlation between the structures observed in the (6.1 MeV,3^-) and (7.1 MeV, 2^+1^-) channels at 128° but, in view of density of peaks in the latter channel, this correlation might well be fortuitous. More plausible is the explanation that what we observe are indeed compound nuclear fluctuations (corresponding to $\Gamma/D \gg 1$) superimposed on an intermediate structure resonance. Such a picture is also consistent with the fact that the amplitude of the observed structures seems to increase near the maximum of the intermediate structure peak. This situation is very reminiscent to the observation of fine structure within the giant dipole resonance[16] and within analog resonances.[17] In this context it is interesting to check whether the envelopes of the measured exitation functions are consistent with the presence of one intermediate structure resonance. In order to test this, calculations have been performed for the elastic and the

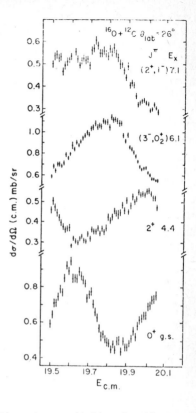

Fig. 7 Fine step excitation functions.

(6.1 MeV,3^-) channel. The calculations assumed an isolated resonance with $J^\pi = 14^+$ superimposed on an energy independent background. Furthermore, the background was also assumed to be constant in angle over the acceptance of the detector. Model calculations[11] employing, among other things, the background amplitude used in the phase shift analysis showed that under these assumptions, the resonance position and total width can be determined reliably, while for the partial widths, only order of magnitude estimates are possible.

The results of these resonance calculations are shown in Figs. 8 and 9. As is obvious from these figures, the general features of these data are consistent with one resonance. However, the resonance parameters, determined independently for the elastic and inelastic channel, are not consistent with each other as is evident from Table III. Especially distressing is the fact that neither position nor total width agree to better than, ±100 keV. On the other hand, this inconsistency is not too surprising considering the fact that (from the phase shift analysis) $|S_{res}^{el}| \lesssim 2\Gamma_{el}/\Gamma \sim 0.2$ while Hauser-Feshbach calculations[15] predict that the elastic S-matrix fluctuates with a standard deviation of $\sqrt{<|S_{fl}^{el}|^2>} \sim 0.05, -0.01$. Under these conditions, and keeping in mind that the width of the intermediate structure resonance is only about 3 times larger than the average coherence widths, it is obviously difficult to make a quantitative resonance analysis.

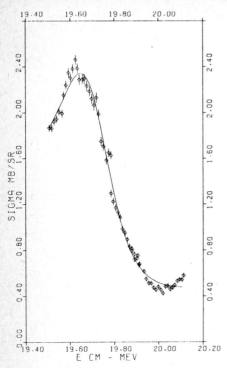

Fig. 8 Resonance analysis of elastic excitation functions.

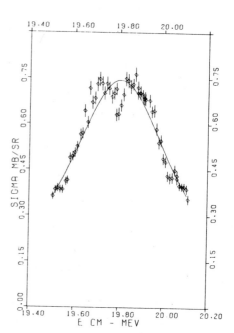

Fig. 9 Resonance analysis of inelastic excitations functions.

Table III

Resonance parameters deduced from the fits to the excitation functions shown in Figs. 8 and 9.

Elastic excitation function

θ_{lab}	χ^2	Γ_{el} MeV	E_R MeV	Γ MeV	ϕ	σ_o mb/sr	Γ_{el}/Γ
26°	6.8	0.040	19.86	0.213	-3°	0.88	0.19
38°	5.6	0.110	19.71	0.329	-39°	0.73	0.33

$^{16}O(3^-)$ excitation function

θ_{lab}	χ^2	$(\Gamma_{el}\Gamma_{3-})^{1/2}$	E_R	Γ	ϕ	σ_o	$(\Gamma_{el}\Gamma_{3-})^{1/2}/\Gamma$
26°	1.4	0.040	19.83	0.43	-153	0.30	0.13
38°	4.2	0.090	19.80	0.59	178	0.003	0.15

$[J_{res} = 14^+]$.

SUMMARY AND CONCLUSION

Summarizing the results presented in this paper, we have shown that the structure near 19.7 MeV in the $^{16}O+^{12}C$ system is with a high probability (>>99%) of nonstatistical origin.

A quasi phaseshift analysis of elastic scattering angular distributions yielded strong evidence that this structure indeed is an isolated resonance in the sense of a bound state embedded in the continuum with quantum numbers, $J^\pi=14^+$ and $\Gamma_{el}/\Gamma \approx 0.12\pm 0.03$. Furthermore, high resolution excitation functions have revealed the presence of substructure which are consistent with compound nucleus fluctuations superimposed on the intermediate structure resonance. The ever presence of these fluctuating components in the S-matrix, however, seem to rule out a quantitative and precise resonance analysis of the data.

As to the dynamical origin of these states, no consistent picture has yet been found. However, the fact that the data in many, but not all, ways exhibit some statistical features lets one wonder whether the intermediate structure itself might not bear some statistical aspects, too. In particular, because of the many, strongly coupled exit channels, one can, in general, not expect to find a pure resonance which couples only weakly to other states. Even if one restricts oneself only to elastic and inelastic scattering, i.e., neglects completely all rearrangement channels and treats the compound nucleus only in an average way, than at any given energy, there are, due to the strong coupling, several active entrance channel angular momenta (see, e.g., Fig. 1 or Ref. 14, p. 410). From this picture, one can easily imagine that adding several more inelastic and rearrangement channels might sufficiently complicate the sequence of levels observed in the excitation function so that their overall features can effectively be described by statistical techniques. Under these circumstances, the observed 'fluctuations' are mainly due to 'molecular levels' with $\Gamma/D \lesssim 1$ and not due to compound nuclear levels with $\Gamma/D >> 1$ although the cross section due to the latter is also not negligible.

The data discussed in this paper have been taken in collaboration with G.M. Berkowitz, R.H. Freifelder, J.S. Karp and T.R. Renner.

APPENDIX

In this appendix we would like to show that, because of unitarity, the effective number of <u>independent</u> channels which governs the size of fluctuations, is much less for the total reaction cross section than might be inferred from the number of <u>open</u> reaction channels. We first treat the case of the total reaction cross section and, at the end, briefly discuss the modifications due to direct reactions.

Fluctuations in the total reaction cross section

$$\sigma_T = \pi\lambda^2 \sum_\ell (2\ell+1)(1-|S_\ell|^2) \tag{1}$$

arise because the elastic S-matrix fluctuates about its mean; the fluctuating component is defined as

$$S_\ell^{fl} = S_\ell - \langle S_\ell \rangle, \tag{2}$$

where $\langle\rangle$ denotes energy average. The notation in eq. (1) is for the case of entrance channel spin 0, for simplicity. The effective number of independent channels is obtained[18] from the variance of the distribution of σ_T, i.e.,

$$\frac{\langle \sigma_T^2 \rangle - \langle \sigma_T \rangle^2}{\langle \sigma_T \rangle^2} = \frac{1}{N_{eff}}. \tag{3}$$

Equation (3) can be evaluated by inserting the definition (2) into eq. (1) and making use of formula (9.1) of Ref. 19 to evaluate the fourth moment of S_ℓ^{fl}.

Using the definition of the transmission coefficient

$$T_\ell = 1 - |\langle S_\ell \rangle|^2 \tag{4}$$

and the Hauser-Feshbach expression for the second moment of S_ℓ^{fl}, i.e.,

$$\langle |S_\ell^{fl}|^2 \rangle = 2 \cdot \frac{T_\ell \cdot T_\ell}{\sum_{\alpha L} T_L(\alpha)}, \tag{5}$$

N_{eff} can be expressed in terms of optical model transmission coefficients. The final result is then

$$N_{eff} = \frac{\left\{ \sum_\ell (2\ell+1)\left[T_\ell - \frac{2T_\ell^2}{\sum_{\alpha L} T_L(\alpha)}\right]\right\}^2}{\sum_\ell (2\ell+1)^2 \left[\left(\frac{2T_\ell^2}{\sum_{\alpha L} T_L(\alpha)}\right)^2 + 2(1-T_\ell) \frac{2T_\ell^2}{\sum_{\alpha L} T_L(\alpha)}\right]}. \tag{6}$$

Note that eq. (5) corresponds to the strong absorption limit discussed in Ref. 19 with an elastic enhancement factor of 2.

We have evaluated eq. (6) for the case of $^{16}O + ^{12}C$ at $E_{cm} \sim 20$ MeV using the Hauser-Feshbach program STATIS[20] with a level density description as in Ref. 21. This yields a value of about 500-600 for N_{eff} for the total reaction cross section. In general, we have, of course, to take into account the damping of the fluctuations due to direct reactions. Since the main contribution to eq. (6) arises from ℓ-values close to ℓ_{gr}, where the direct reaction part is largest, this damping is expected to be important. It can simply be estimated by replacing the transmission coefficients in eq. (6) by the diagonal elements of the penetrability matrix, i.e.,

$$T_\ell = P_\ell + \frac{\sigma_{Direct}(\ell)}{\pi\lambda^2(2\ell+1)}, \tag{7}$$

where $\sigma_{Direct}(\ell)$ is the direct reaction contribution to partial wave ℓ. Model calculations employing eq. (7) indicate that damping by approximately an order of magnitude is not unreasonable.

REFERENCES

* Supported in part by the National Science Foundation.

1. For a recent compilation of experimental results in the field of heavy ion resonances, see W. Henning, Europhysics Conf. on the Dynamics of Heavy Ion Collisions, Hvar, 1981.
2. See, e.g., R. E. Malmin, J. W. Harris and P. Paul, Phys. Rev. C18 (1978) 163.
3. R. Stokstad, D. Shapira, K. Chua, P. Parker, M. W. Sachs, R. Wieland and D. A. Bromley, Phys. Rev. Letters 28 (1972) 1523.
4. E. R. Cosman, A. Sperduto, T. M. Cormier, T. N. Chiu, H. E. Wegner, M. J. LeVine and D. Schwalm, Phys. Rev. Letters 29 (1972) 1341.
5. P. Sperr, D. Evers, A. Harasim, W. Assmann, P. Konrad, K. Rudolph, A. Denhoefer and C. Ley, Phys. Letters 57B (1975) 438.
6. J. J. Kolata, R. M. Freeman, F. Haas, B. Heusch and A. Gallmann, Phys. Rev. C19 (1979) 408.
7. P. Taras, A. R. Rao and G. Azuelos, Phys. Rev. Letters 41 (1978) 840.
8. R. E. Malmin, R. H. Siemssen, D. A. Sink and P. P. Singh, Phys. Rev. Letters 28 (1972) 1590.
9. At this energy, a γ-ray particle angular correlation study suggests the presence of a resonance with $J^{\pi}=12^+$, see C.M. Jachcinski, P. Braun-Munzinger, G. M. Berkowitz, R. H. Freifelder, M. Gai, R. L. McGrath, P. Paul, T. R. Renner and C. D. Uhlhorn, Phys. Letters 87B (1979) 354.
10. G. R. Satchler, Phys. Reports, C55 (1979) 183.
11. H. W. Wilschut, Thesis, Stony Brook 1981; H. W. Wilschut, et al., to be published.
12. W. S. Freeman, H. W. Wilschut, T. Chapuran, W. F. Piel, Jr. and P. Paul, Phys. Rev. Letters 45 (1980) 1479.
13. Y. Kondo, D. A. Bromley and Y. Abe, Phys. Rev. C22 (1980) 1068 and refs. quoted therein.
14. O. Tanimura and T. Tazawa, Phys. Reports 61 (1980) 253, Phys. Rev. Letters 46 (1981) 408.
15. P. Braun-Munzinger and J. Barrette, Phys. Rev. Letters 44 (1980) 719.
16. E. M. Diener, J. F. Amann and P. Paul, Phys. Rev. C7 (1973) 695.
17. See, e.g., E. Lanke, H. Genz, A. Richter and G. Schrieder, Phys. Letters 58B (1975) 289.
18. H. L. Harney and A. Richter, Phys. Rev. C2 (1970) 421, also T.E.O. Ericson, Ann. Phys. (N.Y.) 23 (1963) 390.
19. D. Agassi, H. A. Weidenmüller and G. Mantzouranis, Phys. Reports 22 (1975) 145.
20. Statis, A Hauser-Feshbach Computer code, R. G. Stokstad, Yale Report #52, 1972.
21. F. Pühlhofer, Nucl. Phys. A280 (1977) 267.

CORRELATION MEASUREMENT SEARCHING FOR RESONANT ^{12}C-^{12}C STATES INDUCED BY THE ^{12}C(^{16}O,α) REACTION

K.KATORI, T.SHIMODA*, T.FUKUDA*, H.OGATA**, I.MIURA** and M.TANAKA+

Laboratory of Nuclear Studies, Osaka University, Osaka, 560, Japan
* Department of Physics, Osaka University
** Research Center for Nuclear Physics, Osaka University
+ Kobe Tokiwa Junior College

Great attension has currently been paid on whether ^{12}C-^{12}C molecular resonances appear as final states in the ^{12}C(^{16}O,α)^{24}Mg reaction[1]. In order to distinguish ^{12}C-^{12}C molecular resonance states from structure caused by inelastic excitation of ^{16}O, one decisive way is to excite those states in ^{24}Mg by the ^{12}C(^{16}O,α)^{24}Mg reaction and to measure ^{12}C-decay from those states. Natural carbon target of 780 μg/cm^2 thick was bombarded at E_{lab}(^{16}O)= 142 MeV with the ^{16}O^{5+} beams from the 230-cm cyclotron at Research Center for Nuclear Physics. Detector for α particles was a conventional telescope and fixed at θ_α(lab)=-6°. A telescope consisting of ΔE and position sensitive detectors was used for heavy-ion detection. Measurements were made at ten laboratory angles from 8° to 32.5°. Particles from ^{6}Li to ^{20}Ne coincident with α-particles were identified.

For the α-^{12}C energy correlation, events corresponding to three final states, ^{12}C(gs)-^{12}C(gs)-α, ^{12}C(gs)-^{12}C(2+)-α and ^{12}C(2+)-^{12}C(2+)-α were clearly separated in the three-body Q-spectrum. Fig.1 shows triple differential cross sections in the ^{12}C(^{16}O,α^{12}C)^{12}C reaction as a function of relative kinetic energy between α and ^{12}C(gs). Several peaks line up at the same relative kinetic energy of α-^{12}C from θ(^{12}C)=8° to 15°, indicating sequential α-decay of ^{16}O to ^{12}C(gs) in the inelastic excitation of ^{16}O projectile on ^{12}C target. The 11.3, 12.8, 14.5 and 15 MeV states of ^{16}O are strongly excited with the yield of 0.5-1.0 mb/sr^{2}MeV. The triple differential cross sections at θ_{lab}(^{12}C)=27.5°, 30.0° and 32.5° are shown in Fig.2 as a function of ^{12}C(gs)-^{12}C (gs) relative kinetic energy. Even at the backward angle of θ_{lab}(^{12}C)=32.5°, the yield coming from the inelastic excitation of ^{16}O at E_x=14-22 MeV may remain, so that the yield coming from ^{12}C(gs)-^{12}C(gs) molecular resonance (E_x(^{24}Mg)~51-55 MeV(18+)) may be masked. Fig.3 shows the triple differential cross sections at θ_{lab}(^{12}C)=27.5°, 30.0° and 32.5° as a function of relative kinetic energy of two carbons corresponding to the mutual 2$^+$ excitation. Energy spectra between 24 and 33 MeV appear to resemble at three differential angles, whose energies correspond to the ^{12}C(2+)-^{12}C(2+) molecular resonance(18+)[2].

To estimate the intensity of the sequential decay of ^{16}O projectile for the energy range of $E_x \sim 51-55$ MeV of $^{24}Mg^*$, a separate experiment was performed at $\theta_\alpha = -18°$ and $\theta(^{12}C) = 10°$. Assuming that angular dependence of the inelastic cross sections of $^{12}C(2+) + ^{16}O^* \to ^{12}C(2+) + ^{12}C(2+) + \alpha$ was similar to those of $^{12}C(gs) + ^{16}O^* \to ^{12}C(gs) + ^{12}C(gs) + \alpha$, the observed cross section was larger by a factor of 10 than the estimated cross section of $^{12}C(2+) + ^{16}O^* \to ^{12}C(2+) + ^{12}C(2+) + \alpha$ at $\theta(^{12}C) = 27.5°$. This indicated an existence of the resonant $^{12}C(2+) - ^{12}C(2+)$ states in ^{24}Mg. Thus, we can claim that at the excitation corresponding to the $^{12}C(2+) - ^{12}C(2+)$ molecular resonance (18+) the states of ^{24}Mg were able to be observed at the backward angles as final states.

Whether the humps in the α-singles spectrum really correspond to the ^{24}Mg states with the $^{12}C(2+) - ^{12}C(2+)$ configuration which was observed in the present correlation experiment, depends on the intensity. Thus, ^{12}C angle-integrated cross section at $E_x(^{24}Mg) = 51-55$ MeV was estimated from the yield at $\theta(^{12}C) = 32.5°$ to be an order of 100 μb/sr, which was only a few % of the cross section of the corresponding hump[1]. Therefore, the statement that all yield of the hump in the inclusive α-spectrum is ascribed to the $^{12}C(2+) - ^{12}C(2+)$ resonance states, which have a major cross section of the $^{12}C(2+) - ^{12}C(2+)$ resonance (18+)[2], is hard to be supported.

1) K.Nagatani et al.,Phys.Rev.Letters 43 (1979) 1480.
2) T.M.Cormier et al.,Phys.Rev.Letters 40 (1978) 924.

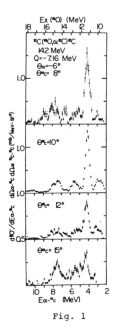

Fig. 1

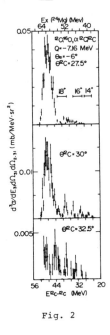

Fig. 2

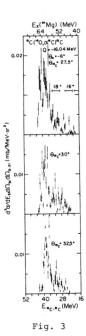

Fig. 3

RESONANCES IN ^{26}Al, ^{29}Si and ^{30}Si: ARE THEY ENTRANCE CHANNEL DEPENDENT?

S.T. Thornton
Department of Physics, University of Virginia
Charlottesville, Virginia 22901
U.S.A.

One of the most interesting and fundamental challenges to arise from the study of heavy-ion collisions at bombarding energies well above the Coulomb barrier is that of understanding and describing the nature and properties of the compound nucleus. Indeed, the validity of the classical conception of an equilibrated, long-lived nucleus in which all nucleons share the available energy and angular momentum must be questioned in the regime of high excitation energies (E^*) available to experimentalists today. The occurrence of resonances in many lighter heavy-ion systems has served to focus a great deal of attention on this question since it is not altogether clear whether resonances are a signature of compound nucleus formation or depend mostly upon the entrance and/or exit channels. The work presented in this paper addresses the compound nucleus versus entrance channel effect aspect of heavy-ion reactions by searching for resonances correlated in different entrance channels leading to the same compound nucleus.

We chose to address experimentally three questions pertaining to the resonance phenomenon:

1) Can resonances exist at high excitation energies (two to three times the Coulomb barrier) in the compound nucleus? It is thought that at very high excitation energies, many more channels are available for decay and that any nonstatistical structure may be averaged out.

2) Are there more non-α-conjugate systems which exhibit resonance behavior? Based on calculations in Ref. 1, there are such systems which are likely candidates due to the possibility of populating isolated states in a region of low level density. Resonances have been observed in several non-α-conjugate systems but some of the results are disputed.

3) How strongly is resonance behavior dependent upon the entrance channel? Correlations between exit channels have been well documented[2] for some reactions but correlations between entrance

channels have not been widely observed.

We have pursued these questions by measuring excitation functions for $^{17}O + ^{12}C$, $^{16}O + ^{13}C$, $^{16}O + ^{10}B$, and $^{17}O + ^{13}C$ over a range of high excitation energies. While questions 1) and 2) are quite obviously addressed by these measurements, question 3) is probed by making measurements on: a) the $^{17}O + ^{12}C$ and $^{16}O + ^{13}C$ systems over the same E^* (^{29}Si), b) the $^{16}O + ^{10}B$ system over the same E^* (^{26}Al) covered in an earlier experiment on $^{14}N + ^{12}C$ (Ref. 3) and c) the $^{17}O + ^{13}C$ system over the same E^* (^{30}Si) previously measured for the $^{18}O + ^{12}C$ and $^{16}O + ^{14}C$ entrance channels.[4,5,9]

Our typical experimental procedure has been to use a magnetic spectrometer to momentum analyze reaction α-particles with high resolution and measure excitation functions to many states in the residual nucleus. We also use solid state detectors in the scattering chamber for monitors and/or to measure elastic and inelastic scattering. The work discussed here was done at the Max-Planck-Institut für Kernphysik in Heidelberg using the Q3D magnetic spectrometer. The energy ranges and angles of investigation are given in Table I. Collaborators for this work are R.L. Parks, K.R. Cordell, C.-A. Wiedner, R. Gyufko, and G. Köhler.

TABLE I. Experimental parameters for the four reactions studied.

Reaction	Bomb. Energy (MeV)	Ex. Energy[a] (MeV)	Energy Step (MeV)	$\theta(\alpha)$[b] (deg.)	$\theta_1(el)$[c] (deg.)	$\theta_2(el)$[d] (deg.)
$^{17}O + ^{12}C$	30.0-57.0 (lab) 12.4-23.6 (c.m.)	33.5-44.7	0.50 0.21	10.0 14.0	35 89	41 110
$^{16}O + ^{13}C$	29.5-52.6 (lab) 13.2-23.6 (c.m.)	33.5-43.9	0.30 0.13	10.0 13.9	33 75	48 114
$^{17}O + ^{13}C$	30.0-57.0 (lab) 13.0-24.7 (c.m.)	39.8-51.5	0.50 0.22	10.0 13.4	35 84	44 110
$^{16}O + ^{10}B$	42.4-60.1 (lab) 16.3-23.1 (c.m.)	35.8-42.6	0.30 0.12	10.0 14.6	33 94	38 114

[a] Excitation energies refer to the compound nucleus for each of the four reactions i.e., ^{29}Si, ^{29}Si, ^{30}Si, ^{26}Al.

[b] $\theta(\alpha)$ is the angle at which the α-particles were detected.

[c] $\theta_1(el)$ is the angle at which elastic particles were detected directly.

[d] $\theta_2(el)$ is the angle for elastic particles which corresponds to detecting the recoil particles at angle $\theta_1(el)$.

A typical set of excitation functions is shown in Fig. 1 for the $^{10}B(^{16}O,\alpha)^{22}Na$ and $^{16}O + ^{10}B$ elastic reactions. The solid lines indicate the location of possible nonstatistical structures. We have adopted a combination of four tests to locate nonstatistical structure in a set of fluctuating excitation functions as discussed in detail in Ref. 6. These tests are the deviation function D(E), the cross-correlation function C(E), the distribution of maxima test, and the sum of excitation functions. The distribution of maxima test is particularly useful because it gives the probability of finding nonstatistical structure. We have adopted the convention of calling an event with a distribution of maxima probability of less than n^{-1} (where n is the number of points in each exitation function, typically $n^{-1} \simeq 0.01$) nonstatistical based on the simple statistical consideration that one such event would be likely out of n points. Others[7] have also chosen a probability of 0.01 as a cutoff but it is obvious that the smaller the probability, the more confident we are that the corresponding structure is nonstatistical.

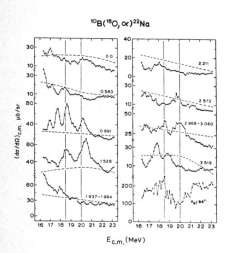

Fig. 1. Excitation functions for the $^{16}O + ^{10}B$ reaction for the elastic ($\theta_{c.m.} = 94°$) and α-particle ($\theta_{c.m.} = 14.6°$) exit channels. The excitation energies of the residual states in ^{22}Na are indicated. States with isospin T=1 (0.657 and 1.952 MeV) are not populated. The solid lines show the location of nonstatistical structure. The dashed curves are Hauser-Feshbach calculations as described in the text.

The results for the three compound nuclei considered here (^{26}Al, ^{29}Si and ^{30}Si) are shown in Figs. 2-4 and listed in Table II. A careful examination shows that for each compound system, the structures seen in the various entrance channels do not all occur at the same excitation energies.

The lack of correlation between the various entrance channels is somewhat surprising since they were chosen specifically because they bring in very similar amounts of maximum angular momentum, J_{max}, at a given excitation energy (see Fig. 5). Therefore, we should be populating

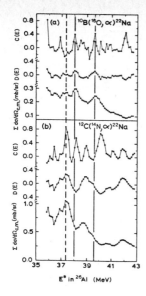
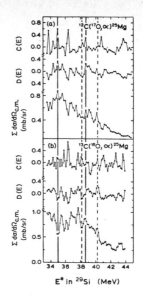

Fig. 2. Results of the sum, C(E) and D(E) tests for (a) the $^{10}B(^{16}O,\alpha)^{22}Na$ reaction and (b) the $^{12}C(^{14}N,\alpha)^{22}Na$ reaction (Ref. 3). The solid (dashed) lines indicate the location of nonstatistical structure in the $^{16}O + ^{10}B$ ($^{14}N + ^{12}C$) system.

Fig. 3. Results of the sum, C(E) and D(E) tests for (a) the $^{12}C(^{17}O,\alpha)^{25}Mg$ reaction and (b) the $^{13}C(^{16}O,\alpha)^{25}Mg$ reaction. The solid (dashed) lines indicate the location of nonstatistical structure in the $^{17}O + ^{12}C$ ($^{16}O + ^{13}C$) system.

these compound nuclei at nearly the same E^* and J. If this is the case and the structures we see are compound nuclear in origin, then they should occur at the same E^* for these different entrance channels.

We do have some evidence that nonstatistical structure occurs for the higher excitation energies in these non-α-conjugate systems. The probability for the structures in $^{17}O + ^{12}C$, $^{16}O + ^{13}C$, and $^{16}O + ^{10}B$ is not overwhelmingly convincing, but in $^{17}O + ^{13}C$ (and in one previously seen in $^{14}N + ^{12}C$) the probability is low enough to suspect such structure. We can conclude, however, that the nonstatistical structure is less likely to be observed at higher energies for reasons already discussed. Resonant contributions to single level Breit-Wigner resonances are also damped by $1/(2i+1)(2I+1)$ where i,I are the intrinsic spins of the projectile, target. This will, for example, damp the observation of resonances in $^{17}O + ^{13}C$ compared to $^{16}O + ^{14}C$ and/or $^{18}O + ^{12}C$ where they have been observed.

We have little or no evidence that we are populating the same nonstatistical structures at the same excitation energies in the compound

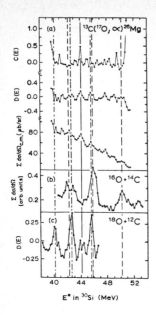
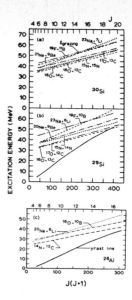

Fig. 4. (a) Results of the sum, C(E) and D(E) tests of the $^{13}C(^{17}O,\alpha)^{26}Mg$ reaction. The solid line indicates the location of a nonstatistical maximum. (b) Results from the sum of elastic scattering at eight angles for the $^{16}O + ^{14}C$ reaction where the dashed lines indicate the resonances reported in Ref. 4. (c) Deviation function for elastic scattering at eight angles for the $^{18}O + ^{12}C$ reaction where the dot-dash lines indicate the resonances reported in Ref. 5.

Fig. 5. Excitation energy in the compound nucleus versus $J(J+1)$ for spin states J in (a) ^{30}Si, (b) ^{29}Si and (c) ^{26}Al. The solid lines are calculated yrast lines while the patterned lines show the grazing angular momentum for various entrance channels (plotted versus $\ell(\ell+1)$).

TABLE II. Candidates for nonstatistical structure in the four reactions studied.

Reaction	$E_{c.m.}$ (MeV)	E^* (MeV)	Probability	Γ_{exp} (keV)	Γ_{st}^a (keV)	$\tau(10^{-21}\,sec)$	$t_{coll}^b\,(10^{-22}\,sec)$
$^{17}O + ^{12}C$	13.9	34.9	2×10^{-3}	~400	195	1.6	5.9
	17.6	38.7	3×10^{-3}	~400	240	1.6	5.6
$^{16}O + ^{13}C$	17.8	38.1	1×10^{-3}	~250	230	2.6	5.4
	19.9	40.2	6×10^{-3}	~300	260	2.2	5.3
$^{17}O + ^{13}C$	17.3	44.1	7×10^{-4}	~500	240	1.3	5.3
$^{16}O + ^{10}B$	18.5	38.0	5×10^{-3}	~400	290	1.6	5.2
	20.0	39.5	7×10^{-3}	~500	310	1.3	5.1

[a] Expected width for statistical structure from Ref. 8.
[b] Approximate collision time.

nucleus in this experiment. However, in the compound nucleus ^{30}Si structure does appear for the ^{18}O + ^{12}C and ^{16}O + ^{14}C reactions at the same excitation energies. This is rather confusing, because we observe one prominent nonstatistical structure in the ^{17}O + ^{13}C channel. Kolata et al.[9] have not observed structure in the ^{17}O + ^{13}C exit channel correlated with structure in the ^{16}O + ^{14}C or ^{18}O + ^{12}C exit channels.

A number of implications arise from the fact that we are not populating the same nuclear states through similar entrance channels. One possibility is that the nonstatistical structures we see are semi-isolated, high-spin, compound nuclear states which can only be excited by a <u>particular</u> combination of E^* and J^π. Another possible reaction mechanism which is consistent with our results is that of a quasimolecule. It has been shown that some observed resonances represent quasistable states formed by the colliding ions. These states may subsequently decay into a compound nucleus, elastic channel, inelastic channel or some other exit channel. The existence of such states seems likely in the reactions considered here because of the high total angular momenta involved. The number of available states in the compound nucleus is very small for large J so the system is almost forced into a highly deformed quasimolecular configuration which is capable of absorbing the angular momentum. The decay of such a system is similarly governed by the large J in that it may only decay into channels which can carry away large angular momenta. Clearly the inelastic, elastic and α-particle exit channels are good candidates for decay.

If quasimolecular states are being formed, we may expect to see them in the α-particle exit channel. If actual compound nucleus formation is not very favorable due to the large J, then a large fraction of the events leading to the α channel would be due to formation and decay of a quasimolecule. Additional support for this idea is seen in Table II where the widths of the states we see are all greater than the predicted width for normal statistical (compound nuclear) fluctuations. Moreover, the lifetimes associated with these widths are roughly equal to the time for one rotation of the pertinent molecule. We also note that the theoretical predictions of Heenen and Baye[10] indicate that quasimolecular resonances can exist in non-α-conjugate systems. They did a multichannel generator-coordinate study which predicted, for example, that ^{16}O + ^{14}C and ^{14}C + ^{14}C should resonate.

Experiments on the ^{16}O + ^{14}C system[4,9] are in agreement with this prediction. However, the likelihood of quasimolecular resonances existing for the present entrance channels is reduced because of the low lying excited states for the nuclei involved.

Based on the evidence at hand, it is not altogether clear whether the nonstatistical structures we see are quasimolecular or compound nuclear. In fact, the question may not even be relevant for the high excitation energies and spins involved here. We know that, if one is formed at all, the compound nucleus must be highly deformed in order to exist with such large J-values. Perhaps we are approaching a spin-energy regime in which an entrance channel quasimolecule and a highly deformed compound nucleus are one and the same. The understanding of the nonstatistical structure is by no means understood in terms of these simple models.

References

1. S.T. Thornton, L.C. Dennis and K.R. Cordell, Phys. Lett. 91B, 196 (1980).

2. E. Almqvist, D.A. Bromley and J.A. Kuehner, Phys. Rev. Lett. 4, 515 (1960).

3. K.R. Cordell, C.-A. Wiedner and S.T. Thornton, Phys. Rev. C23, 2035 (1981).

4. K.G. Bernhardt, H. Bohm, K.A. Eberhard, R. Vandenbosch and M.P. Webb, in Nuclear Molecular Phenomena, edited by N. Cindro, (North-Holland, Amsterdam, 1978), p. 367.

5. M.P. Webb, R. Vandenbosch, K.A. Eberhard, K.G. Bernhardt and M.S. Zisman, Phys. Rev. Lett. 36, 779 (1976).

6. L.C. Dennis, S.T. Thornton and K.R. Cordell, Phys. Rev. C19, 777 (1979).

7. R.A. Dayras, R.G. Stokstad, Z.E. Switowski, and R.M. Wieland, Nucl. Phys. A265, 153 (1976).

8. R.G. Stokstad, Proc. Int. Conf. on reactions between complex nuclei, Nashville, Tennessee, vol. 2 (North-Holland, Amsterdam, 1974), p. 327.

9. J.J. Kolata, C. Beck, R.M. Freeman, F. Haas, and B. Heusch, Phys. Rev. C23, 1056 (1981).

10. P.-H. Heenen and D. Baye, Phys. Lett. 81B, 295 (1979).

GROSS STRUCTURE IN MISMATCHED CHANNELS*

Peter Paul

Department of Physics
State University of New York
Stony Brook, New York 11794

ABSTRACT

A summary is given of recent results of an ongoing program to search for gross structure in the $^{16}O+^{16}O$ system in the region above the Coulomb barrier, in inelastic channels which differ widely in their kinematic conditions from the elastic channel. Various models which explain the gross structure either as shape resonances or as diffractive L window effects, are severely tested under these conditions. The symmetric ^{16}O system offers a variety of single and double nuclear excitation channels which we have studied over the range of c.m. energies between 25 and 35 MeV. Strong energy correlations are observed between gross structure peaks in a strong well matched channel and several severely mismatched single excitation channels. A diffraction model appears to be unable to account for the correlation. A recent coupled-channel calculation explains the data well and indicates strong effects of the channel coupling on the ion-ion optical potential. A measurement of the double excitation also shows gross structure peaks which are correlated among each other. However, the double excitation peaks appear to be anti-correlated with the single excitations. This surprising result is not yet understood.

---

*Supported in part by the National Science Foundation.

I. INTRODUCTION

At energies above the Coulomb barrier one observes regularly spaced peaks about 2 MeV wide in the c.m. system, in the elastic and many inelastic scattering excitation functions of many symmetric light heavy ion systems.[1] The energy sequence and the widths of these gross structure peaks suggests naturally an explanation in terms of standing waves, or shape resonances, in the pockets of the real part of the ion-ion potential.[2] As such, these resonances extend the molecular bands which are observed below the Coulomb barrier. Observation of the resonances requires a weakly absorbing potential for the partial waves near the grazing angular momentum. In asymmetric systems these resonances are not evident because the spacing of the peaks is about equal to their widths since odd and even partial waves are allowed.

However, it has been pointed out[3] that a much less interesting explanation is possible for these regular peaks, in a strong absorption model. In this model the incoming wave is diffracted by the strong imaginary part of the potential near the outer barrier. Peaks arise from the kinematical matching from the angular momentum windows in the incoming (elastic) and outgoing (inelastic) channels. The two models are schematically depicted in Fig. 1. Only in the resonance model does the incoming wave sample the interior of the real potential exciting the shape resonance in the elastic channel. If a resonance, of appropriate spin, occurs near the same energy in the potential of the inelastic channel a double resonance occurs producing a strong peak in both the entrance and the exit channel.[2]

The two models depend on the kinematic conditions in the two channels in different ways. Thus a comparative study of the gross structure in inelastic channels which are either well matched or badly mismatched, can provide a test between the diffraction model and the resonance model. Typically, the resonance model predicts peaks in

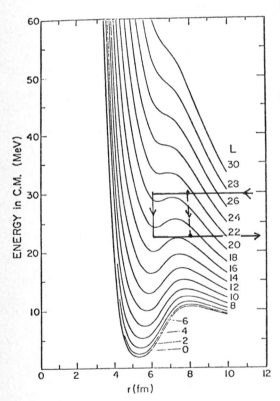

Fig. 1 Real part of the Gobbi potential[4] between two ^{16}O nuclei. Paths are shown for a diffractive and resonant process.

well matched and the mismatched channels which are correlated in energy, while the diffraction model predicts anti-correlated structures. Using the ^{16}O beams from the Stony Brook tandem we have investigated the region around 30 MeV (c.m.), i.e., around a grazing angular momentum of L≃20ℏ, in the ^{16}O+^{16}O system.

II. INELASTIC CHANNELS IN THE ^{16}O+^{16}O SYSTEM

The ^{16}O+^{16}O system has been one of the first in which gross structure has been found in the elastic and strong, well matched inelastic channels.[5,8] The spectrum of low-lying levels in ^{16}O, shown in Fig. 2, makes it also a prime case for the study of mismatched channels because it offers a number of bound, or quasi-bound states of low spin at relatively high excitation energy. Of particular interest are the 0$^+$ state at 6.05 MeV, and the 2$^-$ and 3$^+$ states at 8.87 and 11.08 MeV, respectively. Although energetically open to α decay the latter states are essentially bound because of their unnatural parity. In addition to these single-nucleus excitations one has, of course, the whole spectrum of double-nucleus excitations available. Even when the single excitation is well matched the double excitation can be mismatched since the matching condition depends on the momentum, i.e., $\sqrt{E}$. It is instructive to look at the physical meaning of the channel matching in the different models. In a semiclassical picture it is a matching of grazing angular momenta in the entrance and exit channels.

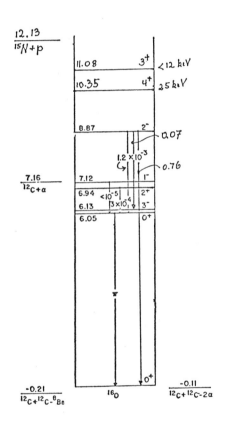

Fig. 2 Level spectrum of ^{16}O.

Figure 3 gives the matching condition for several inelastic channels, relative to the elastic one, from such a semiclassical calculation. The plot spans the range of bombarding energies from 30 to 70 MeV for which we will present data. The grazing angular momenta were computed with a radius parameter of r_o=1.4 fm which corresponds roughly to the location of the outer barrier of the optical potential shown in Fig. 1. All inelastic curves are for the maximally aligned case, i.e. L'=L-I where L,L' are the orbital angular momenta in the elastic and inelastic channels, respectively, and I is the spin of the intrinsic excitation. The crossing of the 6.1-MeV 3$^-$ line with the elastic line near L=20 indicates perfect matching; the double excitation with I=6 and 12.3 MeV is matched only for L=22, almost 10 MeV later.

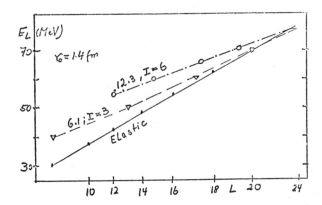

Fig. 3 Matching conditions for grazing angular momenta in $^{16}O+^{16}O$.

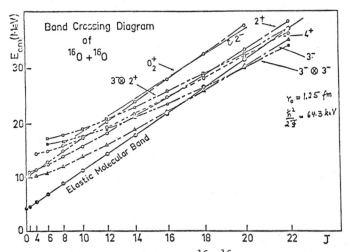

Fig. 4 Band crossing diagram of $^{16}O+^{16}O$ indicating matching of shape resonances in different inelastic channels.

Another description of the matching condition is provided by the band crossing model (BCM) of Abe, Kondo and Matsuse[2], in terms of the energy match of the molecular eigenstates, or shape resonances, in the ion-ion optical potentials in the elastic and inelastic channels. Figure 4 shows the sequence of shape resonances obtained from a realistic real optical potential.[6] For the elastic channel they form a rotational band. Its moment of inertia corresponds to a radius parameter r_o=1.25 fm, which is the radius of the potential pocket. The inelastic bands are obtained by assuming uncoupled intrinsic and orbital motion. Again, only the maximally aligned cases are shown, L'=J-I, where J is the resonance spin determined by the incoming L. Because of the different slopes of the bands in Fig. 4 and 5, the matching conditions are not quite the same in the two cases. The 6-MeV 3⁻ excitation is now matched earlier, near L=18. The 6-MeV 0⁺ and the 8.9-MeV 2⁻ excitation are never matched over the entire energy region (only one unit of h can be removed from the orbital motion for the 2⁻ excitation). Several double excitations are also shown. The BCM predicts that, at the crossing points, eigenstates in the entrance and exit channels are close enough to couple efficiently, and a strong resonance is observed in the inelastic channel. Conversely, one might not expect correlated resonances in the severely mismatched case.

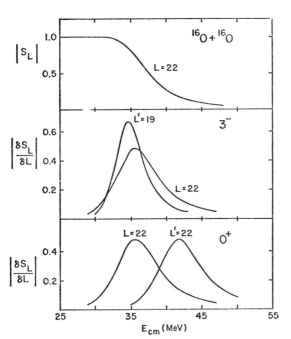

Fig. 5 Channel matching in the diffraction model.

Finally, we discuss how the matching condition expresses itself in the strong absorption or diffraction model.³ Figure 5 gives the example of the L=22 partial wave. In this model the kinematic conditions are important at the outer barrier. The inelastic cross section is proportional to the derivatives of the S matrices for the elastic and inelastic channels, $(dS/dL)^{1/2}$ $(dS'/dL')^{1/2}$, according to their respective angular momenta. These derivatives are shown in Fig. 5 as a function of energy, for the elastic channel and the 6-MeV 3⁻ and 0⁺ excitations, respectively. Obviously, near 35 MeV, the 3⁻-channel L-window has almost maximal overlap with that in the elastic channel which produces a large peak in the cross section. A much smaller peak can be expected in the 6-MeV 0⁺ excitation around 40 MeV. Thus, we expect that the matched peaks will be almost maximally out of phase with the mismatched ones. This prediction was the basis for the first experiment described below.

III. THE 6.05-MeV 0⁺ EXCITATION FUNCTION

The energy degeneracy of the excited 0⁺ state and the 3⁻ state offers the possibility to compare directly two very different matching conditions. Because the energy separation of the two states is only 80 keV, it is difficult to observe the 0_2^+ excitation directly in the particle spectrum. Instead, we chose[7] to measure the total angle integrated population cross section through observation of the nuclear pair transition to the ground state. This is possible because the feeding of the state from higher excited states is negligible (see Fig. 2), and since the side feeding from the fusion-4 α evaporation and from direct transfer to ^{20}Ne with subsequent α decay to this state in ^{16}O can be estimated also to be very small. A specially designed pair detector with large solid angle was then used to measure the excited-0^+ excitation function from 25 to 40 MeV (c.m.). Results are shown in Fig. 6 and compared to the 3^- excitation function[8] obtained by detection of the 6-MeV γ decay, and with the 90° elastic scattering data.[5] Although the 0^+ cross section is only 1/20 that of the 3^- excitation, the 0^+ excitation function displays clear and regular peaks. In addition, they are essentially in phase with the peaks observed in the 3^- curve (this curve contains the sum of single and double 3^- excitations). Two causes are responsible for the small 0^+ cross section: the bad mismatch, and the small E0 matrix element (which amounts to only 10% of the monopole sum rule).

In order to test now the strong absorption model against these data, we make a quantitative fit to both excitation functions using the formalism of Hahn as outlined by Phillips et al.[3] However, we use a different and, as we believe, more realistic parametrization for the S-matrix. The Ericson parametrization[9]

$$S_L = [1+\exp(i\gamma+(L_g-L)/\Delta(E))]^{-1}$$

provides a suitable analytic form if the grazing angular momenta L_g and the fall-off width $\Delta(E)$ are obtained from fits to a realistic optical potential, in this , case the Gobbi potential[4], with an increased absorptive part to simulate the strong absorption inherent in the model. The 3^- and 0^+ excitation strengths were computed using collective form factors with the deformation lengths taken from electron scattering. This procedure produced very reasonable fits to the 3^- curve (see Fig. 7), but the 0^+ strength had to be reduced by a factor of five. The result that the 3^- oscillations are well reproduced is, of course, expected on the basis of the earlier discussion. However,

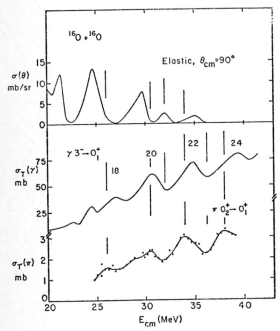

Fig 6 Comparison of the 0_2^+ excitation function (ref. 6) with the 3^- excitation function (ref. 8) and 90° elastic scattering (ref. 5).

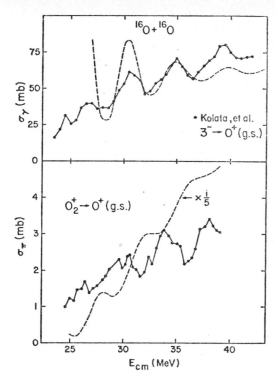

equally expected, the calculation predicts oscillations in the 0^+ cross section which are completely out of phase with the 3^- oscillations, and which are at variance with the data. Thus the simple diffraction model appears to be ruled out, at least in this case. We conclude then that the interior wave must plan an important part in forming the resonance peak. In fact, it can be shown that even with the interior wave no peak is obtained unless the potential has a pocket.

The systematic correlation of the two curves over an extensive energy range is indicative of strong coupling between the two inelastic channels. A realistic coupled channel calculation involving the 3^-, 0^+ and the elastic channels, was recently performed by Tanimura and Mosel.[10] This calculation was able to achieve excellent fits to both curves as shown in Fig. 8, with very few free parameters.

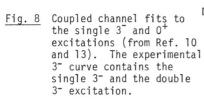

Fig. 7 Diffraction model calculations (dashed lines) to the 3^- excitation (ref. 8) and the 0_2^+ excitation function.

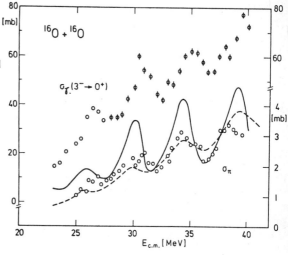

Fig. 8 Coupled channel fits to the single 3^- and 0^+ excitations (from Ref. 10 and 13). The experimental 3^- curve contains the single 3^- and the double 3^- excitation.

Most importantly, they obtain a complete correlation of peak positions in both curves. More systematic studies will be necessary to test the universality of this result, because its implications are most interesting. This becomes apparent as one considers in more detail how the correlation is, in fact, achieved. Although one starts with diagonal potentials which are displaced in each channel by the channel Q value, the coupling interaction strongly affects the resulting effective diagonal channel potentials. As can be seen in Fig.9, in the end all potentials become essentially degenerate, for the partial waves in question (J=18 and 20), all of the way out to the top of the barrier. Thus, because of the strong interaction between channels, even the initially mismatched channels become matched at the outer barrier, and at the same time the shape resonances in the interior are moved close to each other in energy. Inspection of the channel wave function shows that it is the shape resonance which produces the peaks in the cross sections.

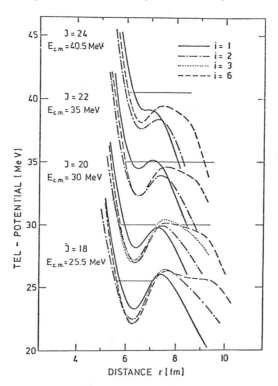

Fig. 9 Equivalent diagonal potentials emerging from a coupled channel calculation (from Ref. 10). Channels 1,2 and 6 refer to the elastic, aligned 3^- and excited 0^+ excitation. For J=18 and 20 these 3 channels have almost degenerated potentials.

This result requires further work; for instance it should be shown experimentally that the spins of the correlated peaks in the 3^- and 0^+ curves are, in fact, the same. If confirmed, it would be highly significant for our knowledge of optical potentials in excited channels. The phenomenological potential which is obtained for the ground states by fits to the elastic scattering cross section contains, of course, all couplings to excited states and is, in this sense, the correct effective potential. However, this potential would be incorrect for the excited channels because these are modified differently by the coupling. Such effects were perhaps already implied in earlier fits using the BCM in which in many cases the potentials had to be shifted arbitrarily to obtain agreement with the data.

IV. HIGHER SINGLE EXCITATIONS

An immediate check of the general validity of these effects if offered by a study of the higher lying, mismatched inelastic channels. The 2^- excitation at 8.88 MeV is the most interesting one because its matching conditions are so similar to those of the 0^+ excitation. Additional interest in this state comes from the fact that it can be excited only in a two-step process. In a collective description the excitation proceeds most efficiently by an octupole-quadrupole sequence through the 6-MeV 3^- state.

As can be seen from Fig. 2, all the higher states are sufficiently separated from each other, so that the respective excitation probability can be obtained directly from the ^{16}O recoil spectrum. In the present experiment[11] both reaction partners were detected in kinematic coincidence using two wide-angle detectors with Z identification in one arm. Position sensitive detectors were used so that kinematic corrections could be applied over an angle range from 25 to 40° (lab) thus providing a somewhat angle-averaged excitation probability. A kinematically corrected Q spectrum is shown in Fig. 10. The spectrum shows all single and double excitations which one expects from the ^{16}O level diagram. The resolution of all peaks is limited by the recoil broadening from γ emission. The higher excitations, especially the double excitations, are quite strong relative to the well matched 3^- single excitation. We note that this is the first experiment which detects the 3^- state directly rather than through its γ decay, and thus separates the single 3^- from the double[8] 3^- excitation, in the $^{16}O+^{16}O$ system.

Figure 11 gives the preliminary excitation functions for all single excitations. The data were taken in 250 keV (c.m.) steps, from 22 to 35 MeV. They show a great deal of narrow structure which must eventually be averaged over to obtain the gross structure. But even in this preliminary presentation gross structure is readily apparent in the 3^- curve, centered at 27.5, 30.5 and 34 MeV. The gross structure in the 2^+ channel tracks these peaks quite accurately. More importantly, the 2^- curve also correlates strongly with the 3^- curve, although its cross section is generally only about 1/10 of the 3^- one. As in the case of the 0^+

excitation, the weak cross section is a combination of the large mismatch and the weak (two-step) excitation probability. As a group, then, these additional single excitations seem to confirm the conclusions reached above, and indicate strong coupling between channels. The calculations which include these additional channels, remain to be done.

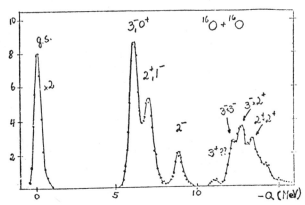

Fig. 10   Kinematically corrected Q spectrum of recoiling ^{16}O ions in kinematic coincidence obtained over an angle range from 25° to 40°, from inelastic scattering of ^{16}O on ^{16}O, at E_{lab}=66 MeV.

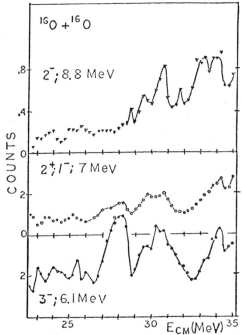

Fig. 11   Excitation functions of several single-nucleus excitations. The data are preliminary and average over an angle range from 25° to 40°.

V. THE DOUBLE EXCITATIONS

Finally, we show in Fig. 12 the excitation functions of the double excitations as they appear from our preliminary analysis. Again these curves show much structure which must be averaged out to obtain the gross structure peaks. The relative cross sections within the group appear to follow the pattern suggested by the matching conditions: the $3^-\times 3^-$ excitation is strongest, the $2^+\times 2^+$ excitation the weakest. But again, only an actual calculation will show how much this ratios are also influenced by the matrix elements. Relative to the 3^- single excitation shown in Fig. 11 one notes that the $3^-\times 3^-$ excitation is about half as strong. This indicates good matching, in agreement with the BCM predictions given in Fig. 4.

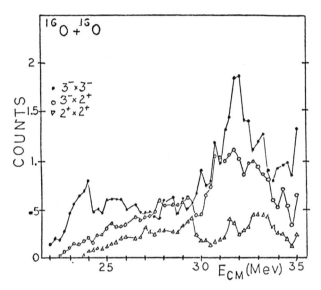

Fig. 12 Excitation functions of the major double excitations from inelastic scattering of ^{16}O on ^{16}O. These data are preliminary.

The energy range spanned by these data is not sufficient to cover several gross structure peaks, and the data are presently being extended to higher energies. However, a peak centered at 32 MeV is quite prominent in both the $3^-\times 3^-$ and the $3^-\times 2^+$ curves. Thus, at least, these two channels are again correlated. However, a closer inspection of the single-excitation curves and the double excitation curves show that the two groups, over the range studied here, are, in fact, anti-correlated. This is a surprising result since it upsets the systematic behavior that seemed to emerge from the earlier studies. In an independent experiment, Wells et al.[12] have recently measured the same double excitations by a γ-γ coincidence technique. Their preliminary results span a much larger energy range, up to 40 MeV, and show several additional gross structure peaks. In the region of overlap the two sets of data are in agreement. In particular, the correlation between peaks in the $3^-\times3^-$ and $3^-\times2^+$ excitations is present also in these (more fully angle-integrated) data. We show in Fig. 13 the strongest double excitation from the preliminary data of Wells et al.[12], up to 40 MeV. The curve is a average through the actual data points. In the same plot the 3^- γ data of Kolato et al.[8] are shown for comparison. The vertical scale is not to scale for the two curves. The anticorrelation between the two curves, up to ∼38 MeV is obvious.

The correlation between peaks in the different double-excitations, on one hand, and the anticorrelation between single and double excitation on the other hand, is a real puzzle. There appears to be sufficient systematic evidence to suggest that this behavior has a deeper reason. But the cross sections are too small relative to the total reaction cross sections to invoke depletion of one channel by another due to unitarity. Thus only detailed coupled channel calculations will be able to provide some insight.

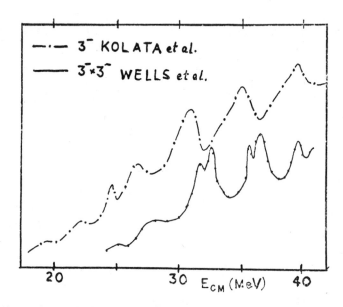

Fig. 13 Comparison of the double-3^- excitation function with the single 3^- excitation function. The 3^- data from Kolata et al. (Ref. 8) include the double excitation. The double excitation curve is obtained form the preliminary data of Wells et al. (Ref. 12). In both cases the curves were hand-drawn through the data.

VI. CONCLUSIONS

It has been shown that a consistent analysis of gross structure in well-matched and mismatched channels produces significant insight into the nature of the gross structure observed in light heavy ion collisions. In particular, it permits a differentiation between a purely diffractive and a resonance explanation. It appears that in the cases studied so far, the explanation of gross structure in terms of shape resonances in an effective optical potential is strongly favored. These resonances would then serve as doorway states for intermediate structure. In addition, there is evidence for strong coupling between at least some channels, perhaps those which involve surface excitations (although this is pure speculation at this point). The diagonal potentials are significantly affected by such a coupling. This makes a quantitative calculation involving several channels, tedious and expensive. Although all single excitations which we have studied in the $^{16}O+^{16}O$ system follow the pattern of strong correlation of peaks, the double excitations as a group seem to be anticorrelated. This phenomenon is not understood as yet, and it is not clear whether this explanation is of a general nature. It appears, for example, that in the $^{12}C+^{12}C$ system the systematics is quite different and all peaks are correlated. The general result, however, that the resonance character of the gross structure is supported by these more detailed studies, is very satisfying.

ACKNOWLEDGMENTS

The results reported above have been obtained in collaboration with two groups at Stony Brook:

The study of the excited 0^+ state involved W.S. Freeman, T. Chapuran, W. Piel and H.W. Wilschut.

The study of the higher excited states was done with D. Abriola, J.S. Karp, R.L. McGrath, W. Watson, S.Y. Zhu and B. Yang.

We thank Dr. D. Balamuth for the permission to use his results prior to publication.

REFERENCES

1. D. A. Bromley, Nuclear Molecular Phenomena, ed. by N. Cindro, North-Holland, 1978, p. 211.
2. Y. Abe, Y. Kondo and T. Matsuse, Prog. Theoret. Phys. Suppl. $\underline{68}$, 303 (1980).
3. R. L. Phillips, K. A. Erb, D. A. Bromley and T. Weneser, Phys. Rev. Letters $\underline{42}$, 506 (1979).
4. A. Gobbi, R. Willard, L. Chua, D. Shapira, D. A. Bromley, Phys. Rev. $\underline{C6}$, 30 (1973).
5. J. V. Maher, M. W. Sachs, R. H. Siemssen, A. Weidinger, D. A. Bromley, Phys. Rev. $\underline{188}$, 1665 (1969).
6. Y. Kondo, D. A. Bromley and Y. Abe, Phys. Rev. $\underline{C22}$, 1068 (1980).
7. W. S. Freeman, H. W. Wilschut, T. Chapuran, W. F. Piel and P. Paul, Phys. Rev. Letters $\underline{45}$, 1479 (1980).
8. J. J. Kolata, R. M. Freeman, F. Haas, B. Heusch, A. Gallman, Phys. Rev. $\underline{C16}$, 891 (1977).
9. T. E. O. Ericson in, "Preludes of Theoretical Physics," (A. deShalit, H. Feshbach and L. VanHowe, eds.), North-Holland, Amsterdam (1965) 321.
10. O. Tanimura and U. Mosel, Phys. Rev. $\underline{C24}$, 321 (1981).
11. D. Abriola, J. S. Karp, R. L. McGrath, P. Paul, W. Watson, B. Yang and S. Y. Zhu, to be published.
12. W. K. Wells, D. P. Balamuth and D. P. Bybell, to be published.
13. U. Mosel, private communication.

IV. RESONANCE STUDIES IN PARTICULAR REACTIONS -

sd - SHELL NUCLEI AND HEAVIER

RESONANCES IN s-d SHELL NUCLEI

J. P. Schiffer

Argonne National Laboratory, Argonne, IL 60439

and

University of Chicago, Chicago, IL 60637

Introduction

I would like to start out this talk with a mild critique of the field of heavy-ion resonance studies. We have been perhaps so bedazzled by the qualitative richness of phenomena, that we have forgotten some of the basic quantitative criteria that allow us to distinguish experimentally verifiable theories, from more nebulous points of view. I would like to comment on four points.

We are suffering from considerable semantic confusion in this field. There is a tendency to treat terms, such as "molecular resonances", "shape resonances", "rotational bands", "Regge trajectories", "barrier top resonances", "intermediate structures" as if they represented a set of well-defined orthogonal theories. In fact, they represent very similar, closely-linked models and people use the same terms to describe somewhat different concepts, or different terms to describe the same concept. We would all benefit if this nomenclature could be defined a little more clearly, so we may have assurance that when we hear a term we know which concept is meant, and not have to guess at it from the general context of the discussion.

Second, the criteria for identifying a resonance and assigning angular momenta, are not very strict. Often a Legendre polynomial is slapped on top of a few oscillations in an angular distribution and if the maxima line up, an unqualified spin assignment is quoted. I really believe one has to do more than that for the angular distribution and, whenever possible, one should try to follow the phase shift through a resonance. At the very least one should indicate the tentativeness of angular momentum assignments if the data are incomplete. Some examples of such problems are shown in Fig. 1.

Thirdly, the question of what is meant by a reduced width seems to be almost forgotten. Yet this is the only parameter, other than the energy and spin-parity of a resonance, that permits testing against models. The fraction of the single-particle width that is contained in a partial width depends sensitively on energy and angular momentum. In R-matrix theory[1] we have

$$\Gamma_{\lambda c} = \underbrace{2kRP_\ell}_{\text{outside}} \times \underbrace{\gamma^2_{\lambda c}}_{\text{inside}}$$

where the outside term is readily calculable, requiring only the definition of a radius, and all the "inside" structure information is contained in γ^2. This reduced width γ^2 then must be divided by the "Wigner-Teichman Limit"

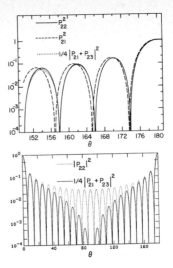

Fig. 1. Hypothetical angular distribution corresponding to a pure L = 22 resonance, alongside an equal coherent mixture of L = 21 and 23 amplitudes. Note that the two are indistinguishable over a limited range of angles, but become more easy to separate if data over an angular interval $\gtrsim \pi/2$ become available.

$$\theta_\lambda^2 \equiv \gamma_\lambda^2 / [3/2 \ (\hbar^2/MR^2)]$$

to give the single-particle fraction. The penetrability factors may be slightly in error because of uncertainties in radii and correction for diffuseness and there are refinements for large reduced widths arising from energy-derivatives of the shift functions--but it is better to have a uniform procedure that may be in error in some systematic way, than to have to rely entirely on qualitative statements.

And finally as my fourth pedantic complaint, I would like to mention the confusion regarding concepts of the mixing width (sometimes called Γ↓) and its connection to the imaginary potential. In the simple optical model,[2] the imaginary potential is introduced to provide the mixing between the simple model degree of freedom (the single particle state in the optical model) and the more complicated states of the nucleus. It simply provides a damping of the model wavefunction on the time scale corresponding to the width of the giant resonance. The quantities of interest are no longer the reduced widths of the individual fine-structure resonances, but the average reduced width or strength function. The reduced width of a whole bump may be extracted without reference to the details of the fine structure underneath. The extent of this mixing on the average is contained in the imaginary potential. But its magnitude does _not_ depend on the fine structure level density as is sometimes assumed in the literature. As was shown many years ago in the estimate of radiative widths by Blatt and Weisskopf,[3] the mixing matrix element _per state_ is proportional to the "complexity" of these states, which in turn is inversely proportional to the density of states. If there is a more complex regime with a higher density of states, _more_ states will be mixed with _smaller_ matrix elements, but the _energy_ spreading width will be unchanged. That is why we have giant resonances with almost the same widths in heavy nuclei as light ones, even though the fine-structure level densities differ by many orders of magnitude. All this, of course, applies to all "simple" states mixing into more complicated ones. It does not matter whether it is a single-neutron state, a multipole giant resonance, or some other model configuration. The behavior of the strength function is a method of model testing that is schematically outlined in Fig. 2.

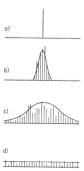

Fig. 2. Schematic representation of model testing by looking for giant resonances in a strength function. The top (a) represents a "perfect model" which puts all the strength into one state (e.g. the shell model for ^{209}Bi g.s.). The next (b) represents a rather good model, such as isobaric analog resonances; (c) is a typical giant resonance, such as the giant dipole resonance or neutron shape resonance; while (d) stands for a model description which may be technically correct, but not useful.

Optical Potential Surface Transparency

For one to see high angular momentum giant resonances in heavy-ion reactions it is a necessary (but not sufficient) condition that the imaginary potential be small in the relevant region. For high angular momentum states this is the surface region of the nucleus-nucleus potential. In Fig. 3(a) we show two such potentials differing only in the surface imaginary part, and Fig. 3(b) shows the dramatic result in the back-angle scattering. Such effects have been known for many years in alpha-particle scattering[4]--where these back-angle structures suddenly disappear as valence nucleons are added in a new oscillator shell. It seems qualitatively plausible that valence nucleons in a new (and larger) shell <u>should</u> suddenly increase the surface absorption, though we have no microscopic theory for imaginary potentials, comparable to the folding model for the real

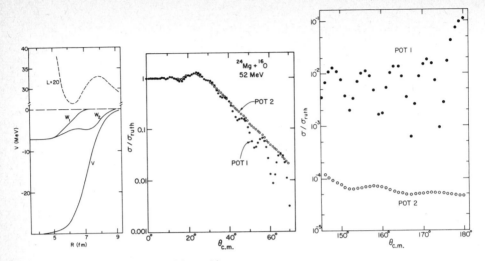

Fig. 3. Elastic scattering for $^{16}O + ^{24}Mg$ calculated for two potentials in which the surface absorption has been modified, as shown in Fig. 3a, in the region critical for the L = 20 partial wave. Potential 1 has small surface absorption, while in potential 2 absorption is increased. The resulting scattering is modified somewhat at forward angles, and is drastically different at back angles.

potential. In heavy-ion reactions similar effects have been found, not only at closed oscillator shells, but in the middle of the s-d shell, where there are dramatic differences between the back-angle scattering of ^{16}O from ^{28}Si and ^{30}Si.[5] Here the $s_{1/2}$ neutron shell is filling, which has a larger surface component and may thus have a strong influence on the surface damping.

But here, as may be seen in Fig. 4, there is also a strong resonance-like behavior in the back-angle scattering. Unfortunately the elastic scattering involves many partial waves and it does not seem very easy to make unique angular momentum assignments to these structures.

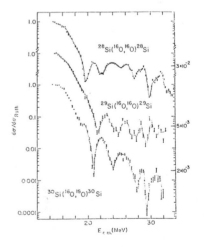

Fig. 4. Excitation functions at 180° for elastic ^{16}O scattering from $^{28,29,30}Si$ from Ref. 5. The average level of back angle yields is shown on the right.

Transfer Reactions

The difficulty with studying resonance effects in elastic and inelastic scattering is the fact, illustrated in Fig. 5 schematically, that there are relatively many partial waves contributing to the process--and the anomalous resonant behavior of one partial wave will not have a unique signature unless one has a very accurate model of the non-resonant processes. There are similar problems in inelastic scattering. The situation is, however, somewhat more favorable for a well-matched transfer reaction.

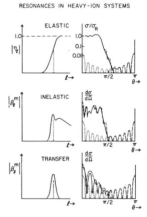

Fig. 5. Schematic illustration of the number of partial waves contributing to elastic or inelastic scattering and a well-matched transfer reaction. The contribution of one anomalous (resonant) partial wave is shown as a dotted line, the coherent sum of the direct and anomalous contribution, as a dashed line.

Such is the case for $^{24}Mg(^{16}O,^{12}C)^{28}Si$, where, as we may see in Fig. 6, there are dramatic effects at 0°, 90° and 180°.[6] The yield at forward angles is well correlated between the ground state of ^{28}Si and the

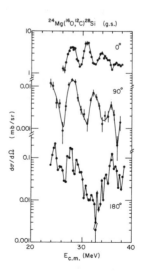

Fig. 6. Excitation functions for the $^{24}Mg(^{16}O,^{12}C)^{28}Si$ reaction at θ_{cm} = 0°, 90°, and 180° from Ref. 6.

1.78-MeV 2^+ state, as well as the 6.9-MeV 3^- and 9.7-MeV 5^- states. In fact, all the states strongly excited in inelastic processes on ^{28}Si seem to show the correlation over the resonances.[7] Detailed angular distributions for the ground-state transition have allowed us to follow the phases through Breit-Wigner circles and assign angular momenta of 20^+, 23^-, and (26^+) to the structures at 27.6, 30.8, and 36 MeV.[6] Please note the critical role played by the excitation functions at 0°, 90° and 180° in these assignments (odd L states <u>cannot</u> appear at 90° and the lack of correlation between 0° and 180° <u>must</u> mean contribution from both odd and even parity states). The reduced widths we can only obtain as a product of the entrance and exit channel widths, since it has not been possible to locate the same resonances uniquely in the elastic channels; the values for $\sqrt{\theta^2_{in}\theta^2_{out}} \approx 1-2\%$. The partial widths for the higher excited states of ^{28}Si can only be estimated roughly because angular distributions have not been measured. It appears that the reduced widths summed over the observed ^{28}Si states and the ^{24}Mg ground state are somewhere between 10 and 40% of the Wigner limit.

Our information on these resonances is still rather limited. We observe a spin sequence that does not fit any simple model. We know as yet nothing of the partial widths for these resonances to excited states of ^{24}Mg or some of the equally well-matched channels in ^{20}Ne + ^{20}Ne. Perhaps they are structurally related to the ground states and strongly-coupled excited states of these nuclei.

The resonances continue up to higher excitation energy as may be seen in the 0° excitation function in Fig. 7. The corresponding excitation function at 180°

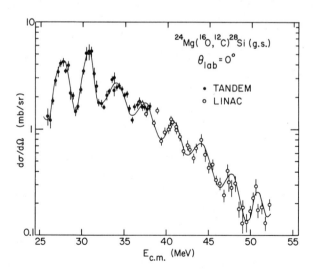

Fig. 7. Excitation function of the ^{24}Mg(^{16}O,^{12}C)^{28}Si reaction at 0°, using the superconducting linac booster at Argonne.

has also been measured and no simple pattern emerges--though some statements about the parities of underlying resonances may be made, as is summarized in Table I.

Table I

E_{cm}	J	π
27.6	20	+
30.8	23	−
36	(26)	+
38.5		−
41.5		(−)
43.5		(+)
46		(−)
50.5		(−)

The general decrease in cross sections is mirrored in the predicted decrease of the direct reaction calculation--but the forward-backward asymmetry seems to remain at roughly 10:1.

Other systems have been investigated in searching for such resonances. These seem much weaker in $^{26}Mg(^{16}O,^{12}C)$ and in $^{28}Si(^{16}O,^{12}C)^{32}S$.

Conclusion

It appears that the system we have studied here, representing ^{40}Ca as the composite nucleus, is perhaps the heaviest one that exhibits strong enough resonances that quantitative measurements may be contemplated. But we have uncovered only a small corner of what is there and even within this system a huge amount of work remains. The work so far represents perhaps 3 man years of research effort. There is easily an order of magnitude more work remaining, unless there is a substantial improvement in detection techniques.

The nature of these resonances is not yet clear. The sequence may perhaps have an explanation that is schematically outlined in Fig. 8, namely that there are several families of quasistationary states in ^{40}Ca, but that the slopes of these families do not necessarily coincide with the slope of the grazing partial waves that provide us with a narrow transparent strip of a window on the underlying structure of the nucleus. We must concentrate a lot of effort and ingenuity in order to maximize the information we gather through this window and only then may we hope to sensibly attempt forming hypotheses about the underlying simple pattern.

That the structures we see are "simple" is clear from the fact that we are in the region of ^{40}Ca where the density of states is higher (by 3-6 orders of magnitude) than the spacing of the observed structures. It is not clear whether the fact that the structures appear primarily in alpha-particle nuclei may have some

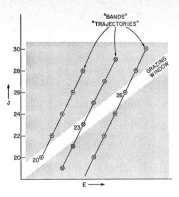

Fig. 8. Schematic representation of a possible relationship between the transparency of the "grazing window" and some underlying ordered structure in heavy-ion scattering or reactions.

special structural significance—or whether it may be a feature of the transparency in the reaction mechanism. A lot of hard work remains but the evidence that there is <u>some</u> relatively simple underlying order appears to be overwhelming.

This work was supported by the U. S. Department of Energy under Contract W-31-109-Eng-38.

References

[1] A. M. Lane and R. G. Thomas, Rev. Mod. Phys. <u>29</u>, 191 (1957).
[2] H. Feshbach, C. E. Porter, and V. F. Weisskopf, Phys. Rev. <u>96</u>, 448 (1954).
[3] J. M. Blatt and V. F. Weisskopf, Theoretical Nuclear Physics, Wiley (1952).
[4] See, for instance, G. Gaul, H. Lüdecke, R. Santo, H. Schmeing, and R. Stock, Nucl. Phys., <u>A137</u>, 177 (1969), or J. S. Eck, W. J. Thompson, K. A. Eberhard, J. Schiele, and W. Trombik, Nucl. Phys. <u>A255</u>, 157 (1975).
[5] P. Braun-Munzinger et al., Phys. Rev. C <u>24</u>, 1010 (1981).
[6] S. J. Sanders et al., Phys. Rev. C <u>21</u>, 1810 (1980).
[7] S. J. Sanders et al., Phys. Rev. C <u>22</u>, 1914 (1980).

HIGH ANGULAR MOMENTUM RESONANCES IN ^{28}Si + ^{28}Si SCATTERING*

R. R. Betts

Chemistry Division, Argonne National Laboratory,
9700 South Cass Avenue, Argonne, Illinois USA

The initial expectation was that resonance behavior in heavy-ion systems would be limited to only a few special systems and to energies not too far above the Coulomb barrier. The past few years have shown us that this behavior has a much wider extent than was previously thought possible, although in many cases the indications of resonance behavior is at best qualitative. In this talk I will present results for the ^{28}Si + ^{28}Si system - the heaviest in which resonance behavior has yet been observed.

Initial measurements[1] of ^{28}Si + ^{28}Si elastic scattering angular distributions (Fig. 1) show little evidence for the surface transparency required for resonance behavior. The angular distributions

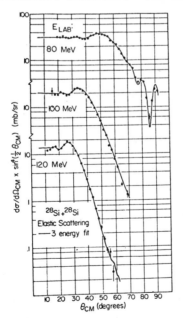

FIGURE 1.

show a Fresnel type diffraction pattern, characteristic of strong absorption. Measurements at large angles[2] and at cross-section levels of about 10^{-4} of the Coulomb cross-section, however, show distinct resonance-like behavior. Elastic scattering cross-sections measured at two angles over a range of energies in the vicinity of twice the Coulomb barrier are shown in Fig. 2. These data do not show the regular behavior observed in lighter symmetric systems but

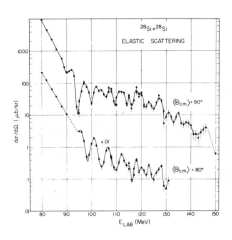

FIGURE 2.

seem to display both structures of width several MeV and indications of much narrower structure, although the step size of 1 MeV (lab) makes the distinction rather difficult. Similar behavior is seen in channels other than the elastic scattering channel as shown in Fig. 3. Of particular interest here is the nature of these inelastic channels. Initially it was thought that a peak observed near 6.5 MeV in the spectrum corresponded to the excitation of the collective 6.89 MeV 3^- state in ^{28}Si. A high resolution study[3] has shown that this initial supposition is incorrect and that this peak corresponds to a mutual excitation of the 1.78 MeV 2^+ and 4.62 MeV 4^+ levels which then appears at an apparent excitation energy of 6.40 MeV. In fact, for these energies and for this angular range, the inelastic scattering spectrum of ^{28}Si + ^{28}Si is dominated by mutual excitations of yrast levels as

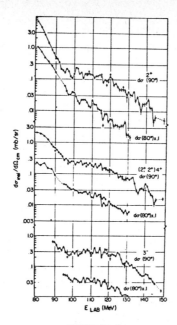

FIGURE 3.

shown in Fig. 4 - a result which can be understood qualitatively in terms of angular momentum matching if the spins of the mutually excited fragments are aligned parallel to one another.

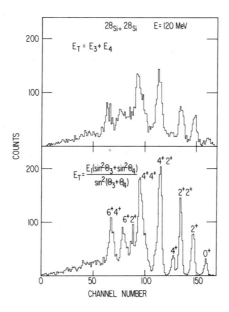

FIGURE 4.

The structures observed in the single angle excitation functions persist when the cross-sections are integrated over a reltively large angular range,[4] $\theta_{CM} \sim 60°-90°$, as shown in Fig. 5, and also appear in the angle integrated yield summed over all final channels with $Q \geq -10$ MeV.

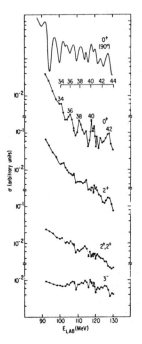

FIGURE 5.

The elastic scattering angular distributions[4] have shapes characterized by single Legendre polynomials squared for each of the broad structures observed in the angle integrated elastic scattering cross-sections. These are shown in Fig. 6. The values of L associated with these Legendre polynomials follow the grazing partial wave rather closely as indicated in Fig. 5. Only in the deep minima between the broad structures do we observe angular distributions not characterized by a single L although the shapes of these are still highly oscillatory.

The appearance of a definite narrow structure near 118 MeV as shown in Fig. 5 led us to a further investigation in which angle integrated cross-sections for elastic and inelastic scattering were measured

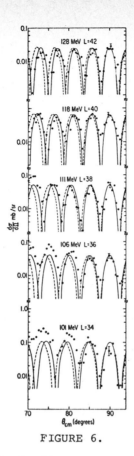

FIGURE 6.

in 100 keV steps over the bombarding energy range 105-121 MeV using a target which was 70 keV thick to the beam.[5] A typical spectrum obtained during this experiment is shown in Fig. 7. Yields for the elastic,

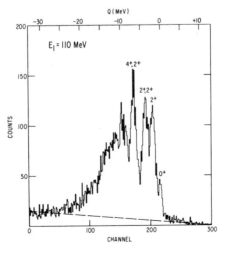

FIGURE 7.

single 2^+, mutual 2^+, mutual 4^+2^+ and the rest of the yield in the spectrum as well as the total yield in the spectrum were extracted. Comparison of repeat points taken during the experiment indicate that the errors associated with these yields are less than 5% with the exception of the elastic scattering channel which has poorer statistics and an error of 10%. These yields are shown in Fig. 8. The data are

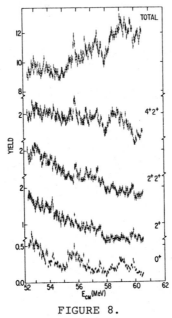

FIGURE 8.

in good agreement with our previous results taken in 1 MeV steps but the finer energy steps reveal a much richer structure than was evident in the earlier data. Structures of width 100-200 keV appear throughout the data for the individual channels as well as in the total yield for the whole spectrum. There appears to be a strong correlation between many of the narrow structures, a feature which is not expected for structures arising from statistical fluctuations.

To put the observed correlations on a quantitative footing we have performed a correlation analysis of the data and compared the results with the expectations for uncorrelated data. As the question of confidence limits on the results of correlation analysis is not well documented, the main features of our analysis are outlined here.

The experimental cross-sections σ were used to generate an average cross-section $\langle\sigma\rangle$ using a Gaussian smoothing function with FWHM = 1500 keV. These quantities were then used to generate

$$y_i = \frac{\sigma_i}{\langle\sigma_i\rangle}$$

where now the broad ($\Gamma \sim 2$ MeV) structures have been removed by the averaging procedure. In principle, the distribution of y can then be compared with the theoretical expectation $P(y)$ based on statistical fluctuations. $P(y)$ is, however, a function of two variables N, the number of channels, and y_D the non-fluctuating or "direct" contribution to the cross-section. N is given by geometrical constraints and may be considered fairly well known but y_D is in general undetermined except by comparison with the experimental results through the relation

$$C(0) = \frac{1}{N_{eff}} = \frac{1}{N}(1 - y_D^2)$$

where $C(0)$ is given by the normalized variance of the data.

$$C_i(0) = \frac{\langle\sigma_i^2\rangle}{\langle\sigma_i\rangle^2} - 1$$

Comparison of the experimental distribution of y with the theoretical expectation is therefore not particularly useful as the latter then has the experimental results factored in. We therefore introduce a new variable

$$X_i = \frac{y_i - \langle y_i\rangle}{\sqrt{\langle y_i^2\rangle - \langle y_i\rangle^2}}$$

where the average is now over the entire range of data. Both experimentally and theoretically X_i are normally distributed with variance unity. For the summed deviation function and normalized cross-correlation

$$D(E) = \frac{1}{N} \sum_{i=1}^{N} x_i$$

$$C(E) = \frac{2}{N(N-1)} \sum_{i>j=1}^{N} x_i x_j$$

and for uncorrelated data we therefore expect $D(E)$ and $C(E)$ to have distributions with mean zero and standard deviations $\frac{1}{\sqrt{N}}$ and $\sqrt{\frac{2}{N(N-1)}}$ respectively. The former of these results is trivial, the latter is approximate but accurate, the exact result requiring numerical integrations. The experimental values for $D(E)$ and $C(E)$ are shown in Fig. 9 together with the expected standard deviations of the theoretical

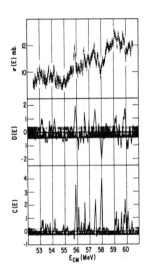

FIGURE 9.

distributions (shaded areas). Most of the narrow structures appear with values of $D(E)$ and $C(E)$ several standard deviations away from zero. The experimental and theoretical frequency distributions are shown in Fig. 10 - the probability that these experimental distributions result from uncorrelated fluctuations is in both cases less than 1 part in 10^5.

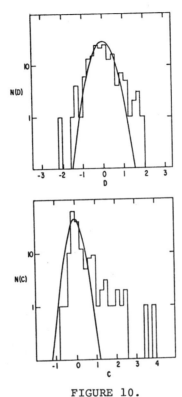

FIGURE 10.

Finally, the energy averaged value of C(E) is expected to be

$$<C(E)> = 0.0 \pm \sqrt{\frac{2}{N(N-1)}} \sqrt{\frac{\pi \Gamma}{\Delta}} = 0.0 \pm .04$$

where $\Delta/\pi\Gamma$ is the number of independent cross-sections (Γ = coherence width, Δ = energy range of the data). This is to be compared with the experimental value of $<C(E)> = 0.27$. We therefore conclude that the narrow structures observed in the data do not arise from statistical fluctuations and thus must be ascribed in true intermediate structure resonances.

On the basis of all the above we conclude that we are dealing with a number of narrow resonances with extremely high angular momentum - of order 40 $\hbar$. If we consider the compound nucleus at these excitation energies and angular momenta using, for example, the rotating liquid drop model to estimate the position of the yrast line, we find level densities which are still several thousand per MeV. This implies a partial width for the average compound nuclear level to decay into the ^{28}Si + ^{28}Si elastic channel of a few eV whereas from the experimental results we estimate values of a few keV. As is the case with the much lighter systems, we are therefore faced with the existence of narrow resonances in a region of high level density which apparently have a strong structural connection with the symmetric entrance channel.

An interpretation in terms of models which utilize a "quasi-molecular" basis is not implausible. Fig. 11 shows the spectrum of

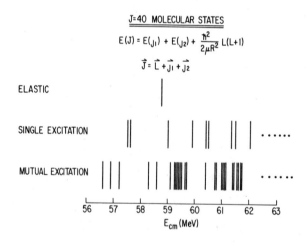

FIGURE 11.

J=40 molecular levels formed by coupling of the excitations of the individual nuclei to the rotations of the dinuclear system with a center-to-center radius equal to the strong absorption radius. The number of such states within the observed width of the gross structures is certainly not inconsistent with the experimental observations. The question of how these basis states are mixed and the value of the background absorption reflecting mixing with more complex excitations are however open ones.

Another interesting speculation is based on the results of calculations of shell structure as a function of deformation and angular momentum for the nucleus ^{56}Ni - the compound nucleus for ^{28}Si + ^{28}Si. These calculations[6] indicate the occurrence of a second minimum at large deformations for a limited range of angular momenta in the vicinity of J=40. Such a second minimum can give rise to shape isomeric states which are expected to decay largely by fission. The connection between these calculations and the present experimental results is tenuous at best, although such fissioning shape isomers may be expected to manifest themselves in the manner observed.

To investigate this possibility we have performed an experiment in which we attempt to observe these resonances by populating the composite system via the ^{16}O + ^{40}Ca entrance channel and looking for decays into ^{28}Si + ^{28}Si. The experiment was performed using a kinematic coincidence arrangement similar to that of Ref. 3 in which the energies and angles of coincident fragments were used to obtain mass identification via two-body kinematics. A mass spectrum obtained at a bombarding energy of 75 MeV is shown in Fig. 12 - the yield of symmetric events is surprisingly large and corresponds to an angle integrated cross-section of several mb. This yield was measured in 250 keV steps over the bombarding energy range 72 to 78 MeV - the target thickness was comparable to the step size. The yield of symmetric events is shown plotted as a function of ^{56}Ni excitation energy in Fig. 13 and is compared with the total yield from the ^{28}Si + ^{28}Si entrance channel which is shown as the solid curve. The ^{28}Si + ^{28}Si data have been averaged so as to correspond to the

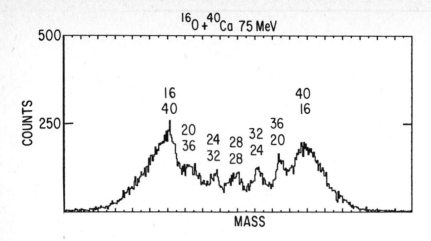

FIGURE 12.

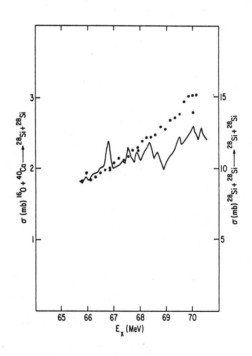

FIGURE 13.

same target thickness as for the $^{16}O + ^{40}Ca$ experiment. The $^{16}O + ^{40}Ca$ data are quite smooth showing none of the prominent structures observed in the $^{28}Si + ^{28}Si$ entrance channel data. The grazing angular momenta for the two entrance channels differ by only 1 ℏ for the same excitation energy and these data would therefore tend to suggest that the observed structures in the $^{28}Si + ^{28}Si$ reactions be described in terms of entrance channel degrees of freedom.

Finally, we address the extent to which resonance phenomena may appear in even heavier systems of which the $^{40}Ca + ^{40}Ca$ system has been considered to be the most likely to show such behavior. Data for the elastic scattering of ^{40}Ca on ^{40}Ca are shown as a function of energy in Fig. 14. These data represent the average differential cross-section over the center-of-mass angular range 77-103°. The solid

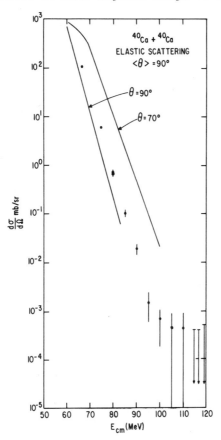

FIGURE 14.

lines show the results of an earlier study by an Orsay group.[7] The cross-sections fall smoothly down to a level of 0.5 μb/sr ($\frac{\sigma}{\sigma_{Coul}} \sim 10^{-6}$) with no hint of any leveling which might be characteristic of resonance behavior. We have also measured the total quasi-elastic and deep inelastic cross-sections for ^{40}Ca + ^{40}Ca in 1 MeV (lab) steps from 170 to 195 MeV. A preliminary analysis of these data indicate no structure at the 3% level and our tentative conclusion is that ^{40}Ca + ^{40}Ca does *not* show resonance behavior.

I would like to acknowledge my collaborators in the work that has been presented here. Namely, S. B. DiCenzo, J. F. Petersen, B. B. Back, S. Saini, W. Henning, I. Ahmad, B. G. Glagola, S. J. Sanders and B. Dichter.

REFERENCES

*This work was performed under the auspices of the Office of High Energy and Nuclear Physics, Division of Nuclear Physics, U. S. Department of Energy, under contract number W-31-109-ENG-38.

1. S. B. DiCenzo, Ph.D. Dissertation, Yale University 1980, (unpublished).
2. R. R. Betts, S. B. DiCenzo and J. F. Petersen, Phys. Rev. Lett. 43, 253 (1979).
3. R. R. Betts, H.-G. Clerc, B. B. Back, I. Ahmad, K. L. Wolf and B. G. Glagola, Phys. Rev. Lett. 46, 313 (1981).
4. R. R. Betts, S. B. DiCenzo and J. F. Petersen, Phys. Lett. 100B, 117 (1981).
5. R. R. Betts, B. B. Back and B. G. Glagola, Phys. Rev. Lett. 47, 23 (1981).
6. M. Ploszajczak (private communication).
7. H. Doubre, J. C. Jacmart, E. Plagnol, N. Poffe, M. Riou and J. C. Roynette, Phys. Rev. C15, 693 (1977).

MICROSCOPIC STUDY OF ELASTIC ^{28}Si-^{28}Si SCATTERING

K. Langanke and R. Stademann
Institut für Theoretische Physik I, University of Münster, Germany

Recently elastic ^{28}Si-^{28}Si scattering has become one of the most exciting heavy-ion systems, since for this system gross and intermediate structure (with widths of 2-3 MeV and around 150 keV, resp.) has been found in the excitation functions for the energy range E = 50-60 MeV [1]. The broad oscillations could be traced back to resonance-like structures in the partial waves ℓ=34-42 which are the highest spin nuclear excitations yet observed directly. In this context ^{28}Si-^{28}Si scattering may become a testcase for the theoretical models (double resonance mechanism[2], band crossing model[3]) which explain the origin of the intermediate structure by a strong coupling of a broad barrier resonance ("virtual state") in the elastic channel to molecular-like states in inelastic channels, since a successful description of the intermediate structure in ^{28}Si-^{28}Si scattering would demand the existence of quasi-molecular states in partial waves up to $\ell\approx 40$ which is clearly higher in spin than calculated until now in microscopic investigations of heavy-ion scattering[4]. The present contribution reports about a microscopic study of elastic ^{28}Si-^{28}Si scattering within the framework of the Generator Coordinate Method. Both fragment nuclei are described by their harmonic oscillator shell model ground states within the jj-coupling scheme. The Brink-Boeker force B1 has been used as effective nucleon-nucleon interaction.

The phase shift analysis exhibit two different kinds of resonant structures: 1) broad barrier resonances with widths of some MeV whose energy positions agree rather well with a spin sequence found by P_ℓ^2-fits to the experimental angular distribution; 2) sharp resonances with widths smaller than a few 100 keV which may be termed "quasimolecular states" due to their life-times. These quasimolecular resonances exist in partial waves up to ℓ=36 which is in contrast to other heavy-ion systems (for example ^{16}O-^{40}Ca). Unfortunately, a microscopic calculation of the inelastic channels is impossible due to computer facilities. But there are good reasons from other theoretical studies to assume that the potentials in the elastic and inelastic channels are rather similar. One may conclude from this that there is evidence from the present microscopic calculation for the existence of quasimolecular states in the elastic and inelastic ^{28}Si-^{28}Si channels up to $\ell \approx 36$.

In the following qualitative analysis it is demonstrated that the double resonance mechanism provides a possible explanation for the intermediate structure in excitation functions, even for the ^{28}Si-^{28}Si system. For simplicity we have restricted this model to its most simple version assuming that
a) the potentials in the elastic and inelastic channels are identical;
b) all quasimolecular states in the inelastic channels occur as resonant structures in the elastic cross section;
c) the energy positions of the assumed resonant structures in the elastic cross section are given by adding the resonance energy in the inelastic channels to the internal excitation energies of the fragment nuclei.

The results of this qualitative analysis are exemplified for the $\ell=38$ barrier resonance which is experimentally located at E = 54-57 MeV. The lowest row of fig. 1 shows the resonance energies of all quasibounds in the inelastic channels (which are assumed to be identical to the present results of the elastic channel) for the energy range E = 40-55 MeV, where E denotes to the relative energy in the respective channels. Through coupling of these quasibounds in the inelastic channel to the first excited 2^+, 4^+ and 6^+ states of the fragment nuclei one can construct 10 resonant structures which are expected to be superimposed on the gross structure of the $\ell=38$-barrier resonance (second row). The third row shows a schematic sketch of the gross and intermediate structure. It is conceivable that the number of resonant structures as well as their energy positions might be slightly changed by using an improved version of the double resonance mechanism. But it should be mentioned that even the simple version of this model predicts an intermediate structure whose spacing ($\approx$ 150 keV) is of the same order of magnitude like the experimental findings.

References

1) R.R. Betts, contribution to this conference and references given there
2) W. Scheid, W. Greiner and W. Lemmer, Phys.Rev.Lett. 25 (1970) 176
3) Y. Kondo, Y. Abe and T. Matsuse, Phys.Rev. C19 (1979) 1356
4) D. Baye, Proceedings of the International Conference on the Resonant Behaviour of Heavy Ion Systems (Aegean Sea, 1980)

Schematic drawing
of the excitation
function including
gross and
intermediate structure

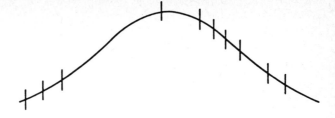

Intermediate structure
of the l = 38 -
barrier resonance

Quasibound states
in the elastic channel
(inelastic channels)

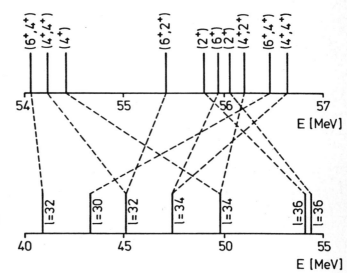

Fig. 1: Schematic drawing of gross and intermediate structure as predicted for the $\ell=38$ barrier resonance using a simple version of the double resonance mechanism.

SEARCH FOR INTERMEDIATE STRUCTURE IN ^{36}Ar VIA THE ^{24}Mg(^{12}C,α)^{32}S REACTION

R. Čaplar[+)], G. Vourvopoulos[*], X. Aslanoglou[*], D. Počanić[+]

[+]Rudjer Bošković Institute, 41000 Zagreb, Yugoslavia
[*]NRC Demokritos, Aghia Paraskevi, Attikis, Greece

The ^{36}Ar composite system, from a number of phenomenological considerations[1,2], appears as a candidate for exhibiting molecular type resonances. Further on, the ^{24}Mg(^{12}C,α)^{32}S reaction is expected (based on the effective barrier arguments[3]), to be a suitable reaction for the observation of possible intermediate resonances of spins up to J~15 i.e. at incoming energies, from, say, the Coulomb barrier up to E_{CM} ~ 24 MeV.

We measured the ^{24}Mg(^{12}C,α)^{32}S reaction from the Coulomb barrier (E_{CM}=11.9 MeV) up to E_{CM}=16.1 MeV in steps of 200 keV and from E_{CM}=15.7 to E_{CM}=19.4 MeV in steps of 330 keV in two independent experiments, using highly enriched ^{24}Mg(99,87%) targets. In the lower energy range the results consist of the excitation functions of the α_0 and α_1 groups at 4 angles (Θ_L=10°, 30°, 50° and 70°) and the angular distribution at selected energies E_{CM}=12.3, 12.9 and 14.3 MeV. In the higher energy range E_{CM}=15.7 - 19.4 MeV the excitation functions were measured at 16 angles (Θ_L=10°, 15°,..., 85°), yielding at the same time the angular distributions.

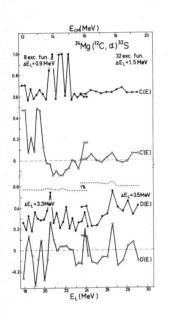

Fig. 1: Results of statistical analysis. D'(E), D(E) C'(E) and C(E) denote the summed deviation function, the summed absolute deviation function, the summed correlation and the summed absolute correlation function, respectively. The dotted line is 1% probability limit.

The measured excitation functions of the ^{24}Mg(^{12}C,α)^{32}S reaction were analyzed using the methods of statistical analysis. For this purpose the deviation and correlation functions (Fig. 1) were calculated and compared with the predictions for the uncorrelated statistical ensemble.

The measured angular distributions of the ^{24}Mg(^{12}C,α_0)^{32}S reaction were fitted with squares of single Legendre polynomials and with squares of the coherent sums of pairs of Legendre polynomials ($\sigma(\Theta) = k \cdot |P_\ell(\cos\Theta) + ae^{i\beta}P_{\ell'}(\cos\Theta)|^2$), Fig. 2. For each angular distribution, ℓ and ℓ' ($\ell' = \ell+1$ or $\ell'=\ell+2$)

[+)] A.v. Humboldt fellow - MPI für Kernphysik, D-6900 Heidelberg

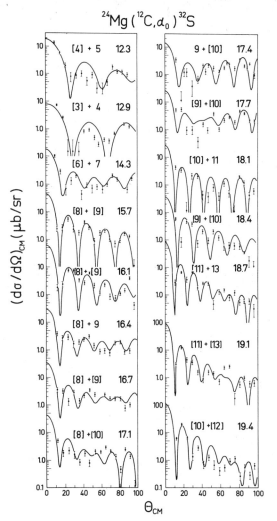

Fig. 2: Angular distributions of the ^{24}Mg(^{12}C,α)^{32}S reaction at centre-of-mass energies indicated in the figure. Solid lines are the minimum χ^2- fits with squares of the coherent sums of pairs of Legendre polynomials. The dominant ℓ-values are plotted without parentheses.

References:

1. N. Cindro and D. Počanić: J. Phys. G.: Nucl. Phys. 6 (1980) 359
2. S.T. Thornton, L.C. Dennis and K.R. Cordell: Phys. Lett. 91B (1980) 196
3. D. Baye, Phys. Lett. 97B (1980) 17
4. P.J. Dallimore and I. Hall: Nucl. Phys. 88(1966) 193

as well as α and β were varied in order to find the best fit to the data.

The results of the measurement and the analysis can be summarized as follows: (i) Structure appears in the excitation functions of both α_0 and α_1 group with peak to valley ratios up to 10 : 1. (ii) The summed absolute deviation function displays maxima at E_{CM} = 12.3, 12.7, 13.7, 14.1, 14.7, 15.1, 16.1, 17.7 and 18.7 MeV but none of them exceeds 1% probability limit. (iii) The summed correlation function (which is normalized) is inside limits ($3\sigma_{C'}$) expected for the uncorrelated statistical ensemble[4] ($\sigma_{C'} \sim$ 0.167 and 0.04 for N=8 and 32 excitation function, respectively) (iv) in most cases two-level fits describe the angular distributions of the α_0 group well. (Only the angular distributions at E_{CM} = 12.9 and 18.1 exhibit pure $P_\ell^2(\cos\theta)$ behavior with ℓ = 4 and 11, respectively). (v) The ℓ-values of the best fits follow closely the ℓ-grazing line.

Thus, the data and the analysis indicate the presence of spin selectivity in the reaction and energy range studied. However, intermediate resonances in the ^{36}Ar composite system, if present, seem to be relatively weak and probably interfere strongly with the statistical background.

MOLECULAR STRUCTURE IN ^{12}C + ^{12}C, ORBITING IN ^{12}C + ^{28}Si, AND FIRST STUDIES OF THE ^{60}Ni + ^{60}Ni INTERACTION*

K. A. Erb, J. L. C. Ford, Jr., R. Novotny[†], and D. Shapira
Physics Division, Oak Ridge National Laboratory
Oak Ridge, Tennessee 37830, U.S.A.

1. INTRODUCTION

The papers presented at this workshop have demonstrated convincingly that molecular resonances are not isolated quirks of nature, but rather, phenomena that occur in a very wide variety of nuclear reactions. Precisely because they are not quirks, and therefore not random, we can hope eventually to understand in detail why and where they appear. We are not yet at this stage. Yet, a very considerable body of data has accumulated, permitting us to begin to attempt to classify the resonances, to determine quantitatively how they relate to each other in a given system, and to study how their properties change from system to system. In the present paper, we discuss some physical implications of a recently proposed classification scheme for the ^{12}C + ^{12}C Coulomb barrier resonances for which the requisite very large body of experimental data is already available. We then present new data[1] suggesting that the back angle resonance-like structure previously observed in quasi-elastic ^{28}Si + ^{12}C reactions[2] reflects the existence of a fully developed, rotating di-nuclear system that governs back-angle yields in many additional exit channels. Finally, we discuss briefly some very recent data for ^{60}Ni + ^{60}Ni scattering in a first look at a previously unstudied region of the periodic table.

---

*Research sponsored by the Division of Basic Energy Sciences, U. S. Department of Energy, under contract W-7405-eng-26 with the Union Carbide Corporation.

[†]Permanent address: University of Heidelberg, Heidelberg, West Germany.

2. CLASSIFICATION OF $^{12}C + ^{12}C$ COULOMB BARRIER RESONANCES

An enormous amount of work has gone into the study of the Coulomb barrier region of the $^{12}C + ^{12}C$ interaction, with the result that approximately forty resonances have been located below E_{cm} = 13 MeV and spin measurements made for at least 28 of these. A portion of the resonance spectrum, based on total reaction cross sections derived[3] from γ-ray measurements,[3,4] is shown in Fig. 1. Perhaps the most striking aspect of the spectrum is its remarkable complexity; if we did not know that many of the structures have greatly enhanced carbon-carbon partial widths, we might be tempted to conclude that no simple, quantitatively accurate description of this spectrum could be achieved. The large partial widths do strongly suggest that a two-body molecular interpretation should be at least approximately correct, however, and a molecular model in which the resonances are treated as quadrupole rotation-vibration excitations has met with qualitative success.[5] A possible analogy with atomic physics phenomena was drawn recently by Iachello,[6] who noted that diatomic molecules dominated by dipole degrees of freedom have the characteristic spectrum

$$E(v,L) = -D + a(v + 1/2) - b(v + 1/2)^2 + cL(L+1), \qquad (1)$$

where v and L denote vibrational and rotational quantum numbers, respectively. Although Iachello used group theoretical techniques, a similar result can be obtained by recognizing that whatever the detailed nature of the attractive forces that bind the nuclei into a molecular configuration, the corresponding potential can be represented in the region of its minimum at $r = r_0$ by the expansion

$$V(r) = V_0 + \frac{1}{2} A(r-r_0)^2 + B(r-r_0)^3 + C(r-r_0)^4 + \ldots$$

The associated spectrum then has the form[7]

$$E(v,L) = \sum_{m,n} A_{mn}(v + 1/2)^m [L(L+1)]^n,$$

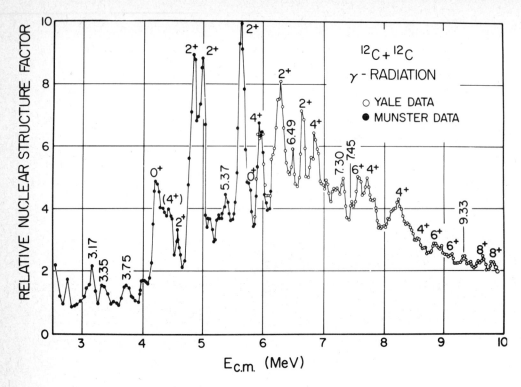

Fig. 1. Total γ-ray yields divided by Coulomb-centrifugal penetrability for the $^{12}C + ^{12}C$ reaction.

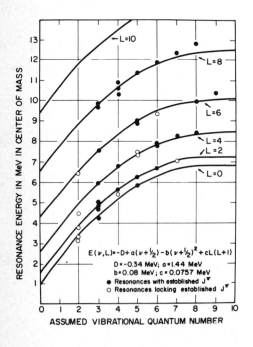

Fig. 2. Classification of known $^{12}C + ^{12}C$ resonances into rotational and vibrational bands.

which simplifies to Iachello's result when the summation is restricted to $m \leq 2$, $n \leq 1$ and rotation-vibration coupling is neglected. We found[8] that all available $^{12}C + ^{12}C$ resonance data below $E_{c.m.} = 13$ MeV can be described extremely accurately, even with this restriction, as is illustrated in Fig. 2. With only four parameters, we reproduce the energies of the 28 correlated resonances whose spins have been determined with an average rms deviation of 44 keV. The remaining 10 resonances, of unknown spin, can also be accommodated comfortably within this scheme by arbitrarily assuming appropriate spin assignments, and these are indicated by means of the open circles in Fig. 2. A more complete discussion of the calculation and a complete list of experimental citations may be found in Ref. 8; in the remainder of this section, we wish to point out some of the physical implications of our approach.

Several important properties of the binding potential corresponding to the theoretical spectrum can be deduced immediately from the parameter values listed in Fig. 2. (Many different parameter sets provide essentially equivalent fits to the data. In what follows, we restrict our discussion to those features of the potential that depend only weakly on the choice of parameter set.)

The rotational parameter, $c = 0.076$ MeV, interpreted in terms of an intrinsic configuration consisting of two point-like ^{12}C nuclei, implies an equilibrium separation of nearly 7 fm and a corresponding minimum in the binding potential at a radius far in excess of that implied by any reasonable $^{12}C + ^{12}C$ optical potential. Thus, this binding potential cannot describe the relative motion of two well-separated ground state ^{12}C nuclei. Chandra and Mosel[9] demonstrated some time ago, however, that the effective mass of the overlapping nuclei can be very much larger than the asymptotic reduced mass, and any such effect would move the deduced equilibrium separation radius toward a more acceptable smaller value. Nevertheless, the 24-nucleon system corresponding to this potential will obviously be very highly deformed, and it will be important in the future to investigate whether nuclear structure calculations can encompass such highly deformed, quasi-stable configurations (shape isomers).

If such a quasi-stable configuration should exist at high excitation in ^{24}Mg, it is difficult to understand why negative parity states should be excluded from its rotational spectrum. None have been observed, but it should be noted that all resonance-sensitive experiments reported to date for this system have involved identical spin-0 bosons (^{12}C nuclei). Thus, negative parity resonances, even if present, could not have been observed. We suspect strongly that the next generation of experiments, involving reactions such as ^{20}Ne(α,^{8}Be)^{16}O, will reveal these states. If observed, and if their spectrum conforms to that found for the positive parity resonances, their presence will provide convincing evidence for a highly deformed molecular shape isomer in ^{24}Mg.

3. MOLECULAR ORBITING IN ^{12}C + ^{28}Si INTERACTIONS

The resonance-like behavior of the large-angle elastic and inelastic scattering yields[2] from collisions involving a variety of 1p and 2s-1d shell nuclei has posed an intriguing puzzle for some time now. While no detailed understanding of the phenomenon has emerged yet, we have very recently discovered that these anomalous quasi-elastic yields represent, at least for the ^{12}C + ^{28}Si[10] and ^{12}C + ^{20}Ne[11] systems, only part of the total large-angle cross section. The fraction of the total reaction cross section contributing to large-angle yields is much larger than had been realized, demonstrating the central importance of whatever processes are responsible and emphasizing that explanations based on the behavior of the elastic yields alone will probably be incomplete. The rather complete ^{28}Si + ^{12}C data to be discussed below provide strong evidence for the formation during the early stages of the collision of a well-developed orbiting di-nuclear molecule that exhausts nearly 25% of the total non-fusion cross section.

Excitation energy spectra for carbon, nitrogen, and oxygen reaction

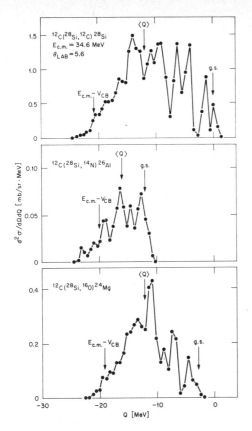

Fig. 3. Energy spectra for C, N, and O emitted at backward angles from the ^{28}Si + ^{12}C interaction.

products observed[10] near $\theta_{cm} = 180°$ from the ^{28}Si + ^{12}C interaction are plotted in Fig. 3. (Experimental details will be suppressed; these, together with a more complete discussion of the following considerations, may be found in Ref. 10.) It is readily seen from the figure that the elastic cross sections, even though remarkably large for such a large center-of-mass scattering angle, are dwarfed by the combined inelastic and rearrangement yields corresponding to more strongly damped reactions. Furthermore, the average Q-values, $\langle Q \rangle$, for a given reaction product were found to remain <u>constant</u> over the entire range of measured angles ($40 \leq \theta_{lab} \leq 20°$ corresponds approximately to $170° \geq \theta_{c.m.} \geq 140°$), as is illustrated for the carbon yields in Fig. 4.

The measured yields were converted to center-of-mass cross sections using the average Q-values (which, again, do not depend on angle in the angular region under investigation), and the resulting angular distributions are plotted in Fig. 5. All rise toward $\theta_{cm} = 180°$ with a $1/\sin\theta$ dependence. Angular distributions measured at different incident energies show essentially identical behavior, except for changes in overall cross section magnitude. Thus, these large-angle yields, which exceed the elastic cross section by approximately two orders of magnitude, appear in a variety of two-body exit

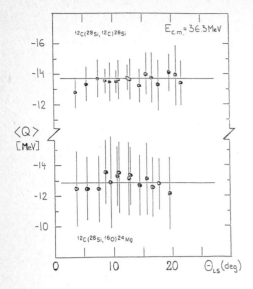

Fig. 4. Average Q-values for C,O yields observed at various angles in the ^{28}Si + ^{12}C interaction.

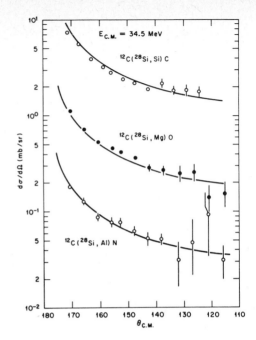

Fig. 5. Angular distributions for the total C, N, and O yields from the ^{28}Si + ^{12}C interaction.

channels, and are strongly damped in energy, are emitted from a <u>long-lived</u> complex formed during the collision.

The bombarding energy dependence of the average Q-values for the various exit channels implies that the long lifetimes can be associated with orbiting. Our measurements, plotted in Fig. 6, show that <Q> increases linearly with beam energy in each case. This behavior is entirely consistent with a simple model in which the total kinetic energy (TKE) is equated with the energy stored in a rotating di-nuclear system:

$$TKE = E_{c.m.} - <Q> = V_{Coul}^{(d)} + V_{Nucl}^{(d)} + Q_{g.s.} + \frac{\hbar^2}{2\mu d^2} \ell_f(\ell_f+1).$$

The experimentally determined slopes and intercepts are reproduced with this equation using a proximity potential for V_{Nucl}, the sticking model to relate

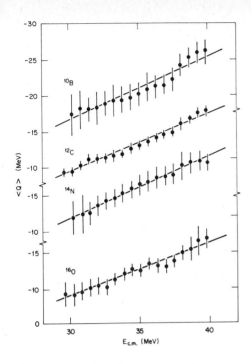

Fig. 6. Bombarding energy dependence of the average Q-values for B, C, N, and O reaction products from the $^{28}Si + ^{12}C$ interaction.

the entrance and exit channel orbital angular momenta (ℓ_i and ℓ_f, respectively), and $\ell_i = (\ell_{gr} + \ell_{cr})/2$ with the grazing and critical ℓ-values respectively determined from elastic[12] and fusion[10] data. Good agreement with the measurements is obtained only when values for the separation coordinate, d, of approximately 7 fm are used, however, showing the peripheral nature of the assumed orbiting mechanism.

Taken together, the above results find a natural interpretation in terms of a rotating di-nuclear molecule. The energy spectra indicate that strong if not complete damping of the energy degree of freedom has occurred, and this perhaps provides appropriate initial conditions for the establishment of a quasi-stable orbiting complex. Clearly, the connection between the back-angle behavior of the $^{28}Si + ^{12}C$ interaction and the deep-inelastic mechanism deserves further investigation. Our findings that $<Q>$ and $d\sigma/d\theta = \sin\theta d\sigma/d\Omega$ remain constant beyond $\theta_{c.m.} = 140°$ for each of the boron, carbon, nitrogen, and oxygen exit channels, show that both energy damping and charge and mass transfer precede the evolution of the di-nuclear system toward back angles. These results would appear to rule out any direct connection between the reaction yields at individual backward angles and the corresponding classical impact parameter. Instead, an interpretation involving an initial fast reaction mechanism followed by orbiting would seem more appropriate. It will be of considerable interest to measure the reaction products over the entire angular range in order to follow the evolution of the system in detail.

The orbiting behavior of a fully developed di-nuclear molecule cannot be described in terms of a single or even just a few partial waves. Thus, resonant structure does not appear in the excitation functions for the total yield. Partial wave summations for individual exit channels that couple to the orbiting complex presumably can be modified by the coupling in an energy-dependent manner, however, leading to resonance-like behavior in such channels. In addition, orbiting reaction amplitudes corresponding to positive and negative angle deflections will be of comparable magnitude near 180°, and their coherent superposition could lead to angular distribution oscillations that depend sensitively on beam energy.

These and many other implications remain to be explored both experimentally and theoretically in the future. Their significance derives from the observation that the cross sections associated with the di-nuclear molecule exhaust at least 25% of the total non-fusion cross section for ^{28}Si + ^{12}C collisions at $E_{c.m.}$ = 40 MeV. If the mechanisms leading to the formation of an orbiting di-nuclear molecule in the ^{28}Si + ^{12}C system are only weakly dependent on the properties of particular channels, as the present data suggest, similar behavior should be a general feature of the collisions of light nuclei.

4. A FIRST LOOK AT ^{60}Ni + ^{60}Ni COULOMB BARRIER INTERACTIONS

Resonant structure similar to that found in lighter systems has been discovered recently in ^{28}Si + ^{28}Si interactions by Betts, et al.[13] It remains an important open question whether collisions involving even heavier nuclei exhibit similar structure. The presence of resonant behavior in these interactions would open new areas for investigation, while its absence would define more sharply the conditions under which resonances exist and may be observed.

We are currently measuring Ni+Ni elastic and inelastic Coulomb barrier scattering ($E_{lab} \sim 200$ MeV) in order to investigate these and related questions,[14] using beams from the new Holifield tandem accelerator at Oak Ridge. We do not know yet whether resonances exist in these systems, but preliminary results reveal for the first time some very significant differences from the behavior observed in the scattering of lighter nuclei.

Fusion is by far the dominant reaction process in collisions of light nuclei at energies up to several times the Coulomb barrier. Broad bumps which may or may not reflect resonances appear in these fusion yields as a function of energy, but <u>all</u> quantitative information concerning resonance spins has been derived from studies of elastic and quasi-elastic exit channels. The latter never contain more than a few percent of the total reaction cross section at energies where resonances are observed.

We have found that the distribution of reaction cross section is entirely different for ^{60}Ni + ^{60}Ni collisions near the Coulomb barrier. Our elastic scattering measurements at E_{lab} = 228 MeV are plotted in Fig. 7, while Fig. 8

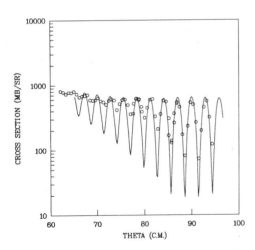
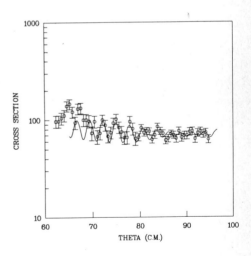

Fig. 7. ^{60}Ni + ^{60}Ni elastic scattering angular distribution at E_{lab} = 228 MeV. The curves are CCBA calculations discussed in the text.

Fig. 8. ^{60}Ni + ^{60}Ni 2$^+$ inelastic scattering angular distribution at E_{lab} = 228 MeV, in mb/sr. The curves are CCBA calculations described in the text.

shows the corresponding differential cross sections for inelastic excitation of the lowest 2^+ level in either target or projectile. For comparison, the results of preliminary coupled-channels calculations using the code PTOLEMY[15] are also shown. A standard light-ion optical potential (V = 40 MeV, W = 15 MeV, r = 1.20 fm, a = 0.54 fm) was used, together with the previously measured B(E2) value, in these calculations. Comparable agreement between theory and experiment was obtained with these parameters at the other energies measured (E_{lab} = 198 and 176 MeV). The quality of the fits can be improved, but even these first-pass calculations lead to some interesting conclusions.

The total 2^+ inelastic cross section accounts for <u>approximately</u> <u>95%</u> of the total reaction cross section at the Coulomb barrier energy of E_{lab} = 198 MeV. Variations in optical model parameters lead to variations in this fraction, of course, but only of the order of a few percent. Thus, very little cross section is left over for fusion; the quasi-elastic channels dominate the ^{60}Ni-^{60}Ni interaction at energies near the Coulomb barrier.

Inspection of the coupled-channels results shows that Coulomb excitation is responsible for these very large inelastic yields. As a consequence, this phenomenon will be a characteristic feature of the collisions of heavy symmetric systems in general.

Our observations have several implications -- as yet mostly unexplored -- for resonance studies in heavy systems. On the one hand, the dominance of the quasi-elastic yields and the corresponding very small fusion cross sections suggest that any resonances present will not damp significantly into the compound (fused) system. On the other hand, entrance channel structures will be masked in the elastic and inelastic exit channels by very strong Coulomb excitation. It thus appears that studies of rearrangement channels offer the best hope for the observation of Coulomb barrier resonances in heavy systems, provided the long-range absorption observed in our work does not preclude the formation of resonances altogether. Alternatively, studies at significantly higher bombarding energies, where the balance between the Coulomb and nuclear interactions will be different, should be undertaken. The new generation of high energy, high resolution accelerators makes such studies possible now.

REFERENCES

1. D. Shapira, et al., to be published.
2. See, e.g., P. Braun-Munzinger, et al., Phys. Rev. C 24, 1010 (1981), and references therein.
3. K. A. Erb, et al., Phys. Rev. C 22, 507 (1980).
4. K.-U. Kettner, et al., Phys. Rev. Lett. 38, 337 (1977).
5. N. Cindro, J. Phys. G 4, L23 (1978).
6. F. Iachello, Phys. Rev. C 23, 2778 (1981).
7. L. I. Schiff, Quantum Mechanics (McGraw-Hill, New York, 1968), p. 453.
8. K. A. Erb and D. A. Bromley, Phys. Rev. C 23, 2781 (1981).
9. H. Chandra and U. Mosel, Nucl. Phys. A298, 151 (1978).
10. D. Shapira, et al., to be published.
11. D. Shapira, et al., Phys. Rev. Lett. 43, 1781 (1979).
12. G. R. Satchler, et al., Nucl. Phys. A346, 179 (1980).
13. R. R. Betts, et al., Phys. Rev. Lett. 47, 23 (1981), and references therein.
14. K. A. Erb, et al., to be published.
15. M. Rhoades-Brown, private communication.

PHASE SHIFT ANALYSIS AND HEAVY ION SCATTERING

C. MARTY

Division de Physique Théorique[+], Institut de Physique Nucléaire,
F-91406, Orsay Cedex

[+]Laboratoire associé au C.N.R.S.

I. Introduction

The heavy-ion systems considered here are the light ones ($^{12}C+^{16}O$, $^{16}O+^{24}Mg$, $^{16}O+^{26}Si$, ...) for which many angular distributions and excitation functions are known in various well defined channels[1]. The backward differential cross sections quite often are large and show oscillatory behaviour with respect to variation of incident energy. It is the reason why the word, if not the idea, of resonance has been put forward about this kind of collisions.

All attempts to fit these data fall into two extreme classes. First one can look at an average description with more or less refined potentials which include surface transparancy[2], odd-even staggering effects[3], superposition of "barrier" and "internal" effects[4]. This approach can be used also at the level of the scattering matrix with appropriate smooth parametrizations S^D [5]. S^D by itself usually is equivalent to standard optical model calculations with form factors of the Woods-Saxon type and strong absorption. It gives cross sections in the backward hemisphere which are by two or three orders of magnitude smaller than the experimental ones. It is well known that any accident to S^D in the vicinity of the grazing angular momentum ℓ_{gr} can qualitatively reproduce the data near 180°. This line of approach has been extensively used in ref.[6,7] under the form for spinless particles (an assumption always kept here) :

$$S = S^D + \delta S \qquad (1)$$

δS contains odd-even effects, resonances, ... The second class has been proposed by Strutinsky[8] is to regard $\delta S_\ell = x_\ell + iy_\ell$ as functions of pairs of random numbers. In his terminology (1) is said to be of the sum (S)-type.

Other forms have been put forward. These fluctuating δS are not connected to given physical mechanisms as in the case of the compound nucleus hypothesis. Qualitatively random δS reproduces the backward effects for systems under consideration. In more recent papers possible origins for these new fluctuations are discussed either phenomenologically[9] or in a more fundamental manner[10].

All the approaches proposed so far give only qualitative reproduction of the data and in such a situation a phase shift analysis as the one recently published by Mermaz[11] can be useful. It is this kind of method, its results and perspectives one wants to discuss here. The first point to be settled is if this approach can give model independent results. As it will be shown in the next section one has some a priori answers to this point. In Section III a specific example will be discussed whereas the possible extensions of the method are proposed in Section IV.

II) Heavy ion phase shift analysis

Though usually discussed and applied in elementary particle physics, for nucleon-nucleon scattering,... this method is not widely used for the analysis of heavy-ion experiments. A phase shift analysis or more properly an amplitude reconstruction is the first step of the so-called inverse problem : given a differential cross section $d\sigma/d\Omega$ at a fixed energy find the corresponding amplitude $f(\theta)$. The second step consists of finding or guessing the dynamics which produces $f(\theta)$ and will be briefly considered in Section IV.

Model independent analysis :

When no long range interactions are present or when one is considering only inelasticity the amplitude can be expanded in a standard way using Legendre polynomials $P_\ell(\cos\theta)$:

$$f(\theta) = \sum_{\ell=0}^{\ell_{max}} B_\ell P_\ell(\cos\theta) \qquad (2)$$

How many sets of $2\ell_{max}+1$ complex coefficients B_ℓ can reproduce $d\sigma/d\Omega$? This quantity

$$d\sigma/d\Omega = |f(\theta)|^2 \qquad (3)$$

is supposed to be known with a good accuracy in the domain $0 \leq \theta \leq 180°$. Barrelet[12] has shown that $2^{\ell_{max}}$ such sets exist. An application has been given to $^{12}C(^{12}C,\alpha_0)^{20}Ne$ [13]. These $\{\delta s_\ell\}$ require no specific assumption on the mechanism underlying the phenomenon. In that sense, though not unique, they are model independent.

In the elastic channel, if a Coulomb field is present, the situation is drastically changed. Charles[14] has shown that, in principle, there is one and only one set of nuclear amplitudes $\{B_\ell\}$ connecting (3) to (2). But to get this unique amplitude one needs : a) To known $d\sigma/d\Omega$ in $4\ell_{max}+3$ angles θ, b) The errors have to be quite small particurlarly in the forward direction. Though the consistency of the method used in [14] has to be discussed more in detail already point a) is not usually fulfilled. In a typical situation as the one which will be discussed in the next Section $\ell_{max} \approx 50$ is a minimal value. Then a) requires 203 experimental points and only 95 have been measured[15]! There is no hope to get the unique $f(\theta)$ and one has just to remember that ideally in the elastic channel the amplitude reconstruction is model independent and requires good measurements also in the Fresnel zone.

Model dependent analysis:

For the elastic channel one is back to the trial and error method which is a very familiar subject. If one starts with some average field one has the standard optical model analysis. Starting instead, at the level of the S matrix, from equation (1) one chooses for S^D some analytical form[5] with parameters fitted on the forward cross section. In an angular window $\underline{\ell} \leqslant \ell \leqslant \bar{\ell}$, the δS_ℓ are found by a minimization procedure, giving some χ^2.

The corresponding amplitudes are obviously non unique since they de-depend at least upon S^D. What can be learned from the δS_ℓ will be discussed in the following Section.

III) The case of $^{16}O + ^{28}Si$ at E_{lab} = 50 MeV as an example

The analysis of a specific case is enlightlening. As in[11] one chooses $^{16}O + ^{28}Si$ at E_{lab} = 50 MeV and for S^D a Frahn-Venter parametrization S^{FV}

$$S^{FV} = g(\ell) + i\mu dg/d\ell \quad ; \quad g(\ell) = [1 + \exp(L - \ell)/\Delta]^{-1}$$

The parameters L, Δ, μ are fitted as said above and the angular window is symmetric around $\ell_{gr} \approx L$. Later[16] on it has been extended up to $\underline{\ell} = 0$ keeping $\bar{\ell} = 30$. With a 10% error bar on every experimental point the corresponding χ^2 is 1.6 with 150 waves for S^{FV}. Mermaz concludes at the existence of a very good fit as can be seen on figure 1 and to the absence of resonance in the sense that no single $P_\ell(\cos\theta)$ fits the backward region in contradiction with most of the theoretical predictions[6,9,10].

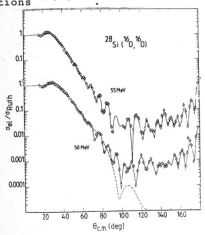

Fig.1 The ratio of the elastic cross section to the Rutherford one (which would be noted here $d\sigma/d\sigma_R$) is given at 50 and 55 MeV. The dots are the experimental points and the full lines the fits. For E_{lab} = 50 MeV the dashed line corresponds to a pure FV scattering matrix.

One can say indeed much more [17]. Figure 1 shows (dashed line) the contribution of S^{FV} alone. One notices a minimum near 95°. From 120° onwards, as expected, δS is responsible for all the scattering amplitude. A look at the Argand diagram of δS_ℓ is not very illuminating except for very large contributions of s and p waves. All other δS_ℓ seem to be distributed more or less randomly around averages values $\bar{x}$ and $\bar{y}$ with standard deviations σ_x, σ_y. All these numbers are given in the first row of table 1 and have been calculated assuming the same weight for every ℓ.

Row Number	Type of waves		$\bar{x}$	σ_x	$\bar{y}$	σ_y
1	$\ell \geq 2$ $\ell \leq 30$	All	.006	.013	.007	.011
2		Even	.007	.015	.004	.009
3		Odd	.006	.011	.009	.012
4	$\ell \geq 2$ $\ell \leq 19$	All	.004	.012	.006	.011
5		Even	.003	.014	.002	.007
6		Odd	.004	.009	.010	.013
7	$\delta S_0 = .108 + i.085$			$\delta S_1 = -.050 + i.011$		

<u>Table 1</u> : $(\bar{x},\bar{y})$, (σ_x,σ_y) are respectively the average values and standard deviations of the pairs $\delta S_\ell = x_\ell + iy_\ell$ with equal weight for all ℓ. Rows 1 to 3 refer to all, even and odd waves for $2 \leq \ell \leq 30$, whereas rows 4 to 6 correspond to the range $2 \leq \ell \leq 19$. δS_0 and δS_1 are along row number 7.

To test this hypothesis sets of random numbers δS_ℓ, $2 \leq \ell \leq 30$, with normal distributions characterized by the values of table 1 have been generated. With the numbers of row 1 unitarity is never violated. If the odd-even effects are taken into account (row 2 and 3) it is violated in 10% of the trials, a percentage which cannot be accepted. The origin of this difference comes mostly from the small changes in σ_x and σ_y between row 1 on the one hand, rows 2, 3 on the other. This shows how careful one has to be in discussing fluctuations which, if present here, are of the Strutinsky S-type. One will return to this point later on.

The modulus of the scattering amplitude δf associated with δS stays at an average constant level over the full angular range but shows about 25 oscillations. Part of them are of "geometrical" origin. They have nothing to do with any dynamical effect but are associated with the asymptotic behaviour of the Legendre polynomials. The number of the oscillations is reduced if δf is split in far-side (δf^+) near-side (δf^-) subamplitudes[18]. The $\delta f^\pm$ are calculated by the formula

$$f^{\pm}(\theta) = (2ik)^{-1} \sum_{\ell=0}^{30} (2\ell+1) e^{2i\sigma_\ell} \delta S_\ell \tilde{Q}_\ell^{\pm} (\cos\theta) \qquad (0°<\theta<180°) \qquad (4)$$

where k is the wave number and σ_ℓ the ℓth Coulomb phase shift. $\tilde{Q}_\ell^{\pm}$ is a Legendre function defined in[18]. Fig.2.A (top) gives $|\delta f^{\pm}|$ which are much less oscillating than $|\delta f|$.

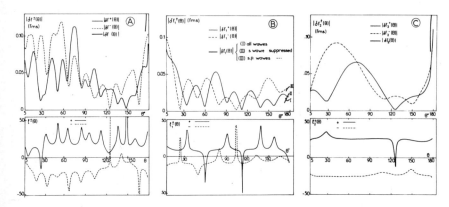

Fig.2 The various amplitudes $|\delta f^{\pm}|$, $|\delta f_1^{\pm}|$ and $|\delta f_2^{\pm}|$, in fermis with a linear scale, are given on the top of columns A, B, C respectively. The corresponding LAM's are given in the bottom part of the figure. Full lines correspond to the farside (+), dashed ones to the near-side (-) amplitudes.

Rather than to plot the phases Arg $f^{\pm}(\theta)$ it is more convenient to used their derivatives $\ell^{\pm}(\theta)$ with respect to θ. $\ell^{\pm}(\theta)$ are the local angular momenta (LAM) associated to each branch[19] and indicate the most active wave at a given angle. If a single Legendre polynomial P_Λ with Λ eventually complex were fitting the backward region one should have in this domain:

$$\ell^{+}(\theta) = -\ell^{-}(\theta) = \text{Re}\,\Lambda \qquad (5)$$

a condition far to be fulfilled in figure 2A (bottom). The backward LAM's are fairly oscillatory and on the average different for the farside and near-side branches.

This is in agreement with Mermaz's second conclusion and even go farther: no single Regge pole is signed from $\delta f(\theta)$. If Regge parame-

---

[+] For large ℓ and θ not too close 0 or π, P_ℓ, $\tilde{Q}_\ell^{\pm}$ respectively behave like $\cos[u(\theta)]/\sqrt{\sin\theta}$ and $\exp\pm i[u(\theta)]/\sqrt{\sin\theta}$ where $u(\theta) = (\ell+\tfrac{1}{2})\theta - \pi/4$. $|P_\ell|$ oscillates with a period π/ℓ whereas $|\tilde{Q}_\ell^{\pm}|$ are smooth. This is the origin of the difference between $|\delta f|$ and $|\delta f^{\pm}|$.

trizations were to be used they should have several singularities and/
or a substantial background$^{(20)}$. However eq.(5) obtained from the asymptotic expression of the $\tilde{Q}_\ell^\pm$ is not valid near 180°. In this extreme domain for the last two maxima of $|f|$ a fit such $b + b'.P_L$ is possible the ratio of the constants b and b' being $|b/b'| \approx .1$. One sees then up to which point the phase shift analysis is in agreement with the theoretical predictions quoted above. It is in a very a narrow domain.

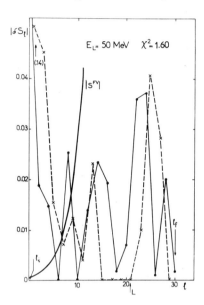

Fig.3 $|\delta S_\ell|$ is plotted as a function of the angular momentum ℓ. The even waves are represented by black points connected by full lines, the odd ones are crosses linked by dashed lines. The smooth curve is $|S_\ell^{FV}|$. The value of $|\delta S_0|$ is .14.

Fig.3 gives $|\delta S|$ as well as $|S^{FV}|$ and reveals the existence of two domains 1 and 2 to which are attributed scattering matrices δS^1 for $0 \leq \ell \leq 19$ and δS^2 for $20 \leq \ell \leq 30$. Domain 2 with its large partial waves is expected to improve the FV fit in the forward direction. For low partial waves (region 1) $|S_\ell^{FV}|$ is extremely small, much smaller than standard optical model values $^{(16)}$. The large magnitudes of s and p waves are probably required to compensate such low values of $|S^{FV}|$. One notes also, in the vicinity of $\ell_{gr} \approx L$ (fig.3) that δS_ℓ vanishes for odd waves for $15 \leq \ell \leq 21$, whereas the even waves in that domain are significantly larger. In this critical region an odd-even effect is clearly seen. It is smeared out in table 1 among the various $\bar{x}, \bar{y}, \sigma_x, \sigma_y$ but it should be more active in region 1 than in region 2 since most of the $\delta S_\ell = 0$ for odd ℓ values are below 20. If in eq.(3) δS^2 is substituted to δS whereas ℓ runs form 20 to 30 one gets amplitudes $\delta f_2^\pm$ whose moduli are given on fig.2C (top). They are quite smooth and as expected, they correct f_{FV} in the forward hemisphere. This is due to δf_2^- in the Fresnel zone where the FV amplitude is entirely near-sided ($f_{FV} \approx f_{FV}^-$) and to δf_2^+ in the Frauenhofer region.

Therefore S^{FV} does not fit accurately the Fresnel region which is an important defect (see Sect.II). $|\delta f_2^\pm|$ rise steeply near 180°. $|\delta f_2|$ is also given there where it reaches a maximum of .2fms which represents almost the whole amplitude due to δS. It is quite clear that a substantial part of the backward rise comes from waves equal to or larger than the grazing one. Their values are around 26 as indicated by the corresponding LAM's which are flat almost everywhere.

Region 1 (Fig.2B) is particularly interesting. The $|\delta f_1^\pm|$ are obtained by substitution of δS^1 to δS in eq.(4), ℓ running for from 0 to 19. They are oscillatory without any apparent correlation between their extrema. The amplitude $|\delta f_1|$ is small near 180°, an effect due to s and p waves. If they are suppressed $|f_1(180°)|$ is multiplied by a factor of nearly two. One can now assert that the region 170-180° is dominated by large partial waves so that all backward effects are of peripheral or "overpheripheral" origin. But as soon as one moves up to 100-110°, δS^1 and δS^2 have the same importance and this angular domain is the place to trace low partial waves in the elastic channel. The small contribution of these waves is well known from α-nucleus scattering[21] and is sometimes attributed to Coulomb randomization [22]. In eq.(4) going from one ℓ to the next one gives a change $2\sigma_\ell - 2\sigma_{\ell-1} = 2tg^{-1}(n/\ell)$ where n is the Sommerfield parameter which is about 10 here. For $\ell \approx 0$ and an increase of one unit in ℓ the phase $2\sigma_\ell$ rotates by a factor which is not far from 180° compensating to a large extent the $(-)^\ell$ due to $P_\ell(-1)$. Therefore Coulomb randomization is not present here, a situation opposite to α-nucleus scattering where n is much smaller. Actually the randomization is of dynamical origin through δS_ℓ^1. This is an incitation to consider the δS_ℓ^1 as random numbers. Table 1 (row 4) gives the characteristics of the normal distributions which have been chosen to generate sets of random δS_ℓ^1. The corresponding χ^2 values lie between 150 and 600 by two or three orders of magnitude larger than Mermaz's value. As at the same time the cross section within the last 15 degrees are quite poor. These are arguments, not too good indeed, against Strutinky S-type of fluctuations. All the problem of stochasticity is actually under study and will be reported later.

With rows 5 and 6 of table 1 an odd-even effect is seen at the level of $\bar{y}$ and σ_y. One has to appreciate more properly the presence or absence of a $(-)^\ell$ in the whole δS since it is not predicted theoretically [23]. As it manifests itself mostly near L one can for instance put all waves amplitudes equal to zero in the region $15 \leq \ell \leq 21$, the remaining δS_ℓ being those of [16]. In doing so, region 2 is practically unaffected and no change is apparent on fig.2C. When the two highest even waves are suppressed in region 1 there are importants effects at backward angles. The most spectacular one is the multiplication by 4 of $|\delta f_1(180°)|$. The odd-even effect as presently studied here acts destructively near 180°.

The various results of this Section can be classified as follows. First one has those which are independent of the FV parametrization : backward effects due solely to δS, absence of Coulomb randomization and also the usefulness of the far-side near-side decompositon. A second type of results depends qualitatively on S^{FV} : no single pole dominance but a possible fit by a dominant P_ℓ in the extreme backward region, side effect common to any S^D going towards zero with ℓ and an odd-even effect at the level of δS. Finally some conclusions rely quantitatively upon the FV choice : backward effects due to waves equal to or larger than ℓ_{gr}, the reduction of the cross section near 180° by the odd-even effect and probably the absence of Strutinsky S-type of fluctuations.

IV) Conclusions

One has now to face the question : is it useful to go on in amplitude reconstruction ? Clearly the answer it positive if δS is not too much model dependent. A brute force method would be to try other S^D on the example of Section III. For instance the modulus and phase of S^{FV} cannot be varied independently and one can use the McIntyre parametrization S^{MI} [5] which is more flexible and also gives a perfectly decreasing cross section near 95° where $|f^{FV}|^2$ has an oscillation as can be seen on fig.1. But S^{MI} also is too small near $\ell = 0$. It may be good to work for S^D with scattering matrix given by standard Woods-Saxon potential like E18. The well known ambiguities at the level of the potential are strongly reduced when one looks at the corresponding S^D.

But before doing so some remarks are appropriate. S^D is supposed to reproduce the direct effects responsible for the forward cross section. One has to fix what is meant by "forward". It surely contains the Fresnel zone but one may wonder if the Frauenhofer domain should be given too. From what has been said in Sect. II one would ask for a very good reproduction of the Fresnel zone only at all energies where data exist. This would fix the energy dependence of the parameters contained or giving rise to S^D.

Then from elementary particle physics one knows that one has to put by hand as many physical information as possible in δS. For instance one can introduce several Regge poles and look at their trajectories or introduce a $(-)^{\ell}$ term as in ref.(7). As the experimental data are scarced and difficult to obtain one is limited in the choice of the phenomena to be included in δS. It is almost impossible to put at the same time resonances <u>and</u> odd-even staggering effect though this has been done for $^{12}C + ^{16}O$ (14). The sounder way is to try separetely the various effects and to see which one is the most important.

The statistical aspects of the collisions one is considering are, if they exist, very interesting. From what is known for reactions with lighter projectiles and assuming the formation of a compound nucleus physical informations can be obtained directly from cross sections. Ericson followed by many others(24) has shown how auto and cross-correlations functions and/or bump counting in angular distributions are related to the average ratio of the width to the spacing of compound nucleus levels. It would be nice in our field to have such relations between physical quantities and cross sections. This is a particular advantage of the phase shift analysis. As the δS_{ℓ} are obtained one has separately the direct cross section $|f^D|^2$, the mixed one $2\text{Re}(f^D.\delta f^*)$ with a remaining term $|\delta f|^2$. If one leaves in δf only the contributions from the δS_{ℓ} which are believed to be random one can study the properties of the fluctuationg part $|\delta f|^2$ in the whole angular range $0 \leqslant \theta \leqslant 180°$. Looking solely to the experimental angular distributions one is limited by the direct effects to trace stochasticity in the backward region $120 \leqslant \theta \leqslant 180°$ which is rather narrow. The various methods which can be used as already mentionned are tested and promising.

In the Introduction it was said that the amplitude reconstruction is just the first step of the so-called Inverse Problem. The second one deals with the dynamics generating the amplitude. The analysis of the data existing for light systems gives some information on this step. There is evidence that a compound nucleus is formed but one is really dealing with a bi-center system. This is backed by experiments near the Coulomb barrier where resonances of molecular type are seen. In deep inelastic collisions one knows also that most of the final products are not far from the initial partners. These situations imply a conservation of the memory of the incident channel much greater than in the compound nucleus picture. The statistical analysis à la Ericson has to be modified. Quantitatively this idea has been discussed by Friedman(10) with reasonable simplifying assumptions. One of his conclusions is the existence in the backward cross section of a $P_\ell^2(\cos\theta)$ angular distribution. This is not in disagreement with the result of Sect.III as it has been seen.

Having all these ideas in mind it is concluded that a preliminary analysis of the example of Sect.III with various S^D is useful. If the δS are not too different one should go on in a systematic way for other systems at several energies as far as the experimental information is large enough to be useful.

All the work reported here has been obtained in collaboration with H. Doubre[17].

REFERENCES

[1] For a review see : J. Barrette and S. Kahana, Comments on Nuclear and particle Physics 9(1980) 67.

[2] S. Kahana, B.T. Kim and M. Mermaz, Phys. Rev. C20(1979) 2124.

[3] D. Dehnhard, V. Shkolnik and M.A. Franey, Phys. Rev. 40(1978) 1549.

[4] S.Y. Lee, Nucl. Phys. A311(1978) 518.

[5] W.E. Frahn, Heavy-ion, high spin states and nuclear structure IAEA Vienna 1975, vol. 1, p. 219.

[6] W.E. Frahn, Nucl. Phys. A337(1980) 324.

[7] W.E. Frahn, M.S. Hussein, L.F. Canto and R. Donangelo, Preprint IFUSP/P-251 (1981), Unv. de Sao Paulo (Brasil).

[8] V.M. Strutinsky, Zeit. Phys. A290(1979) 377.

[9] V.M. Strutinsky , Zeit. Phys. A299(1981) 177 ;
V.M. Strutinsky and S. Kun, Zeit. Phys. A299(1981) 755.

[10] W.A. Friedman, Phys. Rev. C24(1981) 125.

[11] M.C. Mermaz, Phys. Rev. C23(1981) 755.

[12] E. Barrelet, Nuov. Cim. 8A(1972) 331. Cf also ref.[13].

[13] F. Auger , P. Charles, E.F. da Silveira, Nuclear Molecular Phenomena,N. Cindro Ed., North-Holland Pub C°, 1978, p.395.

(14) P. Charles, Deuxième Colloque Franco-Japonais de Physique Nucléaire avec des Ions Lourds, Gif-sur-Yvette, 8-12 Oct. 1979 (unpublished), p.98 ; Thèse d'Etat, Orsay 1981.

(15) P. Braun-Munzinger, G.M. Berkowitz, T.M. Cormier, C.M. Jachinski, J.W. Harris, J. Barrette and M.J. Levine, Phys. Phys. Rev. Lett. $\underline{38}$(1977) 944.

(16) M.C. Mermaz, Private communication.

(17) H. Doubre and C. Marty, Preprint IPNO/TH 81-28.

(18) R.C. Fuller, Phys. Rev. $\underline{C12}$(1975) 1561.

(19) R.C. Fuller and R.J. Moffa, Phys. Rev. $\underline{C15}$(1977) 266.

(20) S. Landowne, Phys. Rev. Lett. $\underline{42}$(1979) 633.
R. Anni, L. Renna and L. Taffara, Lett. Nuov. Cim. $\underline{25}$(1979) 121.

(21) N. Rowley, C. Marty and E. Plagnol, Phys. Lett. $\underline{93B}$(1980) 16.

(22) D. Brink, J. Grabowski and E. Vogt, Nucl. Phys. $\underline{A359}$(1978) 359.

(23) D. Baye, J. Deenen and Y. Salmon, Nucl. Phys. $\underline{A289}$(1977) 511.

(24) For a general review see : A Richter, Nuclear Spectroscopy and Reactions, Part B, J. Cerny Ed., Academic Press New-York 1974, p. 343.

Local-potential description of the bound, quasi-bound and scattering states of the α-nucleus system

R.Ceuleneer and F.Michel,
Université de l'Etat, Faculté des Sciences, B-7000 Mons, Belgium

G.Reidemeister [‡],
Université Libre de Bruxelles, Physique Nucléaire Théorique, CP229,
B-1050 Brussels, Belgium

The description of states of certain nuclei as bound or quasi-bound states of an α-cluster interacting with a core via a local potential has known a revival of interest in the recent years. This approach has proved successfull in the vicinity of the closure of the p-shell [1] and of the sd-shell [2,3]. The possibility of connecting such potentials with the real part of the optical potentials deduced from the analysis of elastic scattering data remains an open problem.

For the α + ^{16}O system the experimental work has concentrated on the resonant states observed in the elastic channel below $E_\alpha \simeq 30$ MeV, in relation to the spectroscopy of ^{20}Ne. The scarcity of the data at higher energies, where the resonant behaviour disappears, has up to now hindered the determination of an unambiguous optical potential. The experimental situation is markedly different for the α + ^{40}Ca system for which numerous angular distributions were measured on the whole angular range up to high incident energies in connection with the "anomalous large angle scattering" puzzle. Subsequent analyses revealed the feasibility of a potential description of the phenomenon [4-7]. Although the familiar discrete ambiguities appear in analyses carried out at low energies [7] (i.e. $E_\alpha \lesssim 50$ MeV), only a single potential family was found to fit the data up to the highest available energies [5,7].

In this contribution we discuss the bound and quasi-bound states associated with the real part of various potentials belonging to this family. In order to take account of the Pauli principle only the physical states allowed by an orthogonality condition type prescription, i.e. those with $N \geq 12$ (where $N = 2n_r + \ell$), were retained. They are found to group into quasi-rotational bands labelled N. The figure shows the states obtained with the real part of the slowly energy-dependent potential A of Delbar et al. [5]; its depth has been fixed to the average value $V_0 = 180$ MeV. Essentially similar results are obtained with the other potentials of the same family. The position

[‡] Chercheur qualifié FNRS

of the head of the N = 12 band is seen to agree within a few MeV with that of the ground state of ^{44}Ti. Moreover both bands terminate at $J^\pi = 12^+$ and qualitatively display the same behaviour. In our approach these states thus appear as α-cluster states. The cluster picture also underlies local-potential calculations recently performed for that system by Pilt [2] and Pál and Lovas [3]; however their potentials were especially constructed to reproduce the g.s. band but are incompatible with the phenomenological optical potential. Our potential also supports excited bands: another positive parity band with N = 14, in the energy region where intermediate resonances have recently been observed [8], and a band of negative parity (N = 13) starting not far from the threshold. It is interesting to compare these results with those of microscopic calculations. One [9] predicts the first band to be of mixed parity and above the threshold. In another [10], bands of different parities are found to be unmixed with energies similar to ours.

Along the same lines we built an optical potential fitting low energy α + ^{36}Ar scattering data, which seems to account for the well-known excited 4p-4h band of ^{40}Ca.

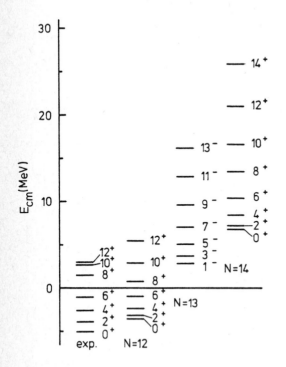

1) B.Buck et al., Phys.Rev. C11(1975)1803
2) A.A.Pilt, Phys.Lett.73B (1978)274
3) K.F.Pál and R.G.Lovas, Phys. Lett.96B(1980)19
4) F.Michel and R.Vanderpoorten, Phys.Rev.C16(1977)142
5) Th.Delbar et al., Phys.Rev. C18(1978)1237
6) H.P.Gubler et al., Phys. Lett.74B(1978)202
7) H.P.Gubler et al., Nucl. Phys.A351(1981)29
8) D.Frekers and R.Santo, to be published
9) H.Friedrich and K.Langanke, Nucl.Phys.A352(1975)47
10) H.Kihara et al., Proc.Int. Conf.Nucl.Struct.,Tokyo, 1977, p.235

Calculation of the internal and barrier wave contributions
to heavy ion elastic scattering made simple

J.Albinski [‡] and F.Michel,
Université de l'Etat, Faculté des Sciences,
B-7000 Mons, Belgium

The semiclassical solution of the three-turning point scattering problem has recently been extended by Brink and Takigawa to the case of complex potentials [1] within the method of comparison equations. Although their method provides valuable information on the physics underlying the optical model description of the data, it has only been applied to a rather limited number of cases, probably because of the difficulties inherent to its programming. Moreover, the method is restricted to analytical potentials, which makes it inapplicable to the study of most folding-model, "model-independent" and numerically-supplied potentials.

This prompted us to investigate the possibility of gaining the same physical information without resorting to a full semiclassical calculation. In this spirit we have implemented an algorithm requiring simple modifications of any conventional optical model code and leading to a very good agreement with the original method [2]. Our

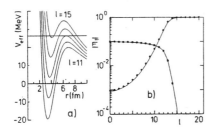

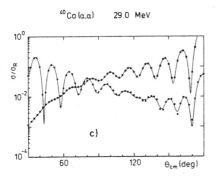

Fig.1.- Comparison of our method with the semiclassical calculation for ^{40}Ca(α,α) at 29 MeV (potential A of ref.4). a) effective potential curves near the grazing angular momentum; b) reflexion coefficients generated by both methods (full line: our method; ● : semiclassical barrier wave contribution; ■ : semiclassical internal wave contribution); c) same as b) for the scattering cross sections.

[‡] Permanent address: Institute of Nuclear Physics, Cracow, Poland

procedure relies on the exponential behaviour of the internal S-matrix coefficients as a function of the absorptive strength; in its simplest version, it only requires three optical model evaluations of the S-matrix coefficients. Results obtained [3] within the frame of the original WKB approximation [1] are compared with those of a slightly more elaborate version of our method in Figs. 1-3 for some representative examples. The very good agreement observed between our method and the semiclassical results makes it a promising tool for systematically investigating the properties of optical potentials deduced from analyses of elastic scattering data.

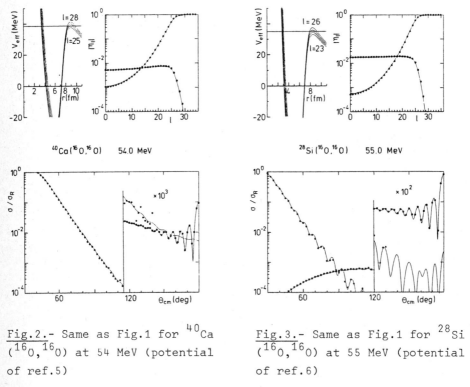

Fig.2.- Same as Fig.1 for ^{40}Ca (^{16}O,^{16}O) at 54 MeV (potential of ref.5)

Fig.3.- Same as Fig.1 for ^{28}Si (^{16}O,^{16}O) at 55 MeV (potential of ref.6)

1) D.M.Brink and N.Takigawa, Nucl.Phys.A279(1977)159
2) J.Albinski and F.Michel, to be published in Phys.Rev.C
3) R.Vanderpoorten, computer code KGB2, Université de l'Etat à Mons (Belgium), unpublished
4) Th.Delbar et al., Phys.Rev.C18(1978)1237
5) N.Alamanos et al., Nucl.Phys.A363(1981)477
6) K.O.Terenetski and J.D.Garrett, Phys.Rev.C18(1978)1944

V. RESONANCE PHENOMENA IN FUSION AND TOTAL REACTION CROSS SECTIONS

Unitarity of the S-matrix and Resonance Phenomena in Nuclear Reaction Cross Sections[*]

I. Rotter
Zentralinstitut für Kernforschung Rossendorf, DDR 8051 Dresden, GDR

1. Introduction

Even the first observation of clearly nonstatistical structure in $^{12}C + {}^{12}C$ induced cross sections about 20 years ago by Bromley et al.[1] created the idea of a formation of quasimolecular states in heavy-ion reactions. In later papers, a doorway mechanism is assumed to play an important role[2] in the resonance phenomena observed experimentally. But the nuclear structure of the doorway state remained an open problem. The suggestion[3], that the doorway state could be a quasimolecular state, meets with difficulties of interpretation on the basis of the traditional conceptions, since a quasimolecular state is not a true compound nucleus state.

The only well understood structures in nuclear reaction cross sections which originate from doorway states are the isobaric analogue resonances in heavy nuclei. These resonances result from nuclear states which have a large decay width and produce a clear intermediate structure in the cross section. The resonances observed in heavy ion reactions are not comparable with this phenomenon.

However, it can be stated some analogy of the resonances in heavy ion reactions to the isobaric analogue resonances in medium heavy nuclei. In this case, fine structure resonances are observed experimentally while the intermediate structure corresponding to the isobaric analogue state appears as one envelope. In contrast to the interpretation of data on heavy nuclei the analysis of the data on medium heavy nuclei raises some problems. These problems are similar to those which are discussed in the analysis of the data obtained from heavy ion reactions: The correlation of the fine structure resonances in the various channels is less and their individuality is larger than expected[4,5]. Structures corresponding to isobaric analogue states can not always well distinguished from other structures which are not connected with any doorway state[5,6,7]. Substructures may appear in the cross section[7,8]. Recently, some correlation of the fine structures is observed not only in the region of an isobaric analogue resonance, as it should be, but also in the analogue free regions[9]. This result is in a clear contradiction to the traditional conceptions. Furthermore, the widths obtained from an analysis of the experimental data for the isobaric analogue states are too small[4] although all corrections in the framework of the traditional conceptions have been considered.

Therefore it seems to be useful to investigate the problem of intermediate structures and doorway states once more in nucleon induced reactions before conclusions on the mechanism of heavy ion reactions can be drawn.
The similarity of the resonance phenomena observed in isobaric analogue resonances on medium heavy nuclei and in heavy ion reactions suggests to look for doorway states also in heavy ion reactions. The isobaric analogue doorway state has a large width due to its simple nuclear structure and a small configurational mixing with the other compound nucleus states due to its isospin.
A quasimolecular state has also a large spectroscopic connection to one of the channels and, furthermore, a small configurational

mixing with the compound nucleus states because of its cluster structure. The last property is lost for the "quasi molecular state" of nucleon induced reactions. In this case, they are the one particle (or shape) resonances the properties of which are very well known. These states mix strongly with the compound nucleus states due to their one-particle structure. An example is the $d_{3/2}$ resonance in ^{16}O the components of which are distributed over a large energy region. Taking into account these differences between shape resonances in nucleon induced reactions and in heavy ion reactions, it is supposed here that the shape resonances (or quasimolecular states) are doorway states in heavy ion reactions.

In the traditional nuclear reaction theories, the interpretation of shape resonances as doorway states is difficult, because a doorway state should be a true compound nucleus state. This problem can be overcome if the cut-off technique of Wang and Shakin[10] for shape resonances is used since it allows to treat the parts of the wavefunction of a shape resonance for $r < R_{cut}$ and for $r > R_{cut}$ separately. The use of this technique in the continuum shell model is formulated by Barz et al.[11] and will be the basic for the considerations of this paper.

It is the aim of the present paper to investigate the resonance phenomena in nuclear reactions in consideration of the unitarity of the S-matrix. The numerical calculations are performed for the ^{15}N + p reaction with realistic wavefunctions for both the compound and the target nucleus. These numerical calculations may be considered as model calculations for the reactions discussed above for which calculations with comparable accuracy cannot be performed.

2. Basic equations of the continuum shell model (CSM)

In the continuum shell model (CSM), the whole function space is divided into the subspace of scattering states and the subspace of discrete states by means of the two projection operators P and Q. The aim of using the projection operator technique is not, as in the Feshbach method[12], the possibility to separate the relevant part from the other part in order to use approximations of different type for both parts. In the CSM, the approximations used in both subspaces (truncation of the number of configurations and and of the number of channels) are comparable. The projection operator technique is used in order to solve the Schrödinger equation with discrete and continuous wavefunctions which both have a very different mathematical behaviour. The division of the function space into the two subspaces is therefore another one in the CSM than in the Feshbach theory. The Q-space of the CSM contains the wavefunctions of all discrete states while the P-space contains the wavefunctions with one particle in a scattering state and the remaining particles in discrete states. The shape resonances belong to the Q-space up to the cut-off radius R_{cut} while the remaining part belongs to the P-space. It is $P + Q = 1$, and the orthogonality condition between the two subspaces is fulfilled by a renormalisation of the wavefunctions[11].

The solution of the Schrödinger equation $H\psi = E\psi$ with discrete and continuous wavefunctions is obtained in the following manner[11].
(i) Solution of the traditional shell model problem (with Woods-Saxon potential)

$$(E - H_{QQ}) \phi_R = 0 \tag{1}$$

with $H_{QQ} \equiv QHQ$. The operator

$$Q = \sum_R |\phi_R\rangle\langle\phi_R| \tag{2}$$

is the projection operator onto the subspace of discrete states. This subspace corresponds to the function space of the traditional shell model due to the cut-off technique used for shape resonances. The states described by the eigenfunctions Φ_R and eigenvalues E_R^{SM} are called QBSEC (quasi bound states embedded in the continuum).

(ii) Solution of the traditional coupled channels equations

$$(E^+ - H_{PP}) \xi_E^c = 0 \qquad (3)$$

with $H_{PP} \equiv PHP$. The operator

$$P = \sum_c \int_{\varepsilon_c}^{\infty} dE \, |\xi_E^c\rangle\langle\xi_E^c| \qquad (4)$$

is the projection operator onto the subspace with one particle in the continuum and the remaining particles in discrete states of the residual nucleus (channels c).

(iii) Solution of the traditional coupled channels equations with source term

$$(E^+ - H_{pp}) \omega_R = H_{PQ} \Phi_R . \qquad (5)$$

The source term describes the coupling of the two subspaces ($H_{PQ} \equiv PHQ$).

The solution ψ is obtained by means of the three functions Φ_R, ξ_E^c, and ω_R:

$$\psi_E^c = \xi_E^c + \sum_{RR'} (\Phi_R + \omega_R) \frac{1}{E - H_{QQ}^{eff}} \langle\Phi_{R'}| H_{QP} |\xi_E^c\rangle \qquad (6)$$

with

$$H_{QQ}^{eff} \equiv H_{QQ} + H_{QP} G_P^{(+)} H_{PQ} . \qquad (7)$$

The operator H_{QQ}^{eff} is that part of the Hamilton operator H which appears effectively in the Q-space when the coupling to the continuum is taken into account. G_P^+ is the Green function for the motion of the particle in the P-space.

The eigenfunctions

$$\tilde{\phi}_R = \sum_{RR'} a_{RR'} \phi_{R'}$$ (8)

and eigenvalues $\tilde{E}_R - \frac{i}{2} \tilde{\Gamma}_R$ of H_{QQ}^{eff} determine the wavefunctions, energies

$$E_R = \tilde{E}_R \ (E = E_R)$$ (9)

and widths

$$\Gamma_R = \tilde{\Gamma}_R \ (E = E_R)$$ (10)

of the nuclear states R.

The solution ψ is the scattering wavefunction modified by the discrete states R:

$$\psi_E^c = \xi_E^c + \sum_R \tilde{\Omega}_R \frac{1}{E - \tilde{E}_R + \frac{i}{2}\tilde{\Gamma}_R} \langle \tilde{\phi}_R^* | H | \xi_E^c \rangle .$$ (11)

The wavefunction of the discrete state R modified by the continuum is

$$\tilde{\Omega}_R = \tilde{\phi}_R + \sum_c \int_{\varepsilon_c}^{\infty} dE' \, \xi_{E'}^c \frac{1}{E^+ - E'} \langle \xi_{E'}^c | H | \tilde{\phi}_R \rangle .$$ (12)

The model is symmetric in the discrete states $\tilde{\Omega}_R$ (modified by the continuum) and the scattering states ψ_E^c (modified by the discrete states). It allows therefore a unified description of nuclear structure and nuclear reaction aspects.

3. The resonance states

Although the general formalism is similar, the concept of a resonance state R defined in the CSM differs from that of the shell model approach to nuclear reactions formulated by Mahaux and Weidenmüller[13]. The energies E_R and widths Γ_R of the Mahaux-Weidenmüller-theory are not determined by eqs. (9) and (10) as has been shown by Lemmer and Shakin[14]. Only the use of the cut-off technique of Wang and Shakin[10] for single-particle resonances allows to solve eqs. (9) and (10) unequivocally. Consequently, the isolated resonances observed in the cross section correspond to the QBSEC introduced by Barz et al.[11] by using the cut-off technique but do not correspond, generally, to the BSEC defined by Mahaux and Weidenmüller[13]. Furthermore, the external mixing of all the resonance states via the continuum is considered explicitely in the CSM instead of the statistical assumptions in the other nuclear reaction theories.

The concept of a resonance state R defined in the CSM corresponds to the concept formulated on the basis of the R-matrix theory (Robson and Lane[15]) since the QBSEC are shell model states with inclusion of the main contributions of the single-particle resonances inside the nucleus. Thus, it is in agreement with the numerous calculations performed successfully with the traditional shell model wavefunctions for many years in analysing different nuclear reactions to get conclusions on the nuclear structure.

A doorway state is defined in the CSM as a state with a simple nuclear structure (i.e. a large spectroscopic connection to one of the channels) as well as a small internal (or configurational) mixing with other resonance states in the neighbourhood. Therefore, external mixing dominates and produces the typical picture of a gross structure at about the energy of the resonance state in the cross section. If the internal mixing would not be small, the "gross structure" would be smeared over a larger energy region due to internal mixing and could hardly be identified.

Doorway states are e.g. isobaric analogue resonance states. Since their doorway properties are known for a long time, they are considered in all nuclear reaction models. Another example of doorway states are the shape resonances. In the CSM, they can be considered as doorway states as discussed above although they may be very broad and a giant resonance in its own right and therefore usually are assumed to belong to the direct reaction part. In heavy ion reactions, the internal (or configurational) mixing of the shape resonances with the other resonance states of more complicated nuclear structure is small because of the different deformation and clustering of both types of states. They may be represented as nuclear states in a second minimum and their interaction with the states of the first minimum (compound nucleus states) can be neglected generally in describing the gross features of the resonance phenomenon. Therefore, the shape resonances in heavy ion reactions are smeared over a smaller energy region than the shape resonances in nucleon induced reactions (e.g. the $d_{3/2}$ resonance in ^{16}O).

Since a doorway state overlaps with N resonance states of the same spin and parity but of a more complicated nuclear structure, one has microscopically to consider N+1 overlapping resonance states. The widths of all N+1 resonance states are changed due to their external mixing with each other. The resonance picture observed may be far from a picture with N+1 resonances due to the interferences and the external mixing[16]. It depends on the ratio Γ/D (Γ-average width, D - average distance of the N+1 resonance states) which is a measure for the degree of overlapping of the individual resonances.

If there are N+1 overlapping resonance states and no one of them is a doorway state, then the resonance picture observed in the cross section is also determined by the ratio Γ/D. The only difference to the doorway case is the fact that all resonance states have widths of the same order of magnitude and the typical picture of a gross structure (as the isobaric analogue resonances in heavy nuclei) does not appear. There is, of course, a lot of examples between these two borderline cases.

4. The S-matrix

In the CSM, the S-matrix is given by the following expression[16]

$$S_{cc'} = S_{cc'}^{(1)} - S_{cc'}^{(2)} \tag{13}$$

with

$$S_{cc'}^{(1)} = \exp(2i\delta_c)\, \delta_{cc'} - 2i\pi \langle \chi_E^{c'(-)} | V | \xi_E^{c(+)} \rangle \tag{14}$$

$$S_{cc'}^{(2)} = i \sum_R \frac{\tilde{\gamma}_{Rc'} \tilde{\gamma}_{Rc}}{E - \tilde{E}_R + \frac{i}{2}\tilde{\Gamma}_R} \tag{15}$$

$S_{cc'}^{(1)}$ describes that part which depends smoothly on energy, while $S_{cc'}^{(2)}$ is the resonance part of the S-matrix. $\chi_E^{c'}$ is the basic wavefunction of the P-space, $V = H - H_0$ the residual interaction.

The functions

$$\tilde{\gamma}_{Rc} = (2\pi)^{1/2} \langle \tilde{\phi}_R^{(-)} | H | \xi_E^{c(+)} \rangle \tag{16}$$

determine the partial widths

$$\gamma_{Rc} = \tilde{\gamma}_{Rc}(E = E_R) \tag{17}$$

in the framework of the CSM.

Although the S-matrix (13) with (14) and (15) has the familiar shape, it differs in some aspects from the usual one:

(i) The values γ_{Rc}, E_R and Γ_R are not parameters but are calculated within the model. Eqs. (9) and (10) have definite solutions since the cut-off technique of Wang and Shakin[10] is used for the single-particle resonances.

(ii) Eqs. (13), (14) and (15) are true for all energies since in eq. (15) not the values γ_{Rc}, E_R and Γ_R but the functions $\tilde{\gamma}_{Rc}$, $\tilde{E}_R$ and $\tilde{\Gamma}_R$ stand. The last two functions depend smoothly on energy with the exception of threshold effects[17].

(iii) It is[18]

$$\Gamma_R \leq \sum_c |\gamma_{Rc}|^2 \qquad (18)$$

where the sign of equality stands only for isolated resonances.

(iv) The function $\tilde{\gamma}_{Rc}$ is complex. It follows from the unitarity of the S-matrix that for overlapping resonance states Im $\tilde{\gamma}_{Rc}$ is a function of all the denominators $(E - \tilde{E}_R)^2 + \frac{1}{4}\tilde{\Gamma}_R^2$. The resonance behaviour of $S_{cc'}^{(2)}$ is therefore determined in this case not by a sum of independent Breit-Wigner terms with energy independent numerators.

(v) The summation over R in eq. (15) includes also the main contributions of the shape resonances for which the cut-off technique of Wang and Shakin[10] is applied.

(vi) All the R dependent terms of eq. (15) include three types of correlations: the internal (or configurational) mixing contained in the Φ_R (eq. (1)), the channel coupling contained in the ξ_E^c (eq. (3)) and the external mixing of the resonance states via the continuum which appears additionally in the eigenfunctions $\tilde{\Phi}_R$ of H_{QQ}^{eff}. Due to all these mixings caused by the residual interaction V, all resonance states R are, generally, correlated.

The inclusion of doorway states in the sum over R in eq. (15) allows to consider the unitarity of the S-matrix for the doorway phenomenon. Suppose there exist N fine structure resonance states overlapped by 1 doorway state. It is

$$\sum_{R} \Gamma_R \approx \Gamma_d \approx |\gamma_{dc_0}^2|^2 \qquad (19)$$

according to the definition of a doorway state. The S-matrix element in the channel c_0 at the energy of the doorway state is

$$S_{c_0 c_0} \approx -1 \qquad (20)$$

while the transmission coefficient is

$$T_{c_0} = \sum_{c'(\neq c_0)} |S_{c_0 c'}|^2 = 1 - |S_{c_0 c_0}|^2 \approx 0 \qquad (21)$$

due to the restraint by the unitarity of the S-matrix.

If there is no doorway state and the fine structure resonance states are more or less isolated then the unitarity condition of the S-matrix can not result in the condition (21). In such a case, the widths of the N resonance states are distributed statistically. The value $|S_{c_0 c_0}|$ differs therefore from unity. Consequently, the transmission coefficient T_{c_0} differs from zero.

That means, the relative contribution which the fine structure resonance states can give to the cross section in a certain channel c_0 depends on whether there is a doorway state or not. The external mixing between the fine structure resonance states and the doorway state results in the reduction of the transmission coefficient. The nucleus becomes more transparent in the channel c_0 at the energy of the doorway state although the fine structure resonances do not lose their individuality. This can be seen also in the cross section, which seems to be reduced in comparison with a cross section calculated from a sum of independent Breit-Wigner resonances. The reduction is connected with the fact that the maximal value of

$$\sigma(E) \propto |S(E) - 1|^2 \qquad (22)$$

is obtained nearly by the doorway state alone. The enhancement of the cross section by the fine structure resonance states is therefore limited.

This result can be understood by means of a consideration of the corresponding phases. The phase shift due to the doorway state is much slower than the phase shifts due to the fine structure resonance states. The phase of the doorway state acts therefore as a background phase which, however, changes with energy. If there is a fine structure resonance at the energy where the doorway state gives its maximal contribution to the cross section then its phase can change only in such a manner that the cross section becomes smaller at this energy and returns back to its maximal value with a width determined by the fine structure resonance. In this manner, the relative phases between overlapping resonance states are changed due to the existence of the doorway state. The relative phases between the resonance states remain unchanged only for nonoverlapping resonances. This phenomenon is similar to the very well known asymmetry of resonances due to their interference with the background. The only difference is that the phase corresponding to the direct reaction background is constant over a larger energy region in contrast to the phase of the doorway state.

The reduction of the cross section due to the overlapping of the resonances will be proved numerically in the next section.

5. The numerical calculations and results

The numerical calculations are performed for the reaction $^{15}N + p$ with 1^- resonances. The wavefunctions of the resonance states are realistic, but the positions are shifted sometimes in order to change the degree of overlapping of the resonances.

Table 1   The configurations taken into account in the shell model calculation (eq. (1)).

№ of the configuration	Occupation of the one-particle states					Number of 1^- states
	$1s_{1/2}$	$1p_{3/2}$	$1p_{1/2}$	$2s_{1/2}$	$1d_{5/2}$	
1	4	8	3	1	0	2
2	4	8	3	0	1	-
3	4	7	4	1	0	2
4	4	7	4	0	1	2
5	3	8	3	2	0	8
6	3	8	3	1	1	6
7	3	8	3	0	2	12
8	3	7	4	2	0	8
9	3	7	4	1	1	18
10	3	7	4	0	2	18

The diagonalisation of the shell model Hamiltonian H_{QQ} for ^{16}O (solution of the shell model problem (1)) is perfomed with basic wavefunctions of 1p-1h and 2p-2h nuclear structure corresponding to $1\hbar\omega$ and $3\hbar\omega$ excitations. The one-particle states taken into account are $1s_{1/2}$, $1p_{3/2}$, $1p_{1/2}$, $2s_{1/2}$ and $1d_{5/2}$. The occupation of the $1d_{3/2}$ state is not allowed in order to avoid the cut-off technique for single particle resonances in these calculations of model type. The nuclear reaction cross section is calculated in such an energy region where the $d_{3/2}$ single particle resonance does not play a role, i.e. where the direct reaction part without cut-off technique for the $d_{3/2}$ waves is smooth (see fig. 1). The number of 1^- states together with their shell model configurations is given in table 1. Altogether, there are 76 states with $J^{\pi} = 1^-$ which are the basic states for the diagonalisation of H_{QQ}. The parameters of the Woods-Saxon potential are of standard type (table 2).

Table 2 Parameters of the Woods-Saxon potential with spin-orbit coupling V^{so}

	Neutrons	Protons
Radius /fm	1.25	1.25
Diffuseness /fm	0.53	0.53
Depth V/MeV		
l = 0	56.82	55.91
l = 1	57.39	57.95
l = 2	54.94	54.36
Depth V^{so}/MeV		
l = 0	0	0
l = 1	9.64	9.89
l = 2	5.27	5.27

The residual interaction is of zero range with spin exchange,

$$V = -500 \text{ MeV fm}^3 \ (0.73 + 0.27 \ P_{12}^{\sigma}) \ \delta \ (\vec{r}_1 - \vec{r}_2). \tag{23}$$

The shell model wavefunctions ϕ_R obtained from eq. (1) are strongly mixed in all the 76 basic states, i.e. also in the isospin (for details see ref.[19]).

The shell model wavefunctions of the target nucleus ^{15}N are calculated in the 1h- configuration space. There are the two states $1/2^-$ and $3/2^-$ according to the ground state and first $3/2^-$ excited state.

The eigenfunctions $\tilde{\Phi}_R$ and eigenvalues $\tilde{E}_R - (i/2)\tilde{\Gamma}_R$ as well as the elastic cross section $^{15}N(p,p)^{15}N_{1/2^-}$ and inelastic cross section $^{15}N(p,p')^{15}N_{3/2^-}$ are calculated with the following number of 1^- resonance states.

(i)   1 resonance state of dominant 1p-1h nuclear structure and dominant isospin T = 1
(ii) 12 resonance states of dominant 2p-2h nuclear structure and dominant isospin T = 0
(iii) All 13 resonance states considered in (i) and (ii).

Table 3   Shell model energies E_R^{SM} used in the coupled-channels calculations with 1^- resonances

No of the state	E_R^{SM} / MeV set I	E_R^{SM} / MeV set II
1	29.1	29.1
2	28.6	28.8
3	28.0	28.5
4	28.2	28.6
5	28.8	28.9
6	29.0	29.0
7	29.6	29.3
8	29.8	29.4
9	30.0	29.5
10	28.4	28.7
11	30.2	29.6
12	29.2	29.1
13	29.4	29.2

The 12 and 13 states, respectively, with $J^\pi = 1^-$ described by the wavefunctions Φ_R (calculated with all 76 basic states) and energies E_R^{SM} are the new basic states for the diagonalisation of H_{QQ}^{eff} (eq. (7)). The parameters of the Woods-Saxon potential as well as the residual interaction V are the same as in the shell model calculation (table 2, eq. (23)). Two channels corresponding to the 1/2⁻ ground state and 3/2⁻ first excited state of ^{15}N are taken into account. The degree of mixing of the eigenfunctions $\tilde{\Phi}_R$ of H_{QQ}^{eff} in the shell model wavefunctions Φ_R (eq. (8)) depends strongly on the degree of overlapping with other resonances. In all cases discussed here the main component is not smaller than $|a_{RR}| = 0.9$.

The calculations (i), (ii) and (iii) are perfomed with two different sets of shell model energies E_R^{SM} of the states (table 3) in order to investigate the influence of overlapping on both the reaction cross section and the widths Γ_R. The wavefunctions used in both types of calculations are the same. Such a type of calculation is possible since the wavefunctions $\tilde{\Phi}_R$ are almost independent on energy.

The state considered in (i) has a large spectroscopic connection to the inelastic channel because of its dominant nuclear structure $(1p_{3/2})^{-1} 1d_{5/2}$ (amplitude: 0.68). The direct reaction part in this channel is small. The influence of overlapping of the other 12 resonance states with dominant 2p-2h nuclear structure and of the resonance state, considered in (i), with dominant 1p-1h nuclear structure on the nuclear cross section can therefore be investigeted easily in the inelastic channel. The results are shown in fig. 1 for the two sets of shell model energies E_R^{SM} (upper part: set I, lower part: set II). The positions E_R and widths Γ_R obtained from eqs. (9) and (10) are listed in tables 4 and 5 and shown in fig. 1 above the two cross sections.

The results (tables 4 and 5) show that the widths of all the resonance states are changed a little only, by external mixing with a doorway state. Such a result differs from the assumptions of all the traditional nuclear reaction theories. The fit of the experimental data is incapable of supporting the traditional doorway concept since the parametrised S-matrix used in fitting the data can identically be rewritten from one interpretation to another [20].

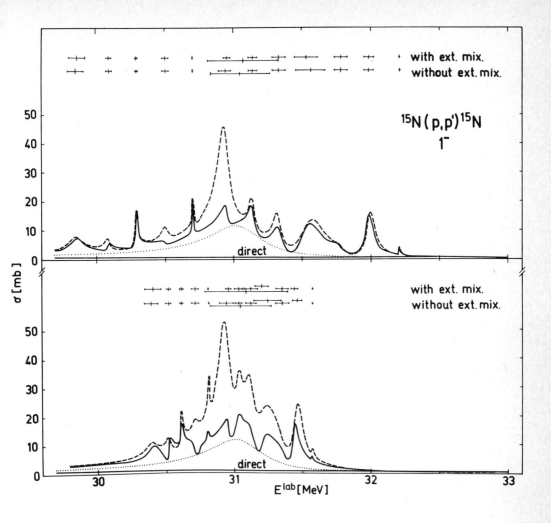

Fig. 1 Cross section of the $^{15}N_{g.s.}(p,p')$ $^{15}N_{3/2^-}$ reaction calculated with the resonance state № 1 (dotted curve) and with all 13 resonance states in consideration of external mixing (full curves) and by neglecting (dashed curves) it. The direct reaction part is denoted by "direct".

Therefore, the analysis of the experimental data must be supplemented by a dynamical calculation[20] as it has been done here on the basis of the CSM. The result obtained means that the fine structure resonances remain a good deal of their individuality in spite of their interaction with the doorway state. Indeed, Kanter et al.[5] have measured a very small life time (15 eV) of a fine structure resonance in the presence of a broad ($\approx$ 200 keV) isobaric analogue resonance in the ^{58}Ni + p reaction. This experimental result could not be explained up to now.

Table 4   Energies E_R and widths Γ_R of the 1^- resonance states with set I of the shell model energies E_R^{SM}

No of the resonance state	Without external mixing E_R/MeV	$\frac{1}{2}\Gamma_R$/keV	With external mixing 13 resonances E_R/MeV	$\frac{1}{2}\Gamma_R$/keV	12 resonances E_R/MeV	$\frac{1}{2}\Gamma_R$/keV
3	29.851	55.9	29.858	58.4	29.858	59.3
4	30.089	21.0	30.090	17.1	30.088	19.8
10	30.287	10.6	30.288	11.1	30.288	11.0
2	30.499	27.1	30.501	25.9	30.499	28.3
5	30.706	6.3	30.701	5.2	30.701	4.3
6	30.938	48.7	30.949	25.0	30.952	46.9
1	31.044	224.5	31.073	257.6	-	
12	31.138	25.2	31.137	28.7	31.137	27.4
13	31.329	37.3	31.331	48.0	31.343	47.0
7	31.564	101.3	31.530	85.0	31.542	91.0
8	31.780	46.3	31.780	43.9	31.781	43.5
9	31.997	34.0	31.984	33.4	31.990	35.4
11	32.213	1.5	32.213	1.3	32.213	1.4

All calculations are performed with (full lines in fig. 1) and without (broken lines in fig. 1) consideration of the external mixing of the resonance states. In the calculations with taking into account the external mixing, the unitarity condition for the S-matrix is fulfilled while this is not the case if external mixing is neglected. In this latter case, the energies and widths of the single resonance states are not influenced by the resonance states lying in the neighbourhood. Only the interference of the resonance states with one another and with the background is taken into account in calculating the cross section while the nondiagonal matrix elements $\langle \phi_R | H | \xi_E^c \rangle \langle \xi_E^c | H | \phi_{R'} \rangle$

describing the external mixing of the resonance states R and R' are neglected.

Table 5   Energies E_R and widths Γ_R of the 1^- resonance states with set II of the shell model energies E_R^{SM}

No of the resonance state	Without external mixing E_R/MeV	$\frac{1}{2}\Gamma_R$/keV	With external mixing 13 resonances E_R/MeV	$\frac{1}{2}\Gamma_R$/keV	12 resonances E_R/MeV	$\frac{1}{2}\Gamma_R$/keV
3	30.385	56.2	30.400	60.5	30.400	62.1
4	30.515	20.8	30.515	13.9	30.514	18.4
10	30.607	10.7	30.608	11.6	30.608	11.6
2	30.713	27.1	30.713	24.7	30.711	29.7
5	30.812	6.2	30.805	5.5	30.803	5.1
6	30.938	48.7	30.954	19.0	30.965	36.0
12	31.031	25.1	31.031	32.1	31.031	31.7
1	31.044	224.5	31.086	305.4	-	
13	31.116	36.7	31.118	48.6	31.162	50.0
7	31.244	100.6	31.199	47.1	31.180	91.8
8	31.353	46.8	31.353	45.2	31.355	43.6
9	31.463	34.6	31.442	23.9	31.450	34.1
11	31.573	1.5	31.573	1.2	31.573	1.2

The most striking result obtained and shown in fig. 1 is the reduction of the nuclear cross section due to the external mixing of the resonances. The reduction increases with increasing degree of overlapping (the lower part of fig. 1 as compared with the upper part).

It should be remarked here that the calculations without consideration of external mixing correspond to the assumptions done in analysing the experimental data with a parametrised S-matrix. A comparison of the results with and without external mixing shows that the individuality of the fine structure resonances is underestimated in such an analysis. The widths of all the resonance states are changed a little only, due to external mixing (table 4). However, their contribution to the cross section seems to be diminished if external mixing is not taken into account in the analysis (for details see ref. [19]).

The CSM calculations are performed without an absorptive part in the potential. Nevertheless, it is possible to do some conclusions on the absorptive part from the results obtained. The absorptive part due to the feeding of the compound nucleus states not taken into account explicitly in an optical model calculation leads, generally, to a reduction of the calculated cross section. However, the contribution which the individual compound nucleus resonance states give to the cross section seems to be reduced in the region of a doorway state if the analysis of the measured cross section is done, as usually, with neglecting the effects of external mixing. Therefore, the absorptive part of the optical model potential should be smaller at the energy of a doorway state than at other energies. With other words: If the analysis of a measured cross section with doorway state is performed without taking into account the effects of external mixing then the cross section has to be multiplied by some enhancement factor at the energy of the doorway state before the analysis can be started. Practically, this means that the absorptive part of the potential must be reduced by division with this factor.

It is very well known that for overlapping resonances [13,18]

$$\Gamma_R \leq \sum_c |\gamma_{Rc}^e|. \qquad (24)$$

This relation should be compared to the reduction of the cross section due to the external mixing of overlapping resonances. In both cases, the quantities observed experimentally, i.e. the total width and the reaction cross section, cannot be represented by a sum of individual terms which are independent from each other, that is the sum of the partial widths in the individual channels, and the sum of the individual cross sections of the isolated fine structure resonances, resp. The total widths and cross sections are reduced due to the overlapping of the resonances producing some stabilisation of the nucleus (enhanced life time and reduced cross section).

6. Conclusions

In this paper, the problem of intermediate structures and doorway states is considered on the basis of the CSM[11] with use of the cut-off technique of Wang and Shakin[10] for single-particle resonances. It is shown that the resonance picture of the cross section is changed by taking into account the external mixing of overlapping resonances via the continuum, i.e. the unitarity of the S-matrix. Generally, the individuality of the fine structure resonances under a gross structure is underestimated if the cross section is analysed by a sum of independent Breit-Wigner resonances, i.e. by neglecting the external mixing between the overlapping resonances. Therefore, problems arise in the interpretation of the other data, e.g. the correlation of the fine structure resonances in the various channels.

The problems of isobaric analogue resonances in medium heavy nuclei can be explained qualitatively by means of the results of the CSM. For example, the widths obtained from an analysis of the experimental data without consideration of the external mixing are too small since the cross section is reduced, generally, at the energy of the doorway state by the external mixing. These problems are discussed in detail in ref. [19].

It is shown further that shape resonances may be considered as doorway states although they are assumed usually to belong to the direct reaction part. The consideration of the external mixing between all the resonance states ensures to fulfill the condition of unitarity of the S-matrix. As a consequence, the nucleus is more transparent at the energy of the doorway state in the corresponding channel than at other energies and in other channels.

The shape resonances of heavy ion reactions (quasimolecular states) have surely doorway properties because of their strong clustering and, consequently, small internal mixing with the resonance states of complicated nuclear structure. They should be seen in the experimental data by an l-dependent absorptive part of the optical potential, large back angle scattering (because of one dominant l-value), large partial width relative to the corresponding channel, and some gross structure in the reaction cross section. The resonance behaviour of the excitation function may be, nevertheless, different for neighbouring nuclei, since the resonance behaviour is determined by the ratio Γ/D, i.e. by the concrete nuclear structure of the compound nucleus, in contrast to the shape resonances which depend more weakly on the concrete nuclear structure of the ion and target nucleus.

These conclusions drawn on the basis of the CSM should be tested by calculations for concrete heavy ion reactions and systematic experimental investigations. The decisive result on the role of external mixing (unitarity of the S-matrix) of a group of strongly overlapping resonance states can be proved not only in heavy ion reactions but also in nucleon induced reactions and in the isobaric analogue resonances on medium heavy nuclei. To prove the general picture of the resonance phenomenon in heavy ion reactions it would be interesting to measure directly the life time of the resonances in a similar manner as it has been done by Kanter et al.[5] for the isobaric analogue resonance case.

References

1) D.A. Bromley, J.A. Kuehner and E. Almqvist, Phys. Rev. Letters 4 (1960) 365; Phys. Rev. 123 (1961) 878; E. Almqvist, D.A. Bromley and J.A. Kuehner, Phys. Rev. Letters 4 (1960) 515
2) H. Feshbach, "Intermediate Structure in Light Ion Reactions", European Conference on Nuclear Physics with Heavy Ions, Caen (1976), J. Phys., supp. C5 - 177 (1976)
3) I. Rotter, Phys. Letters 67 B (1977) 385
4) E.G. Bilpuch, A.M. Lane, G.E. Mitchell, and J.D. Moses, Phys. Reports 28C (1976) 145
5) E.P. Kanter, D. Kollewe, K. Komaki, I. Leuca, G.M. Temmer and W.M. Gibson, Nucl. Phys. A 299 (1978) 230;
6) P. Kleinwächter and I. Rotter, to be published
7) G.H. Terry, H.J. Hausman, and N. Tsoupas, Phys. Rev. C19 (1979) 2155
8) G.M. Temmer, M. Maruyama, D.W. Mingay, M. Petrascu and R.VanBree Phys. Rev. Letters 26 (1971) 1341
9) B.H. Chou, G.E. Mitchell, E.G. Bilpuch, and C.R. Westerfeldt, Phys. Rev. Letters 45 (1980) 1235
10) W.L. Wang and C.M. Shakin, Phys. Letters 32B (1970) 421
11) H.W. Barz, I. Rotter and J. Höhn, Phys. Letters 37B (1971) 4; Nucl. Phys. A 275 (1977) 111
12) H. Feshbach, Ann. of Phys. 19 (1962) 287
13) C. Mahaux and H.A. Weidenmüller, Shell-Model Approach to Nuclear Reactions (North-Holland, Amsterdam, 1969)
14) R.H. Lemmer and C.M. Shakin, Ann. of Phys. 27 (1964) 13
15) D. Robson and A.M. Lane, Phys. Rev. 161 (1967) 982
16) I. Rotter, Annalen der Physik (Leipzig) 37 (1980) 393 and in press
17) I. Rotter, H.W. Barz and J. Höhn, Nucl. Phys. A 297 (1978) 237
18) I. Rotter, Journal of Physics G 5 (1979) 1575
19) P. Kleinwächter and I. Rotter, to be published
20) J.P. Jeukenne and C. Mahaux, Nucl. Phys. A 136 (1969) 49

CHARACTERISTIC RESONANCES AND THE LIMITS TO FUSION IN LIGHT HEAVY-ION SYSTEMS[*]

J. J. KOLATA

Department of Physics, University of Notre Dame

Notre Dame, IN 46556, U. S. A.

ABSTRACT

Comparison of heavy-ion reaction cross section data on several light systems, which have become available in the past few years, has provided strong evidence for the existance of a "characteristic" resonance. The most striking property of this resonance is the fact that, in each case, it is associated with the saturation of the fusion cross section. Equally intriguing is the fact that the "spin" of the resonance, deduced from the critical angular momentum for fusion, appears to place it on or near the extension of the ground-state band of the compound nucleus. Recent work has also shown that in at least two instances this anomaly appears at the same excitation energy in the compound system when populated by different reactions, thus demonstrating that we are not dealing with an entrance-channel phenomenon. Finally, it is shown that the "characteristic" resonances so described appear most strongly in the inelastic scattering and alpha-transfer channels, and that in the heaviest system studied (A = 36) there is little or no evidence for structure in any reaction channel other than these two.

---

[*]Work supported by the U. S. National Science Foundation.

In a recent letter [1] we have shown that the critical angular momentum for fusion of light systems, plotted as a function of the excitation energy of the compound system, displays several characteristic features. Among the most striking of these (Fig. 1) is the existance of a resonance which occurs at the transition between the low-energy behavior of the fusion cross section, which is dominated by the optical-model or "separated nucleus" potentials, and the high-energy behavior which in these light systems appears to be determined by properties of the compound nucleus. The abrupt nature of the onset of saturation in the fusion cross section allowed us to demonstrate a rather precise correlation with these resonances, an example of which is shown in Fig. 2 for the $^{16}O + ^{16}O$ system. It can be seen that the resonance at 26.5 MeV c.m., previously investigated by Singh, et al. [2], dominates the behavior of the ^{20}Ne production cross section (corresponding to alpha-particle transfer). Furthermore, this resonance is clearly correlated in the inelastic-scattering channel to the 3^- state of ^{16}O. Finally, we note the curious asymmetric shape of this anomaly, slowly rising on the low-energy side followed by a rapid falloff once the peak has been reached. This shape also seems to remain the same from system to system, as illustrated in Fig. 3, in which excitation functions for the $^{14}C + ^{16}O$ systems are shown. One notes the very strong characteristic resonances at $E_{c.m.} = 23.5$ MeV, correlated in the $^{14}C(3^-)$ inelastic scattering channel and in the production cross section for ^{18}O, which is formed primarily by alpha-particle transfer. This resonance, previously studied in the elastic channel by Bernhardt, et al. [3], is of some interest since it displays a peak-to-valley ratio of nearly

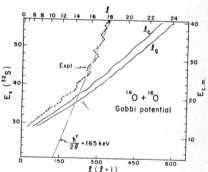

Fig. 1. Critical angular momentum trajectory for $^{16}O + ^{16}O$. The "characteristic" resonance occurs at $E_{c.m.} = 26.5$ MeV.

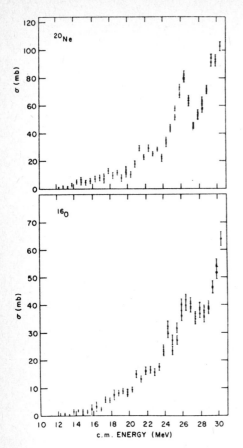

Fig. 2. Alpha-transfer (upper panel) and inelastic scattering (lower panel) cross section for $^{16}O + ^{16}O$, illustrating the correlation of the resonance at $E_{c.m.} = 26.5$ MeV.

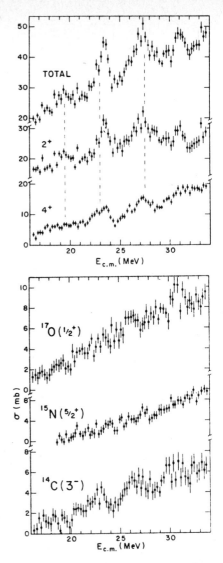

Fig. 3. Production cross section for ^{18}O (upper panel) in the $^{16}O + ^{14}C$ reaction, compared with the inelastic scattering and some single-nucleon transfer cross sections (lower panel). The "characteristic" resonance occurs at $E_{c.m.} = 23.5$ MeV.

one order of magnitude in the elastic data, the largest yet observed in heavy-ion exit channels.

The correlation of these "characteristic" resonances with the saturation of the fusion cross section is shown in Figs. 1, 4, and 5 for several systems. In each case, it can be seen that the limiting angular momentum for fusion, when displayed in an $\ell(\ell + 1)$ plot vs. the excitation energy in the compound system, is characterized by a low-energy and a high-energy behavior. In each of these regimes, the trajectory of the critical angular momentum is approximately a straight line, and the "characteristic" resonance occurs at the intersection of these straight lines in all systems studied. By comparing to the optical-model calculations presented in these Figures, one can see that the low-energy behavior of the fusion cross section is dominated by the optical-model potentials in the entrance channel, ie. the "separated nucleus" potentials. What is more intriguing is the fact that in several systems such as $^{12}C + ^{12}C$ and $^{12}C + ^{16}O$ (Figs. 4 and 5), the high-energy be-

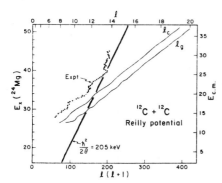

Fig. 4. Critical angular momentum plot for $^{12}C + ^{12}C$. The "characteristic" resonance occurs at $E_{c.m.}$ = 25.5 MeV, at the intersection of the low energy high energy behaviour of the trajectory. Note also the correlation with the extended ground state band of ^{24}Mg.

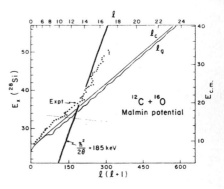

Fig. 5. Critical angular momentum plot for $^{12}C + ^{16}O$. The "characteristic" resonance occurs at $E_{c.m.}$ = 20.5 MeV, at the intersection of the low energy and high energy behavior of the trajectory. Note also the correlation with the extension of the ground state band of ^{28}Si.

havior lies very close to the extension of the ground-state band (gsb) of the compound nucleus. In fact, in the $^{16}O + ^{16}O$ system where the extension of the gsb is not known, a least-squares fit to the high-energy behavior extrapolates back to the ground state of ^{32}S, suggesting similar systematics.

Recent work at the University of Notre Dame has demonstrated the existance of a "characteristic" resonance in the $^{12}C + ^{20}Ne$ system which occurs at the same excitation energy in ^{32}S as the corresponding "characteristic" resonance in $^{16}O + ^{16}O$, even though the fusion Q-values differ by 2.4 MeV. (Fig. 6). This again demonstrates that these structures are a property of the compound system, and not simply an extrance-channel phenomenon, even though their widths are not too different from what might be expected for "shape resonances".

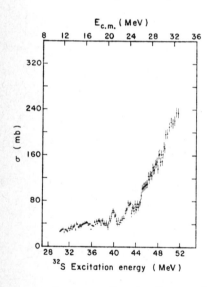

Fig. 6. Cross section for the population of the 1634-keV state in ^{20}Ne in the $^{12}C + ^{20}Ne$ reaction. The structure at $E_x(^{32}S) = 43$ MeV corresponds to a similar anomaly seen in the $^{16}O + ^{16}O$ reaction at the same ^{32}S excitation energy, although the fusion Q-values for the two reactions differ by 2.5 MeV.

Given the association of these resonances with properties of the compound system, it is reasonable to ask whether systematic information on high-energy and high-spin properties of sd-shell nuclei can be deduced from them. Unfortunately, the answer to this question is still rather ambigious. Under the hypothesis that the trajectory of the critical angular momentum for fusion in the high-energy regime extrapolates back to the gsb of the compound nucleus, moments-of-inertia can be deduced for several systems from the location of the resonance, as shown in Fig. 7. It can be seen that the moment of inertia increases gradually from A=24 to A=32,

then decreases at A=36. Because the fractional increase in radius of these light systems with increasing mass number is very large, a more appropriate comparison is in the ratio to the rigid body moment of inertia for a sphere with diffuse outer edge [4]:
$$I_{rigid} = 2/5\, MR^2 + 2Mb^2.$$
Here b is the skin thickness of about 1 fm. The radius R is taken to be [5]:
$$R = 1.28 A^{1/3} - 0.76 + 0.8 A^{-1/3}.$$
With these parameters, the contribution of the diffuse outer edge is about 30% of the total moment of inertia. It can be seen that the experimental moments-of-inertia have qualitatively reasonable values which seem to indicate some effect of deformation. Nevertheless, the correlation with static quodupole moments [6] is not good, and it appears that more information will be necessary before the true value of these resonances as indicators of nuclear structure can be assessed.

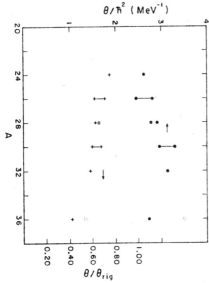

Fig. 7. Moments of inertia deduced from the characteristic resonances in several systems under the assumption that the resonance lies along the extension of the ground state band of the compound nucleus. Also shown is the ratio to the moment of inertia of a rigid sphere with a diffuse surface of width 1 fm.

More important than the search for more examples is an elucidation of the reaction mechanism responsible for the population of these characteristic resonances. The fact that they occur at the transition between the low-energy behavior of the fusion cross section, which is dominated by the "separated-nucleus" potentials, and the high-energy behavior which is apparently determined by properties of the compound system, suggests an analogy to the Landau-Zener effect in molecular physics [7]. One expects enhancements and structure in the energy dependence of cross sections due to this effect at points of crossing between atomic and molecular levels. The nucleus-nucleus analog of this process has been studied by Park, Greiner, and Scheid [8] for $^{12}C + ^{17}O$ and $^{13}C + ^{16}O$ reactions, and by Glas and Mosel [9] for $^{16}O + ^{16}O$. The latter authors find a level crossing to occur at a separation distance of 3.4 fm, which is quite near to the distance of closest approach of two ^{16}O ions moving on the combined Coulomb plus nuclear trajectory calculated with a proximity potential [10], at a c.m. energy and angular momentum corresponding to the "characteristic" resonance (Fig. 8). This calculation does not include the effects of nuclear "friction", so that the actual trajectory in the exit channel will undoubtedly be at a lower c.m. energy and angular momentum. The crucial question is whether the ^{16}O ions can approach so closely without feeling strong frictional forces. In order to answer this question, it is clearly necessary to examine the resonance region with tools that are sensitive to the reaction mechanism.

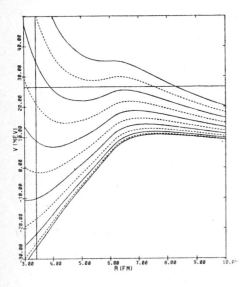

Fig. 8. Total real potential for the $^{16}O + ^{16}O$ system as a function of angular momentum, for $\ell=0$ to $\ell=20$ in steps of two units. The nuclear potential is a modified proximity potential. The horizontal line indicates the location of the characteristic resonance, which can be seen to be at the barrier top for $\ell=16$. The vertical line indicates the location at which the "separated nucleus" and "united nucleus" energy levels first cross in a two-center shell model calculation.

In conclusion, we have established the existance of a "characteristic" resonance in several light systems which occurs at the transition between the low-energy and high-energy behavior of the fusion cross section. We have shown that this resonance can be observed at the same excitation energy in the compound system via two different entrance channels, and also that it is correlated in the inelastic scattering and alpha-particle transfer exit channels. The "spin" of this resonance, deduced from the critical angular momentum for fusion, places it on the extension of the gsb of the compound nucleus for many systems. By extension, we use this property to deduce qualitatively reasonable information on the high-spin properties of the gsb in all systems for which "characteristic" resonances have been located. Finally, we note that the location of this resonance, at the intersection of the low-energy and high-energy behavior of the fusion trajectories suggests an analogy with the Landau-Zener effect in molecular physics. Existing calculations indicate that such an analogy might be apropos, at least for $^{16}O + ^{16}O$, if the "nuclear friction" in this system is small.

References

1. J. J. Kolata, Phys. Lett. 95B (1980) 215.
2. P. P. Singh, et al., Phys. Rev. Lett. 28 (1972) 1714.
3. K. G. Bernhardt, et al., Proceedings of the International Conference on Resonances in Heavy Ion Reaction, Hvar, Yugoslavia, 1977, edited by N. Cindro (North-Holland, Amsterdam 1978), p. 367.
4. G. Sussmann, Z. Phys. A274 (1975) 145.
5. J. Birkelund, et al., Phys. Repts. 56 (1979) 107.
6. R. H. Spear, Phys. Rept. 73 (1981) 369.
7. L. Landau, Phys. Z. Sow. 2 (1932) 46; C. Zener, Proc. R. Soc. A137 (1932) 696.
8. Jae Y. Park, W. Greiner, and W. Scheid, Phys. Rev. C21 (1980) 958.
9. D. Glas and U. Mosel, Phys. Lett. 49B (1974) 301; Nucl. Phys. A264 (1976) 268.
10. J. Blocki, et al., Ann. Phys. (N.Y.) 105 (1977) 427.

Fusion Resonances in $^{12}C(^{16}O,\gamma)^{28}Si$

A.M. Sandorfi and M.T. Collins
Brookhaven National Laboratory, Upton, New York 11973

Recent studies of heavy-ion capture reactions have reported unusual resonances whose structure appears closely linked to low-lying states in the fused nuclei.[1] The $^{12}C(^{16}O,\gamma)^{28}Si$ reaction provides a unique opportunity for investigating the origin of such resonances. Not only has comparatively little structure been observed near the Coulomb barrier of this system,[2] but moreover, the oblate deformation[3] of the low-lying states in ^{28}Si inhibits simple electromagnetic transitions from the inherently prolate heavy-ion channel.

High-energy γ rays from the reaction $^{12}C(^{16}O,\gamma)^{28}Si$ were detected for $^{16}O+^{12}C$ CM energies between 6.4 and 12.9 MeV. The high-energy portion of typical γ-ray spectra are shown in Fig. 7. The decays to the ground and first-excited states of ^{28}Si are clearly evident. At E_{lab}=20.0 MeV, the decay to the fifth excited state at 6.69 MeV appears quite strong. Despite the large backgrounds at lower energies, the areas of the γ_0 and γ_1 peaks are quite well determined.

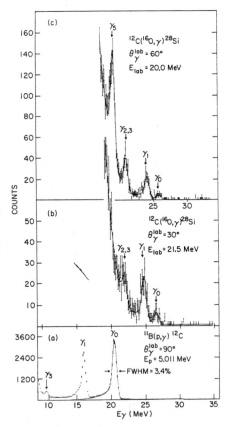

The γ_0 and γ_1 excitation functions are shown in Fig. 2. There are two striking features of these data. First, there are at least 3 relatively narrow (0.2 to 0.5 MeV CM) resonances in the γ_1 yield at about E_{CM}=7.3, 8.5 and 9.8 MeV. Secondly, the cross sections for γ_0 decay are extremely small.

The ground state of ^{28}Si is <u>oblate</u>, and the small $^{12}C(^{16}O,\gamma_0)$ cross sections reflect the difficulty in connecting this intrinsic shape to the inherently <u>prolate</u> heavy-ion entrance channel via an electromagnetic (one-body) operator. With such a simple

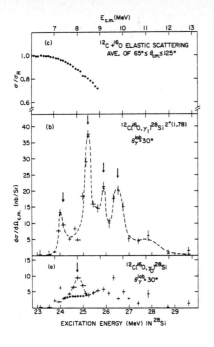

operator, the cross sections will be largest when the initial and final wavefunctions are most similar. This simple picture is borne out in Fig. 1(c) by the very strong γ_5 transitions (~150 nb/sr). The final state for this decay at 6.69 MeV in ^{28}Si is a 0^+ prolate shape-isomer.

The same deformation arguments could be applied to the γ_1 yield since the wavefunctions of the ^{28}Si ground and first-excited states are believed to be very similar.[3] In contrast, the ^{12}C(^{16}O,γ_1) excitation function exhibits several pronounced resonances, and this is quite puzzling. It suggests that the simple model based only on deformations is inadequate, and that the γ_0 yield is small for some other reason. Angular distributions suggest 2^+ assignments for at least the two lowest resonances of Fig. 2. Such resonances could also decay to the ^{28}Si ground state with a comparable rate. Their absence in γ_0 indicates either some as yet unrecognized structural variation among the low-lying levels of ^{28}Si, or an exceedingly complicated structure for the gamma-decaying compound states.

The capture cross sections are proportional to the product of the radiative and elastic-partial widths, $\Gamma\gamma\Gamma_{16_0}$. At least several of the peaks of Fig. 2 cannot be produced by resonances in Γ_{16_0} since none of these have been observed in elastic scattering, Fig. 2(c).[2] The structures seen in radiative capture must result mainly from resonances in $\Gamma\gamma$, and as such must represent special states of ^{28}Si.

Recent measurements of the ^{12}C(^{16}O,γ ^{28}Si) reaction, in which the γ ray backgrounds were dramatically reduced by detecting the recoiling ^{28}Si in coincidence, have revealed substantial cross sections for capture to a large number of bound states in ^{28}Si. We believe the γ ray branching ratios for decay to various bound states in the fused nucleus are essential to a better understanding of heavy-ion radiative capture. These experiments are continuing.

1. A.M. Sandorfi and A.M. Nathan, Phys. Rev. Lett. **40**, 1252 (1978); A.M. Nathan, A.M. Sandorfi and T.J. Bowles, Phys. Rev. **C24**, 932 (1981).
2. H. Spinka and H. Winkler, Nucl. Phys. **A233**, 456 (1974).
3. S. DasCupta and M. Harvey, Nucl. Phys. **A94**, 602 (1967); S.S.M. Wong and G.D. Loughed, Nucl. Phys. **A295**, 289 (1978).

Total Reaction Cross Section of ^{12}C+^{16}O Near the Coulomb Barrier*

E. C. Schloemer, M. Gai, A. C. Hayes, J. M. Manoyan, S. M. Sterbenz, H. Voit†, and D. A. Bromley

Wright Nuclear Structure Laboratory, Yale University, New Haven, Connecticut 06511

The ^{12}C+^{16}O system is one of the few heavy ion systems to date which has shown quasimolecular resonances in the vicinity of the Coulomb barrier. Structures have been observed in various reaction and scattering excitation functions for this system.[1-4] The Coulomb barrier resonances appear typically with widths of approximately 50 to 100 keV (cm). Clearly, the total reaction cross section should exhibit most resonant structures characteristic of the system; we have measured this cross section over the energy range of 7 to 15 MeV in the center of mass of the ^{12}C+^{16}O system.

Gamma-ray yields from recoiling nuclei were measured from the reaction ^{12}C(^{16}O, $x\gamma$)X with energy steps of 43 keV (cm) (from 7 to 11.75 MeV) and 65 keV (cm) (from 11.75 to 15 MeV). Targets were 5 μg/cm^2 ^{12}C evaporated on thick Au backings which slowed the recoil nuclei. Gamma-rays were detected at 90° with a 65 cm^3 GeLi detector 5cm from the target. A liquid nitrogen cold finger was used to limit carbon buildup during the course of the experiment. The data indicate an increase of target thickness of less than 0.5 μg/cm^2 per day. Relative normalization was achieved by monitoring the Rutherford scattering of the oxygen beam from the gold target backing with a surface barrier detector at 90°.

In the figure below, we show the σ_R excitation function over the entire range of our data. The largest fraction of our total yield (about 75%) comes from the 1369 keV gamma-ray of ^{24}Mg. We have integrated yields from ^{27}Al(843, 1014, 2210), ^{26}Mg(1809), ^{26}Al(416), ^{24}Mg(1369), ^{23}Na(438), and ^{20}Ne(1634). Structures are clearly evident over the entire energy region. They appear to be organized into two general types: those below approximately 11.5 MeV with widths of about 75 keV, and those above 11.5 MeV with widths of about 75 keV, and those above 11.5 MeV with widths of about 150 keV. The higher energy structures appear to be superimposed on a gross oscillatory background. Many of these structures have already been studied extensively in other exit channels and resonance energies have been established as indicated. A statistical analysis of our data is under way to help establish those structures which are resonant in nature and those which may be consistent with a statistical interaction mechanism; additional studies are intended to establish J^{π} assignments to as many of the resonances as possible.

---

*Work supported by USDOE Contract No. DE-ACO2-76ERO3074.

†Permanent address: Physikalisches Institut der Universität, Erlangen-Nurnberg, West Germany.

References

1. H. Voit, G. Hartmann, H.-D. Helb, G. Ischenko, and F. Siller, Z. Phys. 255 (1972) 425.
2. W. Treu, W. Galster, H. Fröhlich, H. Voit and P. Dück, Phys. Lett. 72B (1978) 315.
3. J. Schimizu, R. Wada, K. Fujii, K. Takimoto and J. Moto, J. Phys. Soc. Jap. 44 (1978) 7
4. P. Taras, G. Rao, N. Schulz, J.P. Vivien, B. Haas, J.C. Merdinger, and S. Landsberger, Phys. Rev. C15 (1977) 834.

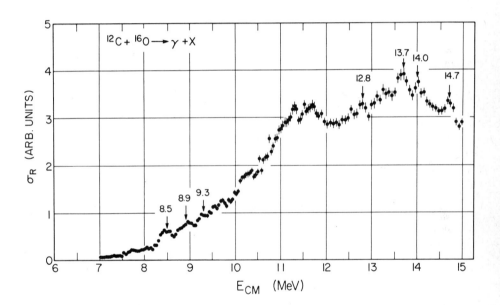

Fig. 1. Excitation function of total reaction cross section. The energies indicated correspond to some of the more familiar resonances in this system.

STRUCTURE IN HEAVY ION REACTIONS INVOLVING ^{14}C

R.M. FREEMAN
Centre de Recherches Nucléaires and Université Louis Pasteur,
67037 STRASBOURG CEDEX, France

I. Introduction :

Over a period of several years during which we have looked for resonances in "light" heavy ion systems using γ-ray techniques we have been able to survey a number of such reactions. In the course of these experiments we have been able to appreciate the specificity of reactions like ^{12}C + ^{12}C, ^{12}C + ^{16}O and ^{16}O + ^{16}O. The question which arose was why does one see structure in many of the channels of these reactions and yet it is so elusive in others. We felt that no theory of the origin of resonances in these systems would be complete without an adequate answer to this question.

We can reply qualitatively by saying that ^{12}C and ^{16}O are both very stable nuclei and reactions between then will take place at relatively lower excitation energies in the compound nucleus where the density of states is smaller. Moreover the number of open channels in these cases will also be smaller and consequently the width of the resonances will be less spread. Calculations relevant to the resonant behaviour of systems below the Coulomb barrier [1] have shown that the density of states and the number of open channels are relatively low in ^{12}C + ^{12}C and ^{12}C + ^{16}O consistent with the experimental observations. In this talk I will deal with measurements which were made above the Coulomb barrier where another consideration comes into play. How well can the system accommodate the high angular momentum brought in by the entrance channel ? Although energetically many channels may be open most of these are for the emission of α particles and nucleons which are unable to carry away much angular momentum. As the energy is increased this angular momentum mismatch between the entrance and exit channels is increasingly felt, especially for the grazing partial waves.

In the light of these ideas it seemed to us that perhaps there remained a certain amount of structure to be discovered in reactions between nuclei with stable nuclear configurations and for energies ranging above the Coulomb barrier. For this reason we turned our attention to the nucleus ^{14}C though it is radioactive and not so readily available either as target or beam. ^{14}C has a stable nuclear configuration because the neutrons are in a closed shell and the protons in a closed subshell. It was considered of particular interest to explore the energy region at about twice the Coulomb barrier and above where the fusion cross section has been observed to saturate in a number of systems. It has been argued that once a certain energy is reached there will be too few states available in the compound nucleus for the higher partial waves to fuse. For this reason the cross section is observed to reach a plateau and eventually to decrease. Without the fierce competition from the fusion processes interesting effects may become visible in the energy dependence of other channels. The $^{12}C + ^{14}C$ reaction had been studied [2,3] in the vicinity of the Coulomb barrier but no conclusive evidence for resonant structure was found. The $^{14}C + ^{16}O$ reaction has been studied [4] at higher energies and strong structure was observed in the elastic scattering and the $^{12}C + ^{18}O$ channel. These findings encouraged us to pursue research in this direction.

At Strasbourg we lacked facilities for ^{14}C studies and have relied on the existing equipment at Munich. The first experiment we tried was $^{14}C + ^{16}O$ using the oxygen beam at Strasbourg and a ^{14}C target procured from Munich. The strongest structure [5] we found for this system was in the channel to ^{18}O. Resonances began to appear at energies superior to $E_{c.m.}$ = 20 MeV. Later we went to Munich where there was a ^{14}C beam available to study the $^{12}C + ^{14}C$ reaction. Over the limited range of bombarding energies studied in this experiment two of the binary channels were found to resonate strongly [6].

However the $^{14}C + ^{14}C$ system was considered to be the reaction "par excellence" for testing the general correctness of these ideas. It was the last of the ^{14}C systems we studied but eventually in 1980 we were able to unite both ^{14}C beam and ^{14}C target at the Munich tandem laboratory and study the reaction over a substantial range of energies. The rest of this talk will be devoted to the results of this experiment. The $^{14}C + ^{14}C$ results are about to be published [7]. We also took the opportunity to extend the $^{12}C + ^{14}C$ to a greater range of energies and these new results will be presented in this talk.

In the months prior to this experiment, the first two experimental studies of the $^{14}C + ^{14}C$ system appeared, both dealing with the elastic scattering [8,9]. The strong periodic structure observed in the 90° elastic scattering data was reminiscent of the $^{16}O + ^{16}O$ system. These results implied that a similar periodic structure should exist for the reaction cross section but it remained to determine how this structure is distributed among the exit channels.

II. The $^{14}C + ^{14}C$ reaction :

The reaction was studied for bombarding energies ranging from E_{lab} = 25 to 65 MeV in steps of 1 MeV. A typical γ-ray spectrum taken at E_{lab} = 40 MeV is shown in

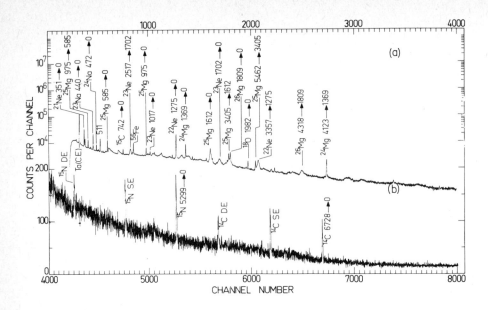

Fig. 1 : γ-ray spectrum recorded during the $^{14}C + ^{14}C$ experiment in the 90° Ge(Li) detector at E_{lab} = 40 MeV. Transitions due to ^{12}C contamination (20 %) are also present.

Fig. 1. Some of the transitions are from the $^{12}C + ^{14}C$ reaction due to a 20 % ^{12}C contamination in the target. By bombarding a similar target of ^{12}C at each energy it was possible to subtract out this contribution and at the same time have another look at this reaction over an extended energy range. Most of the γ-ray intensity comes from fusion-evaporation processes and in the case of the $^{14}C + ^{14}C$ reaction almost all of the fusion cross section was directed into the (αxn) channel to Ne isotopes or into the (xn) channel to Mg isotopes. This subdivision is pertinent to the search for structure as in past experiments it has been found that in the fusion processes it is generally restricted to channels in which there is an α particle in the evaporation chain. In the higher energy region of the γ-ray spectra there are two transitions due to other processes. One is the 6.73-MeV transition from the 3⁻ state of ^{14}C formed by inelastic scattering. The other results from the β-decay of ^{15}C, formed by one-neutron transfer, to the 5.30-MeV level of ^{15}N. The 2s halflife of ^{15}C was negligible compared to the time during which the spectra were accumulated. These two direct channels, in which effects associated with the grazing partial waves should be most evident, were of particular interest.

The excitation functions deduced from the γ-ray spectra are shown in Fig. 2. The results for the Ne isotopes, which account for a large fraction of the total cross section, show a regular series of resonances very like those found in similar channels of $^{16}O + ^{16}O$. The sequence of resonances extends up to the highest energy

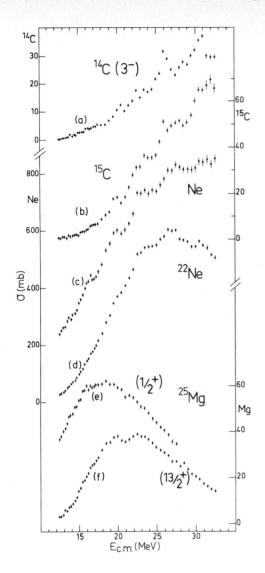

Fig. 2 :
Excitation functions for the $^{14}C + ^{14}C$ reaction deduced from the γ-ray data.
(a) Excitation of the 6728-keV level of ^{14}C.
(b) Production of ^{15}C.
(c) Production of Ne isotopes ($^{21}Ne + ^{22}Ne + ^{23}Ne$).
(d) Yield of the 1275-keV transition of ^{22}Ne.
(e) Yield of the 585-keV transition of ^{25}Mg.
(f) Yield of the 2056-keV transition from the 5462-keV level of ^{25}Mg.

studied E_{lab} = 65 MeV where we were forced to terminate the experiment because of the increasing γ-ray flux from reactions in the Ta backing. Only fragmentary results could be obtained for the fusion cross section to the Mg isotopes. This was not considered a too serious loss as structure has generally been found to be absent from such channels in other experiments. However we were able to compare the cross section to a high spin (13/2) and low spin (1/2) state in the same nucleus ^{25}Mg. To the low spin state the excitation function is smooth but to the high spin state the results contain traces of the same oscillation seen for Ne. This is an additional indication that the structure originates in the higher angular momentum partial waves. The excitation curves for the inelastic and transfer channels are shown at the top of the figure. There is evidence for the presence of the same oscillatory pattern here also though there may be some intermediate structure grafted on to the curves as well. The peak at $E_{c.m.}$ = 26 MeV, for example, is close in energy to a strong anomaly in the elastic scattering data.

III. Comparison of the $^{16}O + ^{16}O$ and $^{14}C + ^{14}C$ systems :

In Fig. 3 the fusion data that was obtained earlier for the $^{16}O + ^{16}O$ reaction [10] is shown to illustrate the similarity with $^{14}C + ^{14}C$. There is, of course, considerable differences in the way in which the outgoing flux is partitioned into different channels. In the neutron rich system $^{14}C + ^{14}C$ the neutron tends to displace the role played by the α particle in $^{16}O + ^{16}O$. But this difference aside the similarities as far as the structure is concerned become very striking. In both cases a regular sequence of resonances is observed in elastic, transfer and fusion channels, and in the fusion channels this structure is most evident in processes where α particles are evaporated from the compound system.

One feature of the $^{16}O + ^{16}O$ reaction was not satisfactorily reproduced in $^{14}C + ^{14}C$. As expected the gross structure of the $^{16}O + ^{16}O$ 90° elastic scattering data was found to be out-of-phase with the resonances of the reaction channels. This was also true for some regions of the $^{14}C + ^{14}C$ data, but not over the entire range of the results. It could be that we do not have enough of the total reaction cross section to make a valid comparison with the elastic scattering. On the other hand the gross structure in the elastic scattering data may be strongly modified by the presence of intermediate structure. But there is no obvious explanation why $^{14}C + ^{14}C$ should differ in this respect.

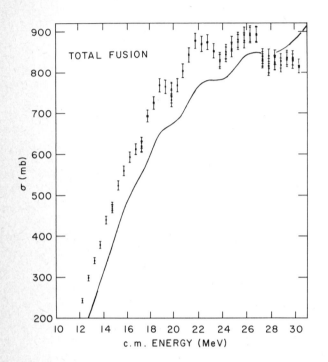

Fig. 3 : Structure in the fusion cross section of $^{16}O + ^{16}O$ deduced from γ-ray measurements.

A number of theories have been put forward to explain the observed structures in $^{16}O + ^{16}O$. Whether these explanations depend on resonant [10-14] or non-resonant [15] models they have one feature in common : each structure in associated with a single even incoming grazing partial wave. There is probably little to be learnt about the detailed mechanism of the reaction from such a regular structure. Any mechanism which highlights the grazing partial waves could lead to a sequence similar to what is observed experimentally. What the experiments do point to is the similarity of the dynamical conditions in both the $^{14}C + ^{14}C$ and $^{16}O + ^{16}O$ systems. There are three conditions which should be fulfilled in order to reproduce the structure observed.

Firstly the system should be composed of identical bosons. The description of the system is thus simplified to the even ℓ partial waves and any structure will be more apparent than in systems where the odd ℓ waves participate as well.

Secondly there should be no direct reaction strongly coupled to the entrance channel. In this respect ^{16}O and ^{14}C are very similar. Both are "spherical" nuclei with no low lying deformed state which would be readily excited in a heavy ion collision. ^{14}C and ^{16}O are the only nuclei (with A > 4) where the gap between the ground state and the first excited state is greater than 6 MeV.

Thirdly there should be a certain degree of surface transparency for these reactions. This is the most difficult of the three conditions to justify. Optical model potentials incorporating a surface transparency have been used to fit the elastic scattering data of some heavy ion systems [16]. This type of potential is apparently required in cases like $^{14}C + ^{14}C$ and $^{16}O + ^{16}O$ to fit the gross structure in the 90° elastic scattering. Weight has been given to the concept of surface transparency by the recent calculations of Haas and Abe [17] who determined the number of channels effectively open for the grazing partial waves of many systems. At energies a few times the Coulomb barrier they found a low minimum in some systems due to the angular momentum mismatch between the entrance and exit channels that was mentioned previously. Among these systems $^{14}C + ^{14}C$ and $^{16}O + ^{16}O$ both show similar behaviour with relatively low minima for grazing partial waves of angular momentum in the vicinity of 20 to 30 $\hbar$.

From the foregoing discussion it seems plausible that the dynamical conditions in the $^{14}C + ^{14}C$ and $^{16}O + ^{16}O$ reactions are rather similar and that this is reflected in the comparison of the experimental results. The regular structures observed in both reactions arise from the stability and spherical nature of the two nuclei. The observation of structure in reactions involving ^{14}C shows clearly that resonances are not confined to systems of nuclei with α-particle substructure like ^{16}O and ^{12}C. In the next section the results for the system where one ^{14}C is replaced by the stable but deformed nucleus ^{12}C will be presented.

The $^{12}C + ^{14}C$ reaction :

The $^{12}C + ^{14}C$ was studied concurrently with the $^{14}C + ^{14}C$ reaction primarily to correct for the ^{12}C contamination of the ^{14}C target, but at the same time we welcomed the opportunity to determine whether the resonant structure we had previously seen extended over a wider range of energies. The ^{12}C + ^{14}C reaction is expected to be even more surface transparent than ^{14}C + ^{14}C. It is however a more complicated system, firstly because the two bosons are no longer identical, and secondly because ^{12}C is a deformed nucleus with its first 2^+ excited state lying lower at 4.43 MeV. This state will be more readily excited in a collision, and the effect of an inelastic process like this, strongly coupled to the entrance channel, will be to disrupt any regular sequence of resonances.

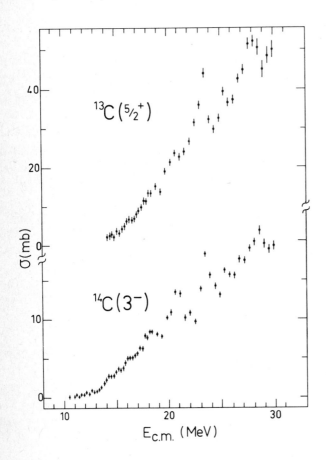

Fig. 4 : Excitation functions for the ^{12}C + ^{14}C reaction for the 3.85-MeV level of ^{13}C ($5/2^+$) and the 6.73-MeV level of ^{14}C (3^-).

It would certainly have been interesting to measure the excitation function for the 4.43-MeV level but this did not prove possible with the γ-ray technique. The lifetime of the state is short and consequently the γ-ray peak is strongly Doppler broadened and difficult to distinguish from the background. There were two other direct channels which were more easily extracted from the γ-ray spectra, and were the channels in which structure had already been found. These are the inelastic scattering to the 3^- state of ^{14}C and the one-neutron transfer to the $5/2^+$ state of ^{13}C. Both states are well matched in angular momentum to the entrance channel and their long lifetimes facilitate determining the intensity of the γ-ray peaks. The excitation functions for both states are shown in Fig. 4. Over the region studied before (18.4 ≤ $E_{c.m.}$ ≤ 26.3 MeV) 1 MeV steps in bombarding energy were taken, twice as large as previously, but at lower energies the reaction was studied in 0.5 MeV steps. In this low energy region the curves have been found to be relatively structureless but at about $E_{c.m.}$ = 20 MeV they break out into the strong structure we had observed in the earlier experiment. This sudden onset of resonant behaviour had also been observed in $^{16}O + {}^{14}C$. It could be interpreted as an effect due to the system becoming increasingly surface transparent as the bombarding energy is increased. When the system is sufficiently transparent for the grazing partial waves resonant effects, whatever their mechanism may be, become possible.

The other transitions visible in the γ-ray spectra were predominately from fusion-evaporation processes. Two examples are shown in fig. 5 for processes involving the evaporation of nucleons only. Consistent with what is generally found the excitation functions for the (pn) channel to ^{24}Na and the (p2n) channel to ^{23}Na are structureless. Both curves correspond to the intensity of the transition from the first excited state though, because of the different level structure of the two nuclei, a smaller proportion of the flux would be channelled towards the first excited state in ^{24}Na than in ^{23}Na. The rise in the ^{23}Na curve at high energies is probably spurious. At the high energies reactions in the Ta backing were beginning to flood the spectra with γ-rays which rendered the measurements increasingly problematic for the low energy transitions.

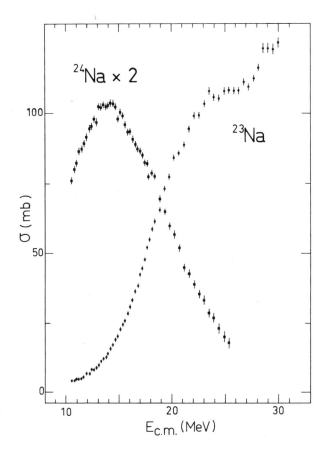

Fig. 5 : Yield functions for the transitions from the first excited state of ^{24}Na and ^{23}Na for the $^{12}C + {}^{14}C$ reaction.

The likelihood of observing structure in fusion processes is always greater when α particles are emitted, and in our earlier experiment it was suspected that there was structure in the (2α) channel to ^{18}O. This structure has been confirmed in the present experiments as shown in fig. 6 where the yield functions for the transitions

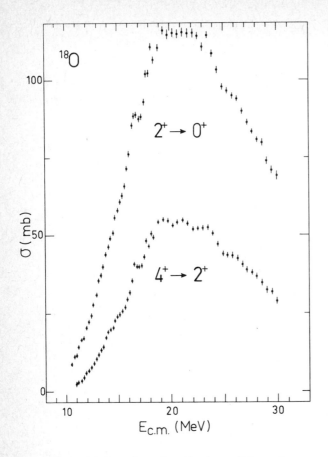

Fig. 6 : Yield functions for the transitions from the first and second excited states of ^{18}O for the $^{12}C + ^{14}C$ reaction

from the first and second excited state of ^{18}O are displayed. In the higher energy part the structure is correlated to some extent with the strong resonances of Fig. 4 and is probably due to the binary reaction $^{12}C + ^{14}C \rightarrow ^{18}O + ^{8}Be$ which cannot be distinguished from the fusion process in these experiments. The structure which appears in the previously unexplored lower energy region is likely to arise from the fusion channel.

In channels for the emission of light particles the strongest structure was observed for ^{22}Ne as shown in fig. 7. This structure being confined mainly to the low energy region was not observed previously. A Hauser-Feshbach statistical model calculation has been included which, though it poorly reproduces the absolute values, predicts the rise at higher energies. This rise is due to channels other than the evaporation of a single α particle, principally (2p2n). There are several other features of this curve which merit comment. At the lowest energies the cross section appears to fluctuate a little. A similar fluctuation was observed [18] in the same channel of $^{12}C + ^{13}C$ and interpreted as a consequence of the limited effective number of open channels in this process. A similar explanation would apply to $^{12}C + ^{14}C$. Another feature is the stepwise nature of the yield function similar to the gross structure often observed in these reactions. A possible explanation for this effect is that the compound nucleus ^{26}Mg is in the s-d shell where positive parity states predominate at low excitation energies. If the difference in the density of positive and negative parity states was still appreciable at the higher excitation energies of the compound nucleus it could lead to a difference between the probabilities

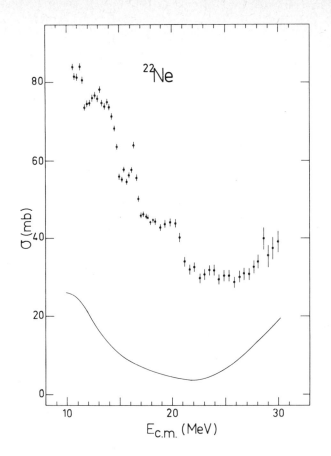

Fig. 7 : Yield function for the transition from the first excited state of ^{22}Ne for the ^{12}C + ^{14}C reaction. The predictions of a statistical model calculation are also shown.

for fusion of the odd and even partial waves and to an oscillation with a period similar to what is observed here. Finally your attention is drawn to the strong anomaly at 16.3 MeV. This appears to be an isolated resonance only about 300 keV wide. Isolated resonances of comparable width have been observed in the ^{12}C + ^{12}C and ^{12}C + ^{16}O reactions.

Conclusions :

In our search for resonances in heavy ion reactions the two factors which appear to be important are the stability of the nuclei and the possibility of a surface transparent region. We have thus been led to study reactions involving ^{14}C. Structure has been found in all the systems we have studied ^{16}O + ^{14}C, ^{12}C + ^{14}C and ^{14}C + ^{14}C, some of which is among the strongest observed in heavy ion reactions. With the γ-ray technique we have used it is not always clear which process is responsible for the observed effects. Further experiments with particle techniques could resolve these ambiguities.

The experiments described in this talk were undertaken in association with coworkers in Strasbourg, F. Haas, B. Heusch and C. Beck. We are indebted to Dr. H.J. Maier of the Universität München for the ^{14}C targets and to Dr. G. Korschinek of the Technische Universität München for the ^{14}C beam. The ^{16}O + ^{14}C reaction was studied in Strasbourg in collaboration with Prof. J.J. Kolata of the University of Notre Dame, Indiana. The ^{12}C + ^{14}C and ^{14}C + ^{14}C reactions were studied in Munich through the courtesy of

Prof. H. Morinaga and Dr. K.A. Eberhard and in collaboration with physicists at the tandem laboratory.

References :

1. D.L. Hanson et al., Phys. Rev. C9, 1760 (1974)
2. M. Feil et al., Z. Phys. 260, 271 (1973)
3. W. Galster et al., Nucl. Phys. A277, 126 (1977)
4. K.G. Bernhardt et al., Nuclear Molecular Phenomena, ed. N. Cindro (North Holland, Amsterdam, 1978) p. 367
5. J.J. Kolata et al., Phys. Rev. C23, 1056 (1981)
6. R.M. Freeman et al., Phys. Lett. 90B, 229 (1980)
7. R.M. Freeman et al., Phys. Rev., to be published
8. D. Konnerth et al., Phys. Rev. Lett. 45, 1154 (1980)
9. D.M. Drake et al. Phys. Lett. 98B, 36 (1981)
10. J.J. Kolata et al., Phys. Rev. C16, 891 (1977) and Phys. Rev. C19, 2237 (1979)
11. W.A. Friedman et al., Phys. Lett. 87B, 179 (1979)
12. Y. Kondo et al., Phys. Rev. C22, 1068 (1980)
13. O. Tanimura and T. Tazawa, Phys. Rep. 61, 253 (1980)
14. R. Vandenbosch and A.J. Lazzarini, Phys. Rev. C23, 1074 (1981)
15. R.L. Phillips et al., Phys. Rev. Lett. 42, 566 (1979)
16. R.H. Siemssen, Nuclear Molecular Phenomena, ed. N. Cindro (North Holland, Amsterdam, 1978) p. 79
17. F. Haas and Y. Abe, Phys. Rev. Lett. 46, 1667 (1981)
18. R.A. Dayras et al., Nucl. Phys. A265, 153 (1976)

Structure in Symmetric Light Heavy-Ion Fusion Cross Sections

N. Rowley, Theory Group, SERC Laboratory, Daresbury, Warrington, UK

N. Poffé, Dept. of Nuclear Physics, Keble Road, Oxford, UK*

R. Lindsay, Dept. of Theoretical Physics, 1 Keble Road, Oxford, UK

The structure observed in the total fusion cross sections of symmetric light heavy-ion systems may be shown to be an entrance-channel effect by comparing the energies E of the 'peaks' with those in the 90° elastic excitation functions $\sigma_E(\frac{1}{2}\pi)$ which lie at higher energies on the same 'rotational band'. In other words the energies for which structure is observed fall on a straight line when plotted against $\ell_g(\ell_g+1)$, where $\ell_g(E)$ is the appropriate grazing angular momentum. The slope of this line is approximately $1/2mR_B^2$, where R_B is the position of the Coulomb barrier. This suggests that the effects are associated with barrier penetration rather than 'molecular resonances' which would be related to a rather smaller radius.

It has been suggested[1] that the fusion structure is a result of surface transparency arising from the small number of open reaction channels and the application of this idea leads to the conclusion that the effect should not be present in the $^{20}Ne + {}^{20}Ne$ system. However, a recent detailed analysis of the evaporation residues in this reaction (performed by one of us (NP) on the Oxford tandem using γ-ray techniques) shows that the structure is quite clearly observable even in this case.

Consider the parametrisation where the fusion transmission coefficients are given by a fermi function in ℓ-space i.e.

$$T_\ell(E) = \left(1 + \exp\left(\frac{\ell - \ell_g(E)}{\Delta(E)}\right)\right)^{-1} \tag{1}$$

Using the Poisson summation formula it is possible to derive the following expression for the total fusion cross section for a non-symmetric system

$$\sigma_f^{NS} \approx \frac{\pi}{k^2}(\ell_g + \tfrac{1}{2})^2 + \frac{8\pi^2 \ell_g}{k^2} \Delta \exp(-2\pi^2\Delta)\sin 2\pi\ell_g, \tag{2}$$

whereas for a system of identical spin-zero nuclei the absence of odd partial waves yields

$$\sigma_f^S \approx \frac{\pi}{k^2}(\ell_g + \tfrac{1}{2})^2 + \frac{8\pi^2 \ell_g}{k^2} \Delta \exp(-\pi^2\Delta)\sin \pi\ell_g \tag{3}$$

Examination of the above formulae shows that the energy-dependent structure in σ_f is appreciably larger for symmetric systems though it still requires a rather small value of Δ to be significant. If the transmission coefficient of eq. (1) comes from barrier penetration a rough approximation to Δ is

$$\Delta(E) \approx \frac{1}{2\pi} \sqrt{\frac{R}{a}} \sqrt{\frac{E-\frac{1}{2}E_B}{E-E_B}} \quad , \tag{4}$$

where E_B is the Coulomb barrier height. This shows that the structure in σ_f may be enhanced by increasing a, the diffuseness of the nuclear potential. For example the oscillations observed for $^{12}C + {}^{12}C$ may be obtained for a $\approx$ 0.8 fm without the introduction of ℓ-dependent absorption[2].

Detailed calculations show that the magnitude of the structure is not significantly changed by strong coupling to inelastic channels though the total cross section may increase and the positions of the peaks be slightly shifted.

(1) F. Haas and Y. Abe, Strasbourg preprint CRN TN80.22 (1980).

(2) O. Tanimura, Nucl. Phys. A334 (1980) 177.

* On leave of absence from the Institut de Physique Nucléaire, Orsay.

VI. SEARCH FOR DIRECT γ-DECAY OF RESONANCES

Search for γ-rays from the Quasimolecular $^{12}C + ^{12}C$ System

V. Metag*, A. Lazzarini, K. Lesko, and R. Vandenbosch
Nuclear Physics Laboratory
University of Washington
Seattle, WA 98195, USA

At this workshop we have heard about very speculative but exciting interpretations of the resonance phenomena in the $^{12}C + ^{12}C$ system. It has been suggested[1] that they are related to rotational bands based on shape isomeric states in ^{24}Mg. We have worked for a long time in the "classical" mass region of shape isomers, the actinide region, where shape isomers are called fission isomers since their dominant decay mode is spontaneous fission. It was therefore of particular interest for us to investigate whether such shape isomers do indeed occur also in other mass regions.

In the actinide region the existence of shape isomers has been associated with secondary minima[2] in the potential energy surface. These structures in the potential energy result from the superposition of shell corrections to the nuclear binding energy onto the macroscopic part of the deformation energy described by the liquid drop model. Since shell corrections are typically only a few MeV their superposition will only give rise to pronounced minima if the macroscopic deformation energy surface is fairly flat. In the actinide region this is achieved by the near cancellation of the shape dependence of the surface and Coulomb energies in the liquid drop model. In lighter nuclei the distorting effect of the Coulomb repulsion is taken over by the centrifugal forces in rapidly rotating nuclei leading again to rather flat deformation energy surfaces at high spins. Strutinsky-type calculations, extended to include nuclear rotation, have been performed by various groups[3,4]. Islands of shape isomerism are predicted to occur at high rotational frequencies not only in the well-studied actinide region but also in the rare earth region, at $Z \simeq 40$, and in the Mg-S region (fig. 1).

---

*Present address: Max-Planck-Institut für Kernphysik, Heidelberg, West Germany.

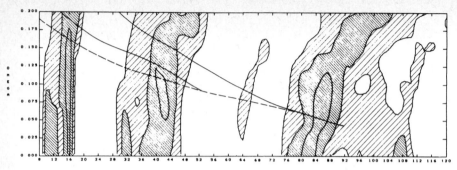

Fig. 1. Proton shell energy contour diagrams versus rotational frequency ω and proton number Z at a deformation of 2:1. The contour line separation is 1 MeV. No contour lines are drawn for positive shell energies (from ref. 3).

A crucial experimental test for these theoretical predictions and for the above mentioned speculative interpretation of heavy ion resonances would be the observation of collectively enhanced E2 γ transitions between these resonances if they were indeed members of the supposed rotational band based on a shape isomer in ^{24}Mg. As in the case of fission isomers it should be possible to derive some information on the shape of the dinuclear system from the collective strength of these transitions. A rough estimate shows that such an experiment is just about feasible.

Provided that the rotational model is applicable, the γ transition strength can be estimated directly from the theoretically predicted quadrupole moment of the quasimolecular 24 nucleon configuration. The groups at Giessen[4] and Lund[5] independently calculate values of about 1.8 b for the charge distribution of the shape isomer in ^{24}Mg. This quadrupole moment exceeds that of the ground state of ^{24}Mg by nearly a factor of 3, a situation even quantitatively analogous to that encountered in the actinide region[6]. One can also make some estimate of the quadrupole moment based on the experimentally determined[7] moment of inertia $2\Theta/\hbar^2 = 10$ MeV^{-1} of the "band". For rigid rotation a lower limit of $Q \geq 1.4$ b is obtained. Assuming that the observed moment of inertia corresponds to 80% of the rigid rotor value, as established for fission isomers[6], a quadrupole moment of 1.9 b is derived. In view of these estimates the theoretical prediction of $Q = 1.8$ b seems quite reliable. Within the rotational model this quadrupole moment gives a B(E2) value for rotational transitions of 270 spu. For typical transition energies of 5-6 MeV this corresponds to a γ width of ≅7 eV which reduces to $\Gamma_\gamma \cong 5.5$ eV if the predicted triaxial shape of the dinuclear system is taken into account. To determine the probability for γ decay one has to divide the radiative width by the total width, which has been determined[8] to run between 200 and 400 keV, of the intermediate structure resonances. Assuming $\Gamma_{total} \simeq 300$ keV (the time associated with a full rotation of the dinuclear system with $2\Theta/\hbar^2 = 10$ MeV^{-1} in the 14^+ state corresponds to a width of 500 keV) a branching ratio of $2 \cdot 10^{-5}$ is expected. This is a very small number but not too small for experimental observation.

In the experiment[9], performed at the tandem accelerator of the University of Washington, we populated the 14^+ resonance of the $^{12}C + ^{12}C$ system at a bombarding energy of $E_{CM} = 25.2$ MeV. The main decay mode of the 14^+ resonance is fission into two carbon nuclei which are either in their ground states or excited to their 2^+ states at 4.44 MeV or higher lying states. A decay of the 14^+ resonance via a quasimolecular γ transition would populate the 12^+ resonance at $E_{CM} = $ 18.5-20.4 MeV which again predominantly decays into the C + C channel. The two outgoing C nuclei were detected in coincidence with two position-sensitive ΔE-E telescopes subtending angles between 70° and 110° in the center-of-mass system (fig. 2). Coincident γ-rays originating either from the searched for "rotational" 14^+-12^+ transition or from the deexcitation of the inelastically scattered carbon nuclei were detected in a 10" x 10" NaI crystal positioned 10 cm away from the target. C-C double coincidences as well as C-C-γ triple coincidences were recorded on tape.

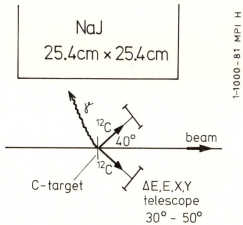

Fig. 2. Detector configuration used in the experiment (from ref. 9).

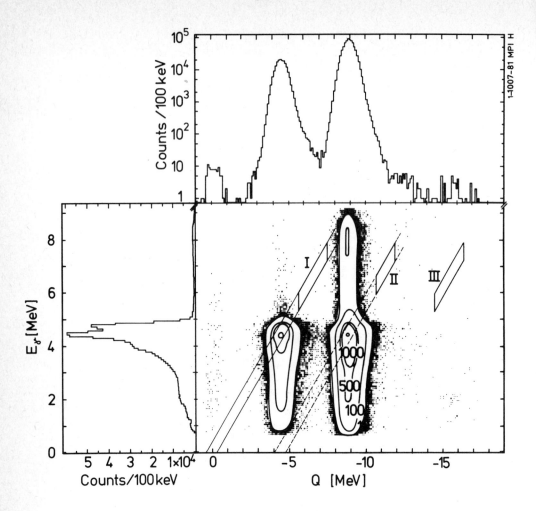

Fig. 3. Scatter plot of γ-ray energy versus the reaction Q value for $^{12}C + ^{12}C$ at a bombarding energy of 50.4 MeV. Small squares correspond to one event, big squares to ≥two events. Contour lines for intensities of 10, 100, 1000, and 5000 counts are indicated. The two straight lines limit the region of events for which the total energy observed in the γ-ray and particle detectors agree within the resolution with the bombarding energy. The dashed lines correspond accordingly to the case that one of the 4.4 MeV γ-rays, emitted from the excited ^{12}C nuclei, escapes detection. The boxes labelled I, II, and III limit the areas where quasimolecular γ-rays are expected, accompanied by no, single, or double excitation of the two C nuclei, respectively (from ref. 9).

The results are summarized in fig. 3 in a scatter plot of events as a function of γ-ray energy and the reaction Q value. The γ and particle spectra resulting from a projection of the coincidence data on either side are also shown. Two groups of events with Q = -4.44 MeV and Q = -8.88 MeV are apparent, corresponding to the excitation of one or both ^{12}C nuclei to the 2^+ state at 4.44 MeV. The strong intensity at E_γ = 4.44 MeV for Q = -8.9 MeV corresponds to the detection of only one of the two emitted 4.44 MeV γ-rays. Events observed close to Q values of -12.1 and -12.7 MeV are attributed to the deexcitation of high-lying states in ^{12}C which decay with known although extremely small branching ratios for γ decay. Including the chance coincidences around Q $\simeq$ 0 and the pileup events at Q $\simeq$ -4.4 MeV, the events shown in fig. 3 appear to be quantitatively understood. Triple coincidences which can be interpreted as decays of rotational shape isomeric states in ^{24}Mg are expected in regions I, II, or III, respectively, depending on whether the supposed quasimolecular γ decay is accompanied by no, single, or double excitation of the C nuclei. No such events are found with γ-ray energies in the range of 5.6-7.5 MeV.

This result can be used to extract an upper limit on the probability P_γ for γ decay of the 14^+ resonance in the ^{12}C + ^{12}C system. It can be shown[9] that, after appropriate corrections for detection efficiencies, the branching ratio for γ decay is determined by the ratio of quasimolecular C-C-γ coincidences to the total number of decays into the C + C channel given by the number of twofold C-C coincidences. Assuming one event as an upper limit we obtain a branching ratio of $\leq 2 \cdot 10^{-6}$, at a level of three standard deviations, $P_\gamma \leq 8 \cdot 10^{-6}$. This limit is more than two times smaller than the branching ratio expected for the decay of shape isomeric states in ^{24}Mg. Although the deduction of this branching ratio from the experimental data is based on certain assumptions we feel that the discrepancy by more than a factor of 2 at least does not support the interpretation of the observed resonances as quasimolecular states. A more definite conclusion can only be drawn, however, if the experiment is repeated with considerably improved sensitivity.

It is illustrative to formulate the present experimental result in terms of the lifetime of the dinuclear system, independent of the rather involved analysis of the widths of the resonance structures. From the estimated partial width for γ decay of $\Gamma_\gamma \simeq 5.5$ eV, corresponding to a partial lifetime of $1.2 \cdot 10^{-16}$ sec, and the experimentally found upper limit of $(\Gamma_\gamma/\Gamma_{tot})_{14^+} \leq 8 \cdot 10^{-6}$ a lifetime of $\leq 1 \cdot 10^{-21}$ sec is deduced for the dinuclear configuration. This is twice the collision time[10] of $5 \cdot 10^{-22}$ sec. In comparison, the time for a full rotation of a nucleus with a moment of inertia of 10 MeV^{-1} and spin 14$^+$ is 1.3·10^{-21} sec, i.e., the colliding nuclei do not find time enough to form a longer-lived quasimolecular configuration.

Blair and Sherif[11] have recently performed a DWBA calculation of the yield of E2 nuclear bremsstrahlung emitted in the collision of two carbon nuclei. They find a ratio of 1·10^{-6} for the bremsstrahlung-to-elastic cross sections, integrated over a γ-ray energy interval of 2 MeV and integrated over the angular range covered by the particle detectors. This value is not in disagreement with the upper limit obtained in the present experiment.

In conclusion, the present experimental limit on the branching ratio of γ-to-particle decay of the 14$^+$ resonance in the ^{12}C + ^{12}C system is fully consistent with the hypothesis that this resonance is associated with a shape resonance in a direct reaction process rather than with a shape isomeric state in ^{24}Mg. An experiment with a considerably improved sensitivity is planned at the Darmstadt-Heidelberg crystal ball which is going into operation this fall.

Acknowledgement

We would like to thank K. Snover for his participation in the early stages of this experiment and appreciate the help of A.G. Seamster in running the experiment. Illuminating discussions with J. Blair, H. Sherif, and L. Wilets are gratefully acknowledged. This work was supported in part by the U.S. Department of Energy.

References

1) E.R. Cosman, invited talk, this workshop.
   E.R. Cosman, R. Ledoux, M.J. Bechara, C. Ordonez, R. Valicenti, and A. Sperduto, preprint 1981.
2) V.M. Strutinsky, Nucl.Phys. A95 (1967) 420; Nucl.Phys. A122 (1968) 1.
3) G. Leander, C.G. Andersson, S.G. Nilsson, I. Ragnarson, S. Åberg, J. Almberger, T. Døssing, and K. Neergård, Proc. Conf. on High-Spin Phenomena in Nuclei, Argonne, 1979, p.197.
4) H. Chandra and U. Mosel, Nucl.Phys. A298 (1981) 151.
5) I. Ragnarsson, private communication, 1981.
6) V. Metag, D. Habs, and H.J. Specht, Phys.Rep. C65 (1980) 2.
7) E.R. Cosman, T.M. Cormier, K. van Bibber, A. Sperduto, G. Young, J. Erskine, L.R. Greenwood, and O. Hansen, Phys.Rev.Lett. 35 (1975) 265.
8) E.R. Cosman, R. Ledoux, and A.J. Lazzarini, Phys.Rev. C21 (1980) 2111.
9) V. Metag, A. Lazzarini, K. Lesko, and R. Vandenbosch, to be published in Phys.Rev. C.

10) A. Gobbi and A.D. Bromley, in Heavy Ion Reactions, ed. R. Bock (North-Holland, Amsterdam, 1979), Vol. I, p.485.

11) J. Blair and H. Sherif, private communication, 1980.

SEARCH FOR DIRECT γ-TRANSITIONS IN $^{12}C+^{12}C^*$

R.L. McGrath, D. Abriola**, J. Karp, T. Renner[†]
and S.Y. Zhu[‡]

Department of Physics
State University of New York
Stony Brook, New York 11794

ABSTRACT

Preliminary results of an experiment to detect γ-transitions between structures known in the $^{12}C+^{12}C$ system at 25.8 MeV and 19.3 MeV (c.m.) are reported. The extracted upper limit for these transitions is evidence against fully collective γ-decay between intermediate width resonances at these energies, but is inconclusive with respect to differentiating gross structure resonances from non-resonant processes.

The C+C molecular states are expected to have extremely large quadrupole γ-transition probabilities. To estimate their size we use the formula for rotational bands built on static deformations, $B(E2) = 5/16\pi \ e^2 Q_o^2 |<J_f 0|J_i 020>|^2$. Using $Q_o=160$ fm^2, the value calculated for molecular states by Chandra and Mosel[1], one finds B(E2) is 180 times the Weisskopf value for $J_i,J_f=14,12$. [Two touching uniformly charged spheres with radius constant 1.25 fm have the comparable moment $Q_o=200$ fm^2.]. The goal of our experiment is to measure the branching ratio Γ_γ/Γ and, thereby, to find the characteristic width over which C+C states are distributed. The catch is, of course, that no matter what the structure, Γ_γ/Γ is very small because the reaction time is orders of magnitude smaller than the mean γ-emission time. Specifically, assuming the relevant width corresponds to intermediate or gross structure width resonances, then $\Gamma_\gamma/\Gamma \sim 2\times 10^{-5}$ or $\sim 3\times 10^{-6}$ assuming $\Gamma_\gamma=8.5$ eV ($E_\gamma=6.5$ MeV) and $\Gamma=0.5$ or 3 MeV, respectively. One expects a still smaller ratio for non-resonant processes such as the one proposed by Phillips et al.[2] Here we estimate $\Gamma_\gamma/\Gamma \sim \tau_{reaction}/\tau_{\gamma-decay}$ and write $\tau_{reaction} = \hbar/2 \cdot d\delta_\ell/dE$. Since the parameterized potential of Phillips gives $\tau_{reaction} \sim \hbar/10$ MeV, we obtain $\Gamma_\gamma/\Gamma \sim 1\times 10^{-6}$ for this mechanism. Similar short reaction times can be found classically for direct reactions involving grazing partial waves ℓ_g where $\tau_{reaction} \Delta\theta/\omega=\Delta\theta\sqrt{2(E-v_c)/\underline{\Omega}}$ and $\Delta\theta$ the "sticking" angle is, say 20°.

Figure 1 shows the schematics of the particular transition searched for here. Cormier et al.[3] reported prominent gross structures centered at about 19.0 and 24.8 MeV (c.m.) with likely spin-parity 12^+ and 14^+. The energy intervals corresponding to the gross widths Γ are hatched. The experiment is sensitive to γ-transitions to the one-MeV region (ΔE) centered at 19.3 MeV. The measurement consists of comparing the number of direct elastic or inelastic decay events to the number of elastic or single inelastic events which have 25.8-(19.3±0.5)=6.5±0.5 MeV "missing" energy.

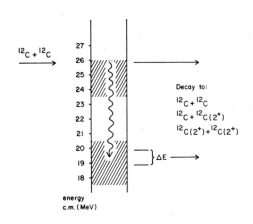

Fig. 1 The $^{12}C+^{12}C$ 25.8 MeV decay processes examined here. The hatched regions show the gross structure regions (FWHM) found in Ref. 3.

Since the γ-ray momentum is only of order 10^{-2} of the ^{12}C momenta, $\Sigma\vec{p}_{12_C} \approx \vec{p}_{beam}$. Thus if it were not for sources of experimental background, e.g. slit scattering, incomplete charge collection, or target contaminants, the γ-transitions could be observed in "singles" ^{12}C energy spectra in the region $Q \sim -6.5$ MeV. We use momentum data from two telescopes to "over determine" the kinematics in order to reduce background. The experimental set up is indicated in Fig. 2. Two ionization counter-silicon position-sensitive detectors comprise telescopes centered at ±40.8° subtending 14°. The intrinsic energy resolution is about 240 keV (FWHM), the angular resolution is about 0.5° which is due primarily to the beam emittance. Slit scattering was reduced by using electropolished masks and collimators. Conventional electronics were used which included pile-up inspection and analog circuits to derive position and Z information. Signals were stored on tape for subsequent even-by-event processing.

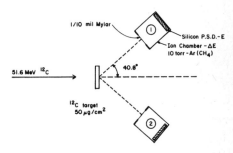

Fig. 2 Experiment schematic. The two telescopes subtend 14°.

The kinematics of all C+C events were tested by computing the apparent net momentum perpendicular and parallel to $\vec{p}_{beam}$ assuming both C's had mass 12 amu. Figure 3 shows the momentum distribution of events collected in a short run with Q-values ~ -8.86 MeV corresponding to the double inelastic $^{12}C(2^+)+^{12}C(2^+)$ reaction. The distribution is consistent with Monte-Carlo simulations based on beam spot size, target thickness and the (assumed isotropic) recoil of the excited ^{12}C nuclei following emission of the 4.43 MeV γ-rays. The figure also shows the momentum distribution of C+C events collected in a short run on a ^{13}C target. Events were selected which have apparent Q-values near -6.5 MeV. It is clear that such contaminant events can be easily distinguished from $^{12}C+^{12}C$ events. Anomalous

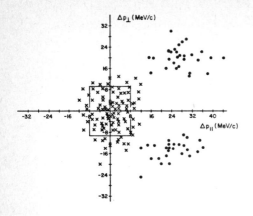

Fig. 3 Momentum distributions of some $^{12}C+^{12}C$ events with $Q \sim -8.86$ MeV (X's) and C+C events from a <u>short</u> run on a ^{13}C target (●'s). The rectangle indicates the FWHM of the distribution.

inelastic events where one or both telescopes give an abnormally small energy signal would be distributed on the right of the momentum plane. In the actual data processing, valid $^{12}C+^{12}C$ events were defined to have net momentum values within the observed full width of the double inelastic distribution indicated by the rectangle.

The resultant background reduction is evident in Fig. 4. The top Q-spectrum is generated from singles data in one telescope; the bottom coincidence Q-spectrum shows all data which satisfy the momentum test.

The latter spectrum is generated by averaging the Q-value computed separately from the data in each telescope since this procedure improves the peak broadening due to the finite beam spot size. The peak associated with the 7.65 MeV 0^+ state suggests the spectrum quality. This particle unstable state has the small radiation branching ratio $(\Gamma_\gamma+\Gamma_\pi)/\Gamma=(4.1\pm0.1)\times10^{-4}$, [4] and the singles inelastic cross section is only 19±2% of the inelastic 4.43 MeV 2^+ cross section. Hence, the ratio of coincident 0^+ events to 2^+ events should be only $(7.9\pm0.8)\times10^{-5}$. The data in Fig. 4 give $(6.1\pm0.7)\times10^{-5}$.

Because of the tails of the strong inelastic peaks it appears that only relatively narrow regions one MeV in width centered at $Q=-6.5$ MeV (elastic decay of the 19.3 MeV intermediate state) and

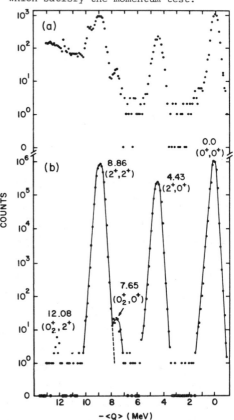

Fig. 4 Part(a), A singles Q spectrum Part(b), A coincidence Q spectrum of all events at 25.8 MeV which satisfy the momentum constraint.

-10.9 MeV (single inelastic decay) might clearly reveal the γ-transition. It is obviously necessary to test to what extent the events in these regions are associated with residual experimental background. This was done by collecting data at 28.8 MeV, an "off-resonance" incident energy.[3] All results are tabulated in Table I. Considering that the "off-resonance" data set has only one third as many events as the 25.8 MeV data set, there is no statistically significant evidence for true γ-decay events in our data. Assuming the "off-resonance" data represent experimental background, we multiply the yield in this run by three to normalize the "on-resonance" run and subtract, getting the net yield 2±5 in both one-MeV intervals. Taking the one standard deviation limit an upper limit on the number of γ-transitions to the number of double inelastic $^{12}C(2^+)+^{12}C(2^+)$ events is $(4\pm7)/4.5\times10^6 = 2.4\pm10^{-6}$.

Table I
Number of Events with Momentum Constraint.

	Final State Q-Value				
Incident Energy (MeV)	4.43 MeV Peak	6.5±0.5 MeV	7.65 MeV Peak	8.86 MeV Peak	10.9±0.5 MeV
25.8	1.15×10^6	8	85	4.48×10^6	8
28.8	$.67\times10^6$	2	31	1.14×10^6	2

The coincidence efficiency is Q-value dependent. The relative efficiencies (based on the c.m. angular interval subtended by telescopes) are 0.7, 1.0, 0.85, 0.7 and 0.5, respectively.

To proceed from this limit to conclusions about γ-decay probabilities we rely on width ratios from the literature. For the gross structures[3,5] $(\Gamma_{(2^+,2^+)}/\Gamma)_P \sim 20\% \sim (\Gamma_{el}/\Gamma)_D \sim 1/2(\Gamma_{2^+}/\Gamma)_D$ where P,D refer to parent, daughter structures. In terms of these ratios we express the ratio R of γ-transition events to direct double inelastic events associated with the parent structure.

$$R = \left(\frac{\Gamma_\gamma}{\Gamma}\right)_P \cdot \left(\frac{\Gamma_{(2^+,2^+)}}{\Gamma}\right)_P^{-1} \cdot \left\{G_{el}\frac{\Gamma_{el}}{\Gamma}\bigg|_D + G_{2^+}\frac{\Gamma_{2^+}}{\Gamma}\bigg|_D\right\} \cdot F.$$

The G's are relative geometrical efficiency factors with $G_{el}=1.4$, $G_{2^+}=0.7$. The factor F is the fraction of the total γ-decay strength contained in the one-MeV intervals examined here. We approximate F by assuming the γ-decay line shape is Lorentzian with width Γ and find $F=(2/\pi)\tan^{-1}(\Delta E/\Gamma)$ ranges from 0.7 to 0.2 for Γ=0.5 to 3 MeV, respectively. From Ref. 3 we estimate that one-half the observed number of double inelastic events are associated with the parent structure. Finally, we obtain an upper limit on $(\Gamma_\gamma/\Gamma)_P$ ranging from $(2 \text{ to } 8)\times10^{-6}$ depending on the assumed line shape width.

This result allows the following conclusions: (a) There is no fully collective γ-transition to the intermediate width resonance at 19.3 MeV (line width 0.5 MeV) because our upper limit 2×10^{-6} is an order of magnitude smaller than estimated for this situation. This may be interpreted either as meaning that the deformed "C+C" states are distributed over a wider energy region than 0.5 MeV, or that the particular parent and daughter structures examined here happen not to comprise a rotational-type band. (b) The upper limit 8×10^{-6} obtained if the broader line shape corresponding to the gross structure width is assumed is comparable in size with general expectations based on framgentation resonant pictures. That is, models where the C+C state is distributed over the gross structure width. (c) The present limit is not good enough to differentiate resonant from non-resonant models.

REFERENCES

*Work supported in part by the National Science Foundation.
**Permanent address: Comisión Nacional de Energia Atómica, Argentina.
†Present address: Nuclear Science Division, Lawrence Berkeley Laboratory, Berkeley, California

1. H. Chandra and U. Mosel, Nucl. Phys. A298 (1978) 151.
2. R.L. Phillips, K.A. Erb, D.A. Bromley and J. Wenenser, Phys. Rev. Letters 42 (1979) 566.
3. T. Cormier, et al., Phys. Rev. Letters 40 (1978) 924.
4. R.G. Markham, S.M. Austin and M.A.M. Shahabuddin, Nucl. Phys. A270 (1976) 489.
5. See also, E.R. Cosman, R. Ledoux and A.J. Lazzarini, Phys. Rev. C21 (1980) 2111.

VII. SPIN ALIGNMENT AND POLARIZATION MEASUREMENTS

MEASUREMENT OF SPIN ALIGNMENT IN $^{12}C+^{12}C$ INELASTIC SCATTERING

W. Trombik

Sektion Physik der Universität München, D 8046 Garching, FRG

I. Introduction

For many years the occurrence of gross and intermediate structures in the excitation function of elastic and inelastic scattering as well as some other exit channels is well known for light heavy-ion systems as e.g. $^{12}C+^{12}C$[1]. In the past the aim of most of the experiments was to search for correlated structures in as many exit channels of a system as possible and to confirm their nature as that of "real resonances". But in spite of this large amount of data, the resonance mechanisms are far from being completely understood, and it seems that new kinds of experiments yielding more detailed information about the reaction processes are necessary.

While the gross structures in the scattering cross sections are usually attributed to an extended surface transparency of the system[2], there are many different models for the explanation of the intermediate structures. The underlying idea for most of the proposed reaction mechanisms can be represented by the generalized doorway model of Feshbach[3]: According to this model the intermediate structures come about by coupling of some specific doorway states with the shape resonance of the entrance channel <u>if</u> the coupling to more complex degrees of freedom of the system respectively the compound nucleus is not too strong.

In $^{12}C+^{12}C$ various configurations are possible doorway states but in most of the models (e.g. the double resonance model[4]) the excitation of the 2^+ (4.43 MeV) state in ^{12}C plays the dominant role as coupling "partner" of the elastic entrance channel. This assumption was - at least to some extent - corroborated by the direct measurement of the inelastic (2^+,4.43 MeV) excitation functions by Cormier et al.[5]. The results are characterized by unexpectedly large cross sections over the whole energy range and by a series of strong broad structures. In addition a wealth of finer structures is apparent in the data especially at lower energies. They are hard to explain by simple non-resonant models.

It was the objective of our experimental work to investigate the spin alignment in the inelastic $^{12}C+^{12}C$ scattering, i.e. the coupling configuration of the relative orbital angular momentum L and the intrinsic spin S of the 2^+ state. We were particularly interested in the correlation of the alignment and the strong gross and intermediate structures of the excitation functions. This question plays a crucial role in many models.

The main predictions for the expected spin alignment are outlined in chapt.II, the experimental method is briefly described in chapt.III and the results are presented in chapt.IV.

II. Theoretical predictions for the spin alignment

a) Non-resonant mechanisms

From the viewpoint of angular momentum matching the aligned configuration (i.e. total angular momentum J=L+2) should be favoured over the whole energy range. If one assumes that the observed structures in the inelastic scattering are predominantly associated with the respective grazing partial waves the J=L (m=o) resp. J=L-2 (m=-2) configurations should be strongly mismatched. Even if these requirements are not strictly valid one would rather expect a general preference of the aligned (m=+2) coupling configuration.

There are non-resonant models which go beyond this simple matching condition. At first I want to quote the diffraction model of Phillips et al.[6] in which the inelastic scattering is parametrized by the Austern-Blair-formalism. This model leads to energy dependent windows of the angular momentum with the consequence that the aligned coupling (m=2) should clearly dominate at the broad maxima of the excitation function, whereas at energies in between both configurations (m=o and 2) should contribute about equally. These conditions for the m-substate population should not depend in the scattering angle. It should be pointed out that this model can, of course, not account for correlations of the alignment with the observed intermediate structures.

Secondly, I want to mention the work of Balamuth et al.[7]. These authors measured particle-gamma angular correlations at the "resonance" energies and tried to reproduce these data on the assumption of different models (isolated resonances, DWBA, coupled-channel-calculations). According to their analysis there is much to be said for the general preponderance of the aligned configuration though it seems questionable whether each gross structure is the consequence of only one partial wave.

b) Specific resonance mechanisms

In many of the proposed resonance models specific coupling configurations are implicitly or explicitly requested or, at least, the results of theoretical calculations reveal implications for the alignment. It is impossible to include all relevant studies here, I rather have to confine myself to some examples.

At first I want to mention coupled-channel calculations performed by Könnecke[8] on the basis of the double resonance model[4]. The preliminary results for 90° (c.m.), achieved with an adiabatic coupling model, show - independently of the gross structures of the inelastic excitation function - a strong dominance of the $m=\pm 2$-configuration with only small fluctuations of the alignment (fig.1).

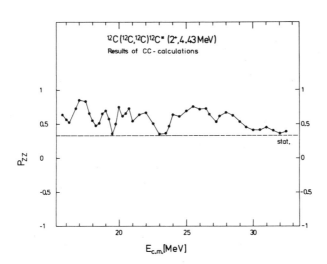

Fig.1: Spin alignment for $\Theta_{c.m.}$ = 90° as resulted from coupled-channel-calculations by Könnecke[8]. For the definition of P_{zz} see eq.(2).

Similar results were obtained by Tanimura[9] in coupled-channel-calculations where the folding procedure was used for the determination of the potential. Also in this case, the intensity of the $m=\pm 2$ substate by far exceeds that of $m=o$ over the whole energy range. It should be underscored that the results of Könnecke and Tanimura are clearly at variance with the predictions of the diffraction model of Phillips et al.

Finally the band-crossing-model of Abe et al.[10] is to be quoted. In this model inelastic molecular bands are calculated for each possible configuration (J=L+2,L,L-2) by analogy with the elastic molecular band. Near the crossing points of different bands the coupling of these two bands should be strong, thus causing a distinct enhancement of the cross sections with pronounced fragmentation of the maxima. The calculations of Abe et al.[10] produce a crossing of the elastic and the aligned inelastic band only around 19 MeV (c.m.). Hence, in this energy range the m=+2-configuration should dominate by far.

Summing up the quoted predictions virtually can be classified into two groups:

(1) General preference for the aligned configuration over the whole energy range with small fluctuations only (e.g. angular momentum matching, calculations by Könnecke and Tanimura)

(2) Strong dependence of the alignment on the gross structures of the inelastic excitation function (e.g. diffraction model).

It should be emphasized that none of the outlined concepts implies drastic changes of the alignment according to intermediate structures.

III. The experimental method

The measurements were performed at the Munich MP-tandem. The principle of the experimental arrangement is schematically illustrated in fig.2: The scattering plane is defined by the beam and the particle detector.

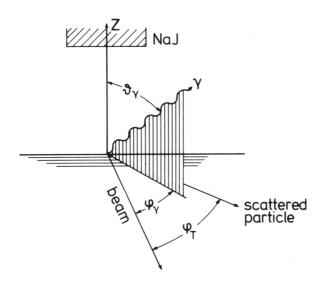

Fig.2: Schematic view of the detector arrangement of the alignment experiments (see text).

The coincident γ-radiation emitted in the direction of the scattering normal (z-axis) is detected with a NaI detector positioned azimuthally symmetric with respect to this axis. This means we select the single inelastic (2^+) scattering events by the particle detectors, and we detect the coincident γ-radiation out of plane. It is very important that the particle spectra are free of background; this could be achieved with the method of kinematical coincidences.

For inelastic particle events identified in this way the total probability for γ-emission is unity. The probability $W_\gamma(m)$ for detecting the γ-ray in the NaI-detector placed according to fig.2 depends, however, on the particular m-substate which was populated. This is a consequence of the different radiation patterns of different Δm transitions and of the chosen symmetry of the detector arrangement. In fig.3 the γ-detection probabilities $W_\gamma(m)$ for the different m-substates are plotted as a function of the NaI aperture angle.

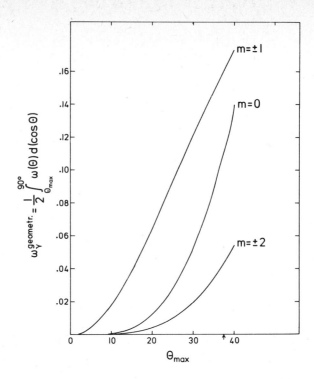

Fig.3: Detection probabilities $W_\gamma(m)$ vs. aperture angle θ_{max} of a γ-detector positioned in the direction of the scattering normal. Here a detection efficiency of 100% is assumed.

In the case of single inelastic scattering the symmetry about the reaction plane requires $P_1 = P_{-1} = 0$ (Bohr's theorem[11]); therefore $P_0 + P_2 + P_{-2} = 1$. From fig.3 one sees that the detection probabilities $W_\gamma(0)$ and $W_\gamma(2)$ differ by factors between 2 and 3. For the absolute $W_\gamma(m)$-values one has to include the detection efficiencies of the real NaI. This was done on the basis of Monte-Carlo-calculations.

Summarizing the procedure, one measures the fraction of γ-radiation detected in coincidence with particles; this fraction is compared with the known γ-detection probabilities $W_\gamma(0)$ and $W_\gamma(2)$:

$$\left(\frac{N_{\gamma\text{-part.}}}{N_{\text{part.}}}\right)_{\exp.} = W_\gamma(0) P_0 + W_\gamma(2) P_{\pm 2} = W_\gamma(0) P_0 + W_\gamma(2) \cdot (1-P_0) \tag{1}$$

From (1) P_0 is deduced.

Usually the alignment is represented in the following form[13] (s=spin of the excited state, here s=2):

$$P_{zz} = \frac{1}{s(2s-1)} \left(\sum_m 3m^2 P_m - s(s+1) \right) = \frac{1}{2} \sum_m m^2 P_m - 1 = 1 - 2P_0 \qquad (2)$$

This convention is used for fig.1 and for all the following figures.

IV. Experimental results

a) First experiment

In the first experiment[12] of this kind we measured the alignment for several scattering angles around 90° (c.m.) at six energies. Three of them corresponded to strong maxima and three to minima of the inelastic excitation function. The results - shown in fig.4 together with excitation functions - proved to be equivocal: At higher energies the alignment is correlated with the maxima but this trend does not continue to lower energies (see e.g. the maximum at 24 MeV). It was concluded that possible reasons for this nonuniform behaviour of the alignment could be:

(1) P_{zz} varies considerably with the scattering angle,
(2) the alignment is - in contrast to the model expectations - strongly connected to the fine structures of the excitation functions,
(3) the assumption that each gross structure can be associated with one single spin value is wrong.

b) Second experiment with improved setup

Because of the results of the first experiment and the open questions we started a new series of experiments with an improved setup. The alignment was systematically measured in finer energy steps and over a wider range of scattering angles. The energy range from 16 to 32.7 MeV (c.m.) was covered in steps varying between 0.1 and 0.4 MeV (c.m.). The energy spread due to the target thickness did not exceed 50 to 80 keV (c.m.); thus there was no averaging over the observed intermediate structure of the inelastic excitation functions. The particle detectors were placed at about 19°, 26.5°, 34°, 41.5°. These angles were slightly changed in order to keep the fourth detector as close to 90° (c.m.) as possible.

For each particle angle the resulting P_{zz}-values are shown in fig.5. The main characteristics are the following:

(1) The energy dependence of the alignment exhibits dramatic fluctua-

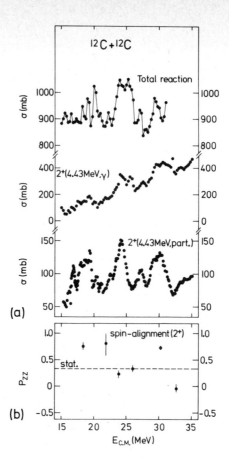

Fig.4: Results of the first alignment experiment[12] together with $^{12}C+^{12}C$ excitation functions.

tions; the P_{zz}-values reach from -1 (pure m=o) to +1 (pure m=±2).
(2) The strength of the fluctuations increases with increasing particle angle. This behaviour could reflect an appreciable contribution of processes with more or less statistical m-substate distribution (e.g. compound processes or simple direct mechanisms at forward angles).
(3) The width and regularity of the observed structures varies strongly, especially at lower energies resembling on that score the excitation functions.

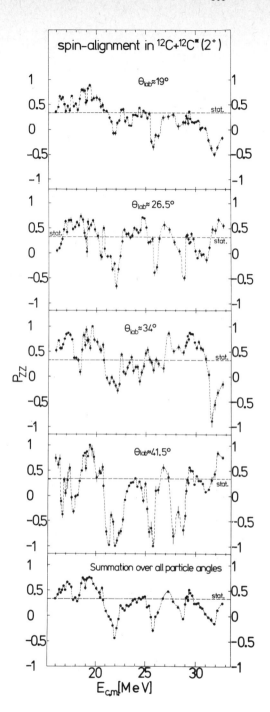

Fig.5:

Spin alignment as derived from the second experiment for different particle angles (lowest curve: summation over all angles). The line labelled by "stat" means $P_o = P_2 = P_{-2} = 1/3$.

(4) It is striking that - in contrast to the three other angles - the alignment for 90° (c.m.) shows relatively regular, strongly pronounced gross structures over the whole energy range. This is clearly at variance with the results of the mentioned coupled-channel-calculations. It is not easy to understand why this angle distinguishes so drastically from the others because for identical particles in the entrance channel the restriction to even partial waves is valid for all angles.

Is the alignment for different angles correlated? This question cannot be replied definitely. From fig.5 one sees that for instance the minima of P_{zz} around 22, 26 and 29 MeV appear at all angles while in the case of the maxima the situation seems more complicated. Nevertheless, averaging the alignment over all angles (lowest part of fig.5) some significant broad structures are left above 22 MeV.

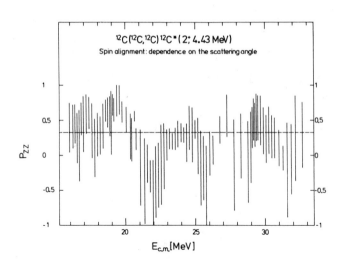

Fig.6: Angle spread of the alignment at respective one energy (without statistical errors).

But in contrast to this tantalizing picture the angle dependence of the alignment is substantial, though. This can be seen in fig.6. Particularly in the regions where the alignment is generally low the angle spread of P_{zz} is strong; this behaviour is clearly inconsistent with the predictions of the diffraction model[6]. On the other hand, the spread of the alignment is reduced in those regions where P_{zz} is relatively high, thus the impression of correlated gross structures is given.

At lower energies - the region of intermediate structures in the excitation functions - the alignment is generally high (one could say in agreement with the band crossing model[10]). A sequence of correlated structures near 17.3, 18.6 and 19.4 MeV seems to emerge. Beyond that it is noticeable that around 19.3 MeV - the energy of the notorious resonances in many exit channels[1] - the preference of $m=\pm 2$ reaches the highest degree within the whole energy range.

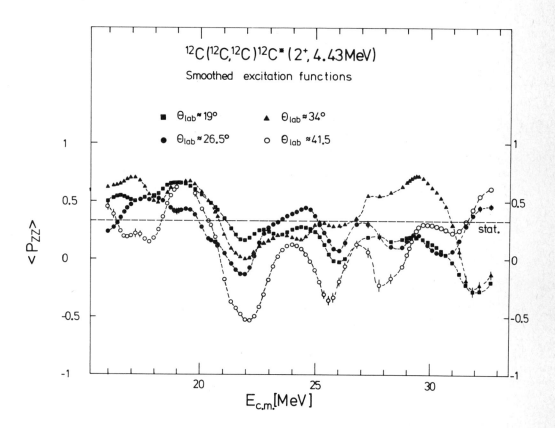

Fig.7: P_{zz}-values for the different particle angles smoothed in regard of energy by a Lorentzian function (see IV.d and Ref.14).

By smoothing the P_{zz} energy dependence by a Lorentzian function (see IV.d) the general trend of P_{zz} is revealed (fig.7). The appearing gross structures are partly correlated (especially between 19 and 26 MeV) but likewise considerable uncorrelations (for instance above 26 MeV) manifest themselves.

c) Comparison with the inelastic excitation functions

Is the alignment correlated with the excitation functions? The answer to this question turns out to be different, depending on whether we see things through rosecoloured spectacles or whether we take a gloomy view (figs.8,9).
In the first case we consider the 90° (c.m.) curves (fig.9). Here the correlation to the observed structures is amazing except at the lowest energies: Each broad maximum of the excitation function corresponds to a peak of the alignment. The maximum at about 27 MeV is particularly worthy of notice: This peak appears as relatively narrow and weak in the excitation function. Until now it was not paid much attention to; no spin assignment[5] was done and it does not emerge in the diffraction model[6]. Nevertheless, the relative maximum of P_{zz} is very pronounced.

The hope, also raised from fig.6, to find extensive correlations between alignment and excitation functions for all angles is disappointed, however. Of course, in figs. 8 and 9 there are some examples of correlation but the cases of disagreement seem to prevail (see e.g. at energies above 26 MeV for Θ_{lab} = 26.5°). This relatively weak degree of correlation should turn up in the excitation function summed up over all angles. In principle this is confirmed (fig.10), although with due optimism one may state that several maxima of the excitation function (e.g. at 24 and 30 MeV) are reflected in the alignment (disregarding the widths and relative strengths, however).

d) Investigation of gross and intermediate structures

For a more definite distinction of gross and intermediate structures we carried out a special procedure: The energy dependence of the cross sections and the alignment was smoothed by a Lorentzian function[14]; the averaging interval was 0.8 MeV (c.m.) which corresponds to a normal running average of about 1.7 MeV (c.m.). After eliminating all finer structures the good correspondence at 90° (c.m.) survives, of course, but also for the other angles the structures of the cross sections are to a certain degree reflected in the alignment (fig.11). For instance, at 19° some maxima (e.g. around 19, 24, 27.3 and 29.8 MeV) appear in the energy dependence of P_{zz} also; at 34° the situation is similar.

With guarded optimism one could see in this respect a confirmation of the concept of kinematical matching. But one should be careful with

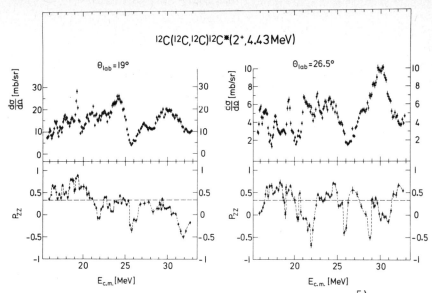

Fig.8: Comparison of the inelastic excitation functions[5] and the alignment for Θ_{lab} = 19° resp. 26.5°.

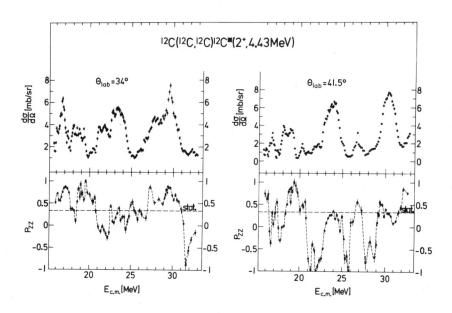

Fig.9: Same as in fig.8 but for Θ_{lab} = 34° resp. 41.5°.

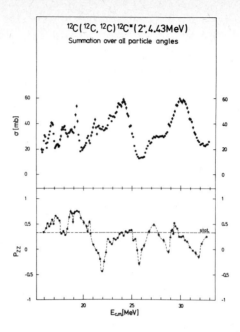

Fig.10: Same as in fig. 8 but summed over all particle angles.

such an interpretation: At first the correlation effects are for the other angles by far weaker than for 90° and, on the other hand, there are several pronounced peaks in $<P_{zz}(E)>$ which should not occur according to the diffraction model (e.g. at 27 MeV for Θ_{lab} = 26.5° and 41.5°). Thus we can conclude that the conditions for the inelastic scattering are evidently more complicated than predicted by relatively simple kinematic models.

To investigate the finer structures we calculated the trend-corrected excitation functions (fig.12), this means the deviations of the actual values from the smoothed values: $\frac{\sigma}{<\sigma>}-1$ and $P_{zz}-<P_{zz}>$. Of course, a perfect agreement of the two trend-corrected excitation functions cannot be expected because a good deal of the observed structures should be statistical fluctuations without any drastic influence on the alignment. Therefore the few "true" correlations should deserve all the more interest and they could be the starting point for further theoretical examination. Without going into detail I refer to the maxima near 16.8, 19.4, 22.8 and 26.8 MeV (Θ_{lab} = 26.5°) as examples.

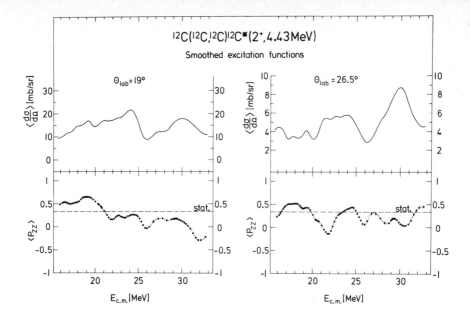

Fig.11: Same as in fig.8 but smoothed by a Lorentzian function (Ref.14, 2I = 0.8 MeV (c.m.)).

V. Summary and conclusions

We measured the spin alignment P_{zz} for the single inelastic (2^+, 4.43 MeV) $^{12}C+^{12}C$ scattering over a wide energy range in fine steps with a γ-detector positioned in the direction of the scattering normal. The obtained P_{zz}-values depend strongly on the particle angle in contradiction with the diffraction model[6].

The energy dependence of the alignment shows very pronounced fluctuations of different widths in disagreement with recent coupled-channel-calculations[8,9]. At 90° (c.m.) only the structures of the excitation function and of $P_{zz}(E)$ are largely correlated whereas for the other particle angles the conformity is limited.

Beyond that some narrower structures (without spin assignments in previous works) in the excitation functions correspond to significant maxima of the alignment. These observations lead us to conclude that simple concepts as e.g. kinematical matching do not justice to the complexity of the reaction mechanism. It is evident from the data that further detailed theoretical investigations are necessary for a deeper understanding of the "resonances".

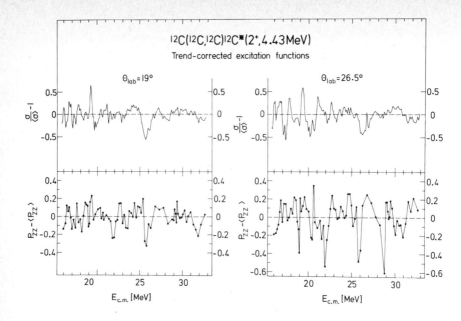

Fig.12: Trend-corrected excitation functions (upper curves: $\frac{\sigma}{<\sigma>}-1$, lower curves $P_{zz}-<P_{zz}>$) for Θ_{lab} = 19° resp. 26.5°.

Acknowledgement

The data presented in this talk have been obtained in collaboration with F.Krug, W.Trautmann, W.Dünnweber, D.Konnerth, K.A.Eberhard, W.Hering, and R.Singh.

This work was supported in part by the Bundesministerium für Forschung und Technologie.

References

1) See e.g. D.A.Bromley, in Nuclear Molecular Phenomena, ed. N.Cindro, North-Holland Pub. Co., Amsterdam, 1978, p.1
2) D.Konnerth, K.G.Bernhardt, K.A.Eberhard, R.Singh, A.Strzalkowski, W.Trautmann, and W.Trombik, Phys.Rev.Lett. 45, 1154 (1980)
3) H.Feshbach, J.Phys.Colloq. 37, C5-177 (1976)
4) W.Greiner and W.Scheid, J.Phys.Colloq. 32, C6-91 (1971)
5) T.M.Cormier, C.M.Jachcinski, G.M.Berkowitz, P.Braun-Munzinger, P.M.Cormier, M.Gai, J.W.Harris, J.Barrette, and H.E.Wegener, Phys.Rev.Lett. 40, 924 (1978)

6) R.L.Phillips, K.A.Erb, D.A.Bromley, and J.Weneser, Phys.Rev.Lett. $\underline{42}$, 566 (1979)

7) D.P.Balamuth, L.E.Cannell, and R.W.Zurmühle, Phys.Rev. $\underline{C23}$, 2492 (1981)

8) R.Könnecke, private communication

9) O.Tanimura, Nucl.Phys. $\underline{A309}$, 233 (1978), and private communication

10) Y.Abe, in Nuclear Molecular Phenomena, ed. N.Cindro, North-Holland Pub.Co., Amsterdam, 1978, p.211

11) A.Bohr, Nucl.Phys. $\underline{10}$, 486 (1959)

12) W.Trombik, W.Dünnweber, W.Trautmann, W.Dahme, K.A.Eberhard, W.Hering, D.Konnerth, and R.Singh, Z.Physik $\underline{A296}$, 187 (1980)

13) R.J.Blin-Stoyle and M.A.Grace, in Handbuch der Physik, ed. S.Flügge (Springer-Verlag, 1957), Vol.42, p.555

14) A.E.Bisson, K.A.Eberhard, and R.H.Davis, Phys.Rev. $\underline{C1}$, 539 (1970)

DWBA ANALYSES OF RESONANCE STRUCTURE IN THE $^{16}\text{O}(^{16}\text{O},^{12}\text{C})^{20}\text{Ne}$ REACTION[*]

Yosio KONDŌ and Taro TAMURA

Department of Physics, University of Texas

Austin, Texas 78712, U.S.A.

ABSTRACT

We demonstrate that the EFR-DWBA is a practical tool to use in analyzing certain resonance phenomena, observed in transfer reactions induced by heavy ions. We show this by taking the $^{16}\text{O}(^{16}\text{O},^{12}\text{C})^{20}\text{Ne}$ reaction as an example. It is seen that the available experimental angular distributions, excitation functions and the nuclear alignment are reproduced rather well this way. In the course of the discussion, it is also shown that the enhancement mechanism embodied in the band crossing model plays a key role in good fit achieved. To our knowledge, this is the first successful analysis of resonance phenomena in heavy-ion reactions, going beyond the elastic and inelastic processes.

---

[*]Work supported in part by the U.S. Department of Energy.

I. Introduction

Since the first observation of three resonances [1] in the sub-Coulomb $^{12}C + ^{12}C$ reaction, resonance phenomena in heavy ion reactions have been studied extensively both theoretically and experimentally. Nogami [2] suggested that the coupling of the ground state to the collective 2^+ excitation of ^{12}C might be playing an indispensable role. Imanishi [3] formulated Nogami's idea using a coupled-channel method and successfully reproduced basic features of the above triplet resonances. Scheid, Greiner and Lemmer [4] extended Nogami's idea to propose a "double resonance mechanism" and applied it to the study of the intermediate structure observed in the $^{16}O + ^{16}O$ elastic scattering. Having in mind these pioneer works and also the concept of potential resonance "bands," suggested by microscopic studies [5,6] of interaction between composite particles, Kondō, Matsuse and Abe proposed a "band crossing model" [7,8]. And it was applied successfully in explaining resonances in inelastic scattering of the $^{12}C + ^{12}C$, $^{12}C + ^{16}O$ and $^{16}O + ^{16}O$ systems [8-11]. With this model, we understand that the crossing of potential resonance bands, i.e., a simultaneous onset of potential resonances in both the entrance and exit channels, is essential in making the resonance conspicuous.

Recent experiments have demonstrated that resonant behavior also takes place in α-transfer reactions, such as $^{16}O(^{16}O,^{12}C)^{20}Ne$ [12,13], $^{20}Ne(^{16}O,^{12}C)^{24}Mg$ [14], $^{24}Mg(^{16}O,^{12}C)^{28}Si$ [15,16] and $^{28}Si(^{16}O,^{12}C)^{32}S$ [17]. In particular for the $^{16}O(^{16}O,^{12}C)^{20}Ne$ reaction, leading to the 4^+ member (E_x = 4.25 MeV) of the ground band of ^{20}Ne, it was shown that excitation functions at several angles have pronounced gross peaks that are correlated in their energies [12,13]. The purpose of the present report is to show that the resonant behavior in this α-transfer reaction and related measurements of the alignment can also be explained on the basis of the band crossing model, combined with the use of the distorted wave Born approximation (DWBA).

In section II, we present the DWBA analysis of the $^{16}O(^{16}O,^{12}C)^{20}Ne(4_1^+)$ reaction, while that of the $^{16}O(^{16}O,^{12}C)^{20}Ne(6_1^+)$ reaction is given in section III. These analyses are summarized in section IV, where works of other groups are also discussed.

II. EFR-DWBA Analysis of the $^{16}O(^{16}O,^{12}C)^{20}Ne(4_1^+)$ Reaction*

There are known several sets of optical model parameters [11,18-20] (see Table I) which successfully reproduce the gross energy dependence of the $^{16}O + ^{16}O$ elastic scattering [18]. This knowledge certainly facilitates our analysis, although it is also interesting to see whether these potentials behave the same or differently when used for transfer reactions involving resonances. As for the exit $^{12}C + ^{20}Ne(4_1^+)$ channel, the knowledge of the optical model parameters to be used is very limited. In the present analysis, we consider two potentials (see Table II); the Vandenbosch [21] and the Gobbi [20] potentials. We shall discuss the implications of these potentials after presenting results obtained by their use.

In Fig. 1, we first show several angular distributions calculated for the $^{16}O(^{16}O,^{12}C)^{20}Ne(4_1^+)$ reaction at $E_{lab}(^{16}O) = 51.5$ MeV, and compare them with the data of Rossner et al. [13]. In this figure, the dashed line shows the DWBA cross section obtained by using the DBA-Vandenbosch potential, i.e., by using the KBA [11] and Vandenbosch [21] potentials for the entrance and

---

*The contents of this section were also reported in Refs. [22,23].

TABLE I. Distorting Potentials for the $^{16}O + ^{16}O$ Channel

1) Maher Potential (J.V. Maher et al., Phys. Rev. **188** (1969) 1665.)

$$U(r) = \frac{-17 - i(0.4 + 0.1 E_{cm})}{(1 + \exp((r-6.8)/0.49))}.$$

2) Gobbi Potential (A. Gobbi et al., Phys. Rev. **C7** (1973) 30.)

$$U(r) = \frac{-17.0}{(1 + \exp((r-R_0)/0.49))} - i \frac{0.8 + 0.2 E_{cm}}{(1 + \exp((r-R_i)/0.15))},$$

$R_0 = 1.35 (16^{1/3} + 16^{1/3})$, $R_i = 1.27 (16^{1/3} + 16^{1/3})$.

3) Chatwin Potential (R.A. Chatwin et al., Phys. Rev. **C1** (1970) 795.)

$$U(r) = (-17.0 - i \frac{0.22 E_{cm}}{(1 + \exp((J-J_{cr})/0.4))}) \frac{1}{(1 + \exp((r-6.8)/0.49))},$$

$J_{cr} = 6.7 (2\mu(E_{cm}-6.7)/\hbar^2)^{1/2}$.

4) KBA Potential (Y. Kondō et al., Phys. Rev. **C22** (1980) 1068.)

$$U(r) = \frac{100}{(1 + \exp((r-3.5)/0.3))} + (-16.0 - 0.014 L(L+1)$$

$$- i \frac{0.3 E_{cm}}{(1 + \exp((J-J_{cr})/0.4))}) \frac{1}{(1 + \exp((r-6.55)/0.5))},$$

$J_{cr} = 6.7 (2\mu(E_{cm}-7.7)/\hbar^2)^{1/2}$.

TABLE II. Distorting Potential for the $^{12}C + ^{20}Ne$ Channel *

1) Vandenbosch Potential (R. Vandenbosch et al., Phys. Rev Lett. **33** (1974) 842.)

$$U(r) = \frac{-17 - i(-0.333 + 0.54 E_{cm})}{(1 + \exp((r-R)/0.57))}, \quad R = 1.35 (12^{1/3} + 20^{1/3}).$$

2) Gobbi Potential (A. Gobbi et al., Phys. Rev. **C7** (1973) 30.)

$$U(r) = \frac{-17.0}{(1 + \exp((r-R_0)/0.49))} - i \frac{0.314 + 0.2 E_{cm}}{(1 + \exp((r-R_i)/0.15))},$$

$R_0 = 1.35 (12^{1/3} + 20^{1/3})$, $R_i = 1.27 (12^{1/3} + 20^{1/3})$.

* In this table, E_{cm} is the center of mass energy of the $^{16}O + ^{16}O$ channel.

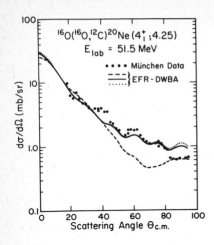

Fig.1 Angular distributions of the $^{16}O(^{16}O, ^{12}C)^{20}Ne(4_1^+)$ reaction at $E_{lab}(^{16}O) = 51.5$ MeV. Data are those of Ref. [13]. The dashed and solid lines, respectively, represent the EFR-DWBA cross sections with the KBA-Vandenbosch and KBA-Gobbi potentials. The post form was used for both calculations. The DWBA cross section with the prior form and the KBA-Gobbi potential is shown by the dotted line.

exit channels, respectively. The predicted angular distribution has a slope which is too steep for $\theta_{cm} \gtrsim 40°$. The solid line shows the result with the KBA-Gobbi potential, i.e., with the KBA and Gobbi [20] potentials for the incident and exit channels, respectively. In this case the experimental angular distribution is reproduced reasonably well. Certainly the weakly-absorptive Gobbi potential is preferred for the exit channel, rather than the strongly-absorptive Vandenbosch potential. The same preference was found when other $^{16}O + ^{16}O$ potentials [18-20] were used for the entrance channel. The DWBA cross sections shown in Fig. 1 were normalized to the data at forward angles. The normalization factors were 17 and 3.5 for the KBA-Vandenbosch and KBA-Gobbi cases, respectively.

All the above calculations were performed by using the same form factor. Radial wave functions of the $\alpha-^{12}C$ and $\alpha-^{16}O$ systems were calculated by the separation energy method using a Woods-Saxon potential with the parameters $r_0 = 1.35$ fm and $a = 0.65$ fm. The post form was used consistently. (A result with the prior form and with the KBA-Gobbi potential is given in Fig. 1 by the dotted line, which demonstrates the practical equivalence of the two forms.) The adopted alpha-particle spectroscopic factors of ^{16}O and ^{20}Ne were those of the SU(3) model [24]. Numerical calculations were performed using an exact-finite range (EFR)-DWBA code, SATURN-MARS [25], which was modified to take account of the symmetrization of the system and to include an angular momentum dependent (J-dependent) imaginary potential [19].

The relative merits of the entrance channel optical potentials of Refs. [11,18-20] were not very clear in predicting the above angular distribution at $E_{lab}(^{16}O) = 51.5$ MeV, but became more noticeable when the excitation functions were considered. The upper and lower panels of Fig. 2 show the excitation functions [13] of the reaction at $\theta_{cm} = 57°$ and 78°, respectively. In obtaining the theoretical results presented in this figure, the optical potential used for the exit channel was fixed to that of Gobbi. The dotted

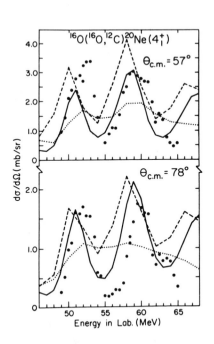

Fig.2 Excitation functions of the ^{16}O(^{16}O,^{12}C)^{20}Ne(4_1^+) reaction at $\theta_{cm} = 57°$ and 78°. Data are those of Ref. [13]. The dotted, dashed and solid lines represent the EFR-DWBA cross sections with the Gobbi-Gobbi, Chatwin-Gobbi and KBA-Gobbi potentials, respectively. (The experimental cross sections given for the excitation functions [13] appear to be too large by about a factor of 1.5, as seen by comparing the cross sections at $E_{lab}(^{16}O) = 51.5$ MeV with those given for the angular distribution [13] at the same energy. The theoretical values were thus multiplied by a factor of 1.5 in this figure.)

lines in Fig. 2 show the DWBA excitation functions with the Gobbi-Gobbi potential. Although some resonant structure was predicted, it was too weak compared with the data. The Maher [18]-Gobbi potential, for which results are not shown in Fig. 2, had a similar difficulty. The dashed lines in Fig. 2 show the excitation functions with the Chatwin [19]-Gobbi potential, and it is seen that the experimental resonant structure was reproduced reasonably well. Note that the Chatwin potential has a J-dependent imaginary part. Since the Gobbi and the Chatwin potentials both have the same real part and predict quite similar excitation functions [20] for the $^{16}O+^{16}O$ elastic scattering, it is often conjectured that the effects of the "surface transparency" in the Gobbi potential and the "J-dependence of the imaginary part" in the Chatwin potential are basically equivalent. However, as seen in Fig. 2, they predict markedly different excitation functions of the transfer reaction. In passing we may note that the KBA potential also has a J-dependent imaginary part.

The solid lines in Fig. 2 show the results obtained with the KBA-Gobbi potential. They are similar to those obtained with the Chatwin-Gobbi potential, but agree somewhat better with experiment. In fact, the energy interval, widths and amplitudes of the gross structure are reproduced rather well, although the peak positions are still somewhat shifted to the lower energy side. This last trouble may indicate that the Gobbi potential, derived by fitting the $^{16}O+^{16}O$ elastic scattering, needs refinements to be used for the $^{12}C+^{20}Ne$ channel.

Excitation functions were also calculated with the KBA-Vandenbosch, the Chatwin-Vandenbosch and the Gobbi-Vandenbosch potentials. In these cases, the resonances were predicted to be too weak.

Absolute squares of the DWBA overlap integrals are shown in Fig. 3 for the KBA-Gobbi case which gave the best fit to the data. As seen, only one or two grazing partial waves are contributing strongly at any chosen energy. It is also seen that the resonances are dominated by the aligned configuration [8], in which the orbital angular momentum L and the channel spin I are coupled to give the maximum total angular momentum J, i.e., in which $J = L + I$. In this case, the entrance channel orbital angular momentum L_i equals J, while the exit channel orbital angular momentum L_f equals $J - 4 (= L_i - 4)$ for the aligned configuration. As seen from the solid lines in Fig. 3, the overlap integrals of the aligned configuration are indeed very large at every peak energy of the calculated excitation functions. At $E_{lab}(^{16}O) = 51$ and 59 MeV, the partial waves with $J = 18$ and 20, respectively, make the dominant contributions. The strong enhancement of the overlap integral of the aligned configuration reflects the simultaneous onset of potential resonances in both the entrance and exit channels, and this is precisely the mechanism embodied in the band crossing model for giving the enhanced structure. As seen from

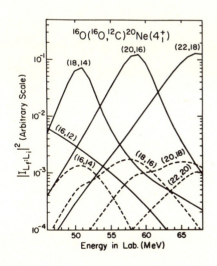

Fig.3 Absolute square of the DWBA overlap integrals, $|I_{L_f:L_i}|^2$, of the $^{16}O(^{16}O, ^{12}C)^{20}Ne(4_1^+)$ reaction for the case with the KBA-Gobbi potential. Attached to each curve is the pair of orbital angular momenta, (L_i, L_f). The solid lines show contributions from the aligned configuration, in which $L_f = L_i - 4$. The dashed lines are those with the $L_f = L_i - 2$ configuration.

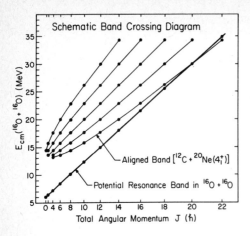

Fig.4 Schematic diagram showing the crossing of the aligned band of the $^{12}C + ^{20}Ne(4_1^+)$ channel with that of the $^{16}O + ^{16}O$ channel. For further details of the band crossing model, see Ref. [8].

the schematic diagram in Fig. 4, a simultaneous onset of potential resonances in both the entrance and exit channels takes place naturally as a result of the crossing of the aligned band of the $^{12}C + ^{20}Ne(4_1^+)$ channel with that of the $^{16}O + ^{16}O$ channel. For further details of the band crossing model, see Ref. [8].

The $L_f = L_i - 2$ configuration, as shown by dashed lines in Fig. 3, contributes with an enhancement at the energies of the potential resonances in both the entrance and exit channels. However, since the potential resonances in the two channels do not occur simultaneously, the enhancement is not as prominent as it was with the aligned configuration. (It should be noted further that, in this reaction, the kinematic matching condition [26] also favors the contribution of the aligned configuration.) Contributions of the other configurations, i.e., $L_f = L_i$, $L_i + 2$ and $L_i + 4$, were found to be much smaller and are not shown in Fig. 3.

An Argand diagram of the DWBA overlap integral, $I_{L_f:L_i}$, for the $^{16}O(^{16}O, ^{12}C)^{20}Ne(4_1^+)$ reaction is shown in Fig. 5. The energy dependence clearly shows the contribution of a resonance term which has its maximum amplitude

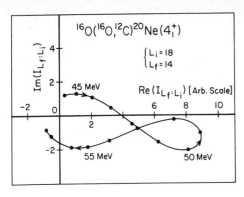

Fig.5 Argand diagram of the DWBA overlap integral for the $^{16}O(^{16}O, ^{12}C)^{20}Ne(4_1^+)$ reaction, with $(L_i, L_f) = (18, 14)$, which plays the dominant role at energies near 51 MeV. Circles were placed at every 1 MeV interval of the incident energy.

at about 51 MeV and a counter-clockwise energy dependence. It is quite legitimate to conclude, from this behaviour of the DWBA overlap integral, that those peaks of the excitation functions are attributed to the onset of resonances.

To our knowledge, two DWBA analyses have been performed for the $^{16}O(^{16}O, ^{12}C)^{20}Ne(4_1^+)$ reaction. Arima et al. [27] performed an analysis at $E_{lab}(^{16}O)$ = 60 MeV using the Buttle-Goldfarb approximation [28]. They emphasized the choice of the transfer potential, as opposed to fitting experimental angular distributions. Anyway, the predicted angular distribution had a strongly forward-peaked pattern, reflecting the use of the optical potentials with a strong imaginary part for both channels. Pougheon et al. [29] performed an EFR-DWBA calculation of the reaction at $E_{lab}(^{16}O)$ = 68 MeV using weakly absorptive optical potentials. Unfortunately, we were unable to reproduce their calculated cross sections.

We used the Vandenbosch potential [21] to demonstrate the failure of using a strongly absorptive potentials in the $^{12}C + ^{20}Ne(4_1^+)$ channel. Note, however, that this does not mean the work of Ref. [21] was unreasonable. Vandenbosch et al. extracted the potential to fit the $^{12}C + ^{20}Ne(0_{gr}^+)$ elastic scattering data [21], which show a much weaker oscillation compared with that in the $^{16}O + ^{16}O$ elastic scattering data [18]. What we have shown is that the effective imaginary potential to be used in the $^{12}C + ^{20}Ne(4_1^+)$ channel can be (or should be) weaker than that in the $^{12}C + ^{20}Ne(0_{gr}^+)$ channel. The purpose of Ref. [21] was to demonstrate that the absorption can be channel dependent, even when two channel energies are chosen so that both result in the same compound system with the same total energy. The $^{12}C + ^{20}Ne(0_{gr}^+)$ channel is more absorptive than in the $^{16}O + ^{16}O$ channel, because the former couples strongly with a larger number of direct reaction channels than does the latter. By the same token, the $^{12}C + ^{20}Ne(0_{gr}^+)$ channel. Therefore, what we have shown above is in line with what was pointed out in Ref. [21], rather than being evidence against it.

III. EFR-DWBA Analysis of the $^{16}O(^{16}O,^{12}C)^{20}Ne(6_1^+)$ Reaction

In this section we report our DWBA analysis of the $^{16}O(^{16}O,^{12}C)^{20}Ne(6_1^+)$ reaction. The 6_1^+ state (E_x = 8.78 MeV) lies above the α-threshold energy of ^{20}Ne and thus the $\alpha + ^{16}O$ system lies in the continuum. In carrying out the calculation, we adopted a bound state approximation, replacing the wave function of the continuum state by that of a bound state. We calculated angular distributions of the reaction at $E_{lab}(^{16}O)$ = 68 MeV using two binding energies, 1.5 and 0.5 MeV, and both with the post and prior forms. The resultant four DWBA cross sections had different magnitudes but nearly the same angular distributions. In the following, we consistently use the binding energy of 0.5 MeV and the post form.

In Fig. 6, the calculated angular distributions for the $^{16}O(^{16}O,^{12}C)^{20}Ne(6_1^+)$ reaction at $E_{lab}(^{16}O)$ = 68 MeV are compared with the data of

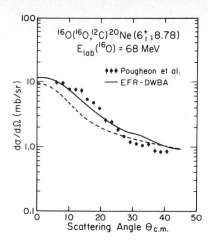

Fig.6 Angular distributions of the $^{16}O(^{16}O, ^{12}C)^{20}Ne(6_1^+)$ reaction at $E_{lab}(^{16}O) = 68$ MeV. Data were taken from Ref. [29]. The solid line represents the EFR-DWBA cross section with the KBA-Gobbi potential. The dashed line shows the angular distribution obtained by using only the resonance term, i.e., only the $(L_i, L_f) = (22,16)$ partial wave.

Pougheon et al. [29]. The solid line shows the EFR-DWBA cross section with the KBA-Gobbi potential, which gave the best fit to the $^{16}O(^{16}O, ^{12}C)^{20}Ne(4_1^+)$ data. The dashed line represents the angular distribution obtained by using only the resonance term, i.e., only the $(L_i, L_f) = (22,16)$ partial wave (see also Fig. 8). The resonance term alone already reproduces the basic features of the experimental angular distribution. The inclusion of other terms further improves the fit (see the solid line).

The upper and lower panels of Fig. 7 show the excitation functions of the reaction at $\theta_{cm} = 60°$ and $87°$, respectively. The solid lines show the DWBA excitation functions with the KBA-Gobbi potential. The data are those of Ref. [13]; they include not only the yield to the 6_1^+ state but also to the 5_1^- state of ^{20}Ne. (No data are available for the excitation function of the reaction to the pure 6_1^+ state of ^{20}Ne.) Therefore, the comparison between the calculation and the data may not be very significant. It may, nevertheless, be worth remarking that the resonant structure of the data is reasonably well reproduced by our calculation.

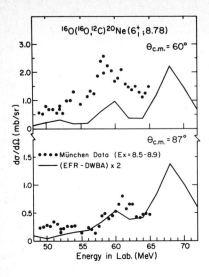

Fig.7 Excitation functions of the $^{16}O(^{16}O,^{12}C)^{20}Ne(6_1^+)$ reaction at θ_{cm} = 60° and 87°. Data are those of Ref. [13]; they include not only the yield to the 6_1^+ state but also to the 5_1^- state of ^{20}Ne. The solid lines represent the EFR-DWBA cross section with the KBA-Gobbi potential.

Absolute squares of the DWBA overlap integrals are shown in Fig. 8 for the KBA-Gobbi potential. Characteristic features of Fig. 8 are almost the same as those of Fig. 3. The strong enhancement of the overlap integrals of the aligned configuration, which are shown by solid lines, reflects the simultaneous onset of potential resonances in both the entrance and exit channels,

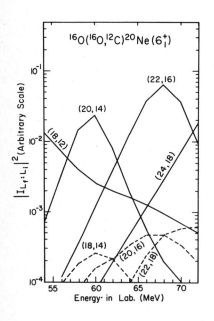

Fig.8 Absolute square of the DWBA overlap integrals, $|I_{L_f:L_i}|^2$, of the $^{16}O(^{16}O,^{12}C)^{20}Ne(6_1^+)$ reaction for the case with the KBA-Gobbi potential. The solid lines show contributions from the aligned configuration, in which $L_f = L_i - 6$. The dashed lines are those with the $L_f = L_i - 4$ configuration.

as is predicted by the band crossing model. It is clearly seen that the aligned configuration of the $(L_i, L_f) = (22,16)$ partial wave has the dominant contribution at $E_{lab}(^{16}O) = 68$ MeV, the energy at which Pougheon et al. measured their data.

Fig. 9 shows the angular dependence of the alignment (i.e. the sum of the populations for the magnetic substates $M = +I$ and $-I$ of ^{20}Ne) at $E_{lab}(^{16}O)$ = 68 MeV. The quantization axis is taken perpendicular to the reaction plane ($\vec{z} = \vec{k}_i \times \vec{k}_f$; the Basel convention). Data were extracted [29] from coincidence measurements of the ejectile ^{12}C and the ^{16}O nucleus which is produced by the decay of the excited ^{20}Ne. The solid line in Fig. 9 represents the EFR-DWBA calculation with the KBA-Gobbi potential, while the dashed line represents the result obtained by using only the resonance term, i.e., only the $(L_i, L_f) = (22,16)$ partial wave. It is clear that the resonance term of the aligned configuration, embodied in our DWBA, is able to reproduce the large alignment of the data. Similar effects were discussed earlier [9,10] in relation to inelastic scattering.

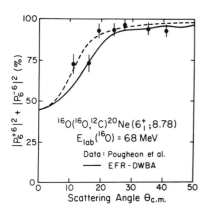

Fig.9 The angular distributions of the alignment (i.e. the sum of the populations for the magnetic substates $M = +I$ and $M = -I$) at $E_{lab}(^{16}O) = 68$ MeV. The quantization axis is taken perpendicular to the reaction plane. Data are those of Ref. [29]. The solid line represents the result of the EFR-DWBA calculation with the KBA-Gobbi potential, while the dashed line represents the result obtained by using only the resonance term, i.e., only the $(L_i, L_f) = (22,16)$ partial wave.

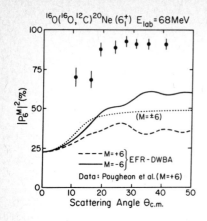

Fig.10 Magnetic substate population of the $^{16}O(^{16}O,^{12}C)^{20}Ne(6_1^+)$ reaction at $E_{lab}(^{16}O) = 68$ MeV. Data are those of Ref. [29]. The solid and dashed lines represent the result of the EFR-DWBA calculation with the KBA-Gobbi potential. The dotted line represents the result obtained by using only the resonance term, i.e., only the $(L_i, L_f) = (22, 16)$ partial wave.

The magnetic substate population of ^{20}Ne at an incident energy of $E_{lab}(^{16}O) = 68$ MeV is shown in Fig. 10. Data are those of Ref. [29]. The solid and dashed lines present the result of the EFR-DWBA calculation with the KBA-Gobbi potential for the M =+6 and -6 substate, respectively. Data show a large difference between the populations of the M =+6 and -6 substates, which indicates the presence of a large polarization of ^{20}Ne. The DWBA calculation was unable to reproduce this large polarization, a fact which is not surprising because the resonance term alone gives rise to exactly the same yield for the M =+6 and -6 substates (see the dotted line). This is seen from the following expression for the angular distribution populating the substate M;

$$|P_I^M(\theta)|^2 \frac{d\sigma(\theta)}{d\Omega} \propto \left| \sum_{\substack{L_i L_f \\ M_i M_f}} (-)^{L_f} (L_i M_i L_f M_f | IM) I_{L_f:L_i} Y_{L_f-M_f}\left(\frac{\pi}{2},\theta\right) Y_{L_i-M_i}\left(\frac{\pi}{2},0\right) \right|^2 .$$

Since the resonance term dominates the yield at this energy, it is unlikely that we would ever obtain a large polarization. A conceivable source of the discrepancy between the data and the calculation is the shift of the resonance energy discussed in connection with Fig. 2. If 68 MeV does not correspond to the center of the resonance, the dominance of the resonance term is reduced, and a larger polarization may result. To see such an effect, we modified the KBA-Gobbi potential slightly so as to shift the potential resonances to higher energies. The magnetic substate population obtained by using this modified potential is given in Fig. 11. It in fact increases the theoretical polarization but not to the extent needed to reproduce the data.

We have shown that we were able to reproduce all the characteristic features of the data except for the large polarization. We remark here, however, that it was stated in Ref. [29] that their procedure to extract the magnetic substate population had a tendency to overestimate the deduced polarization, under certain circumstances. It might be interesting to point out that the magnetic substate populations at $\theta_{cm} = 90°$ of the $^{16}O(^{16}O,^{12}C)^{20}Ne(5_1^-)$ reaction, measured in the same experiment, were such as to result in a large polarization (75% M =+5 and 9% M =-5 substates) [30]. However, the symmetry of the system requires that the populations of the M =+5 and -5 substates should be exactly the same at $\theta_{cm} = 90°$, i.e., no polarization. It may be that the large polarization reported in Ref. [29] is yet somewhat inconclusive.

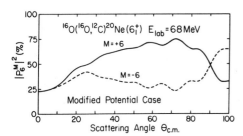

Fig.11 Magnetic substate population. The solid and dashed lines represent the result of the EFR-DWBA calculation with a modified KBA-Gobbi potential, in which the potential resonances are shifted to higher energies.

Bond [31] has suggested a possibility of large polarization in heavy ion transfer reactions using a semi-classical discussion. He further demonstrated it, by using EFR-DWBA, for the $^{16}O(^{16}O,^{12}C)^{20}Ne(6_1^+)$ reaction. However, strongly absorptive potentials were used in his calculation, which may not be appropriate for this particular system.

IV. Summary and Conclusions

In the present article, we have demonstrated a practical and realistic way to analyze resonance phenomena in heavy ion transfer reactions. To our knowledge this is the first successful analysis of resonance phenomena in heavy ion transfer reactions. We have shown that the angular distributions, excitation functions and the alignment of the $^{16}O(^{16}O,^{12}C)^{20}Ne$ reaction are reproduced fairly well by the EFR-DWBA calculations, if the KBA-Gobbi potential is used. The peaks of the calculated excitation functions are attributed to the enhancement of the DWBA overlap integral of the aligned configuration. The enhancement reflects the simultaneous onset of potential reso-ancnances in both the entrance and exit channels, as is predicted by the band crossing model. As emphasized and demonstrated, it was vital to have weakly-absorptive potentials in both channels. It was further shown that the present analysis was even capable of discriminating between weakly-absorptive potentials, as exemplified by the comparison in Fig. 2 of the Gobbi-Gobbi and the Chatwin-Gobbi curves.

Krause et al. [32] have formulated an analysis of the $^{16}O(^{16}O,^{12}C)^{20}Ne$ reaction by describing the whole system as composed of 4 clusters ($^{12}C + ^{12}C + \alpha + \alpha$). This is an interesting model, but seems to be somewhat too complicated to perform realistic calculations, at least at the present time.

Using a coupled-channel method it has already been demonstrated [8-11] that the mechanism embodied in the band crossing model plays an essential role in reproducing resonances of inelastic scattering for the $^{12}C+^{12}C$, $^{12}C+^{16}O$ and $^{16}O+^{16}O$ systems. Recently, Mosel and Tanimura [33] analyzed resonance structure observed in inelastic scattering of the $^{12}C+^{12}C$, $^{14}C+^{14}C$ and $^{16}O+^{16}O$ systems by using one and two step DWBA methods and concluded that the mechanism embodied in the band crossing model plays an essential role in reproducing resonances for <u>well matched</u> channels. Combining these studies with that of our own, given above, we may conclude that the resonance mechanism of the band crossing model does allow one to understand resonance phenomena in a variety of heavy ion reactions.

REFERENCES

[1] E. Almqvist, D.A. Bromley, and J.A. Kuehner, Phys. Rev. Lett. <u>4</u> (1960), 515.
[2] M. Nogami, private communication.
[3] B. Imanishi, Phys. Lett. <u>27B</u> (1968), 267; Nucl. Phys. <u>A125</u> (1969), 33.
[4] W. Scheid, W. Greiner, and R. Lemmer, Phys. Rev. Lett. <u>25</u> (1970), 176.
[5] D. Baye, and P.-H. Heenen, Nucl. Phys. <u>A276</u> (1977), 354.
[6] T. Ando, K. Ikeda, and A. Tohsaki-Suzuki, Prog. Theor. Phys. <u>64</u> (1980), 1608, and references therein.
[7] Y. Kondō, T. Matsuse, and Y. Abe, Proceedings of the INS-IPCR Symposium on Cluster Structure of Nuclei and Transfer Reactions Induced by Heavy-Ions, Tokyo, edited by H. Kamitsubo, I. Kohno, and T. Marumori (The Institute of Physical and Chemical Research, Wako-Shi, Saitama, Japan, 1975), p. 280; Y. Abe, Proceedings of the Second International Conference on Clustering Phenomena in Nuclei, College Park, Maryland, 1975, edited by D. Goldman, J.B. Marion, and S.J. Wallace (National Technical Information Service, Springfield, Virginia, 1975), p. 500; Yosio Kondō, T. Matsuse, and Y. Abe, <u>ibid</u>., p. 532.

[8] Y. Abe, Y. Kondō, and T. Matsuse, Prog. Theor. Phys. Suppl. 68 (1980) 303, and references therein.

[9] T. Matsuse, Y. Abe, and Y. Kondō, Prog. Theor. Phys. 59 (1978) 1904.

[10] Y. Kondō, Y. Abe, and T. Matsuse, Phys. Rev. C19 (1979) 1356.

[11] Y. Kondō, D.A. Bromley, and Y. Abe, Phys. Rev. C22 (1980) 1068.

[12] P.P. Singh, D.A. Sink, P. Schwandt, R.E. Malmin, and R.H. Siemssen, Phys. Rev. Lett. 28 (1972) 1714.

[13] H.H. Rossner, G. Hinderer, A. Weidinger, and K.A. Eberhard, Nucl. Phys. A218 (1974) 606.

[14] J. Schimizu, T. Nakagawa, Y. Fukuchi, H. Yamaguchi, M. Sato, Y. Nagashima, and K. Furuno, 1980 Annual Report of Tandem Accelerator Center, University of Tsukuba, (1981) p. 82; and J. Schimizu, private communication.

[15] M. Paul, S.J. Sanders, J. Cseh, D.F. Geesaman, W. Henning, D.G. Kovar, C. Olmer, and J.P. Schiffer, Phys. Rev. Lett. 40 (1978) 1310.

[16] J. Nurzynski, T.R. Ophel, P.D. Clark, J.S. Eck, D.F. Hebbard, D.C. Weisser, B.A. Robson, and R. Smith, Nucl. Phys. A363 (1981) 253.

[17] J.C. Peng, J.V. Maher, M.S. Chiou, W.J. Jordan, F.C. Wang, and M.W. Wu, Phys. Lett. 80B (1978) 35.

[18] J.V. Maher, M.W. Sachs, R.H. Siemssen, A. Weidinger, and D.A. Bromley, Phys. Rev. 188 (1969) 1665.

[19] R.A. Chatwin, J.S. Eck, D. Robson, and A. Richter, Phys. Rev. C1 (1970) 795.

[20] A. Gobbi, R. Wieland, L. Chua, D. Shapira, and D.A. Bromley, Phys. Rev. C7 (1973) 30.

[21] R. Vandenbosch, M.P. Webb, and M.S. Zisman, Phys. Rev. Lett. 33 (1974) 842.

[22] Y. Kondō and T. Tamura, Bull. Amer. Phys. Soc. 26 (1981) 611.

[23] Y. Kondō and T. Tamura, preprint.

[24] M. Ichimura, A. Arima, E.C. Halbert, and T. Terasawa, Nucl. Phys. A204 (1973) 225.

[25] T. Tamura, and K.S. Low, Comp. Phys. Comm. 8 (1974) 349.

[26] D.M. Brink, Phys. Lett. 40B (1972) 37.

[27] A. Arima, R.A. Broglia, M. Ichimura, and K. Schäfer, Nucl. Phys. A215 (1973) 109.

[28] P.J.A. Buttle, and L.J.B. Goldfarb, Nucl. Phys. 78 (1966) 409.
[29] F. Pougheon, P. Roussel, M. Bernas, F. Diaf, B. Fabbro, F. Naulin, E. Plagnol, and G. Rotbard, Nucl. Phys. A325 (1979) 481.
[30] F. Diaf, F. Pougheon, P. Roussel, M. Bernas, F. Naulin, G. Rotbard, M. Roy-Stephan, and C. Stamm, AIP Conference Proceedings 47 (1978) 750; F. Diaf, Ph.D. thesis (1978), unpublished.
[31] P.D. Bond, Phys. Rev. Lett. 40 (1978) 501; Phys. Rev. C22 (1980) 1539.
[32] O. Krause, B. Apagyi, and W. Scheid, Nucl. Phys. A364 (1981) 159.
[33] O. Tanimura, and U. Mosel, preprint.

VIII. MODELS AND SYSTEMATICS

THEORY OF NUCLEAR MOLECULAR STATES[†]

Detlev Hahn and Werner Scheid
Institut für Theoretische Physik der Justus-Liebig-
Universität, Giessen, West Germany

Jae Y. Park
Department of Physics, North Carolina State
University, Raleigh, USA

1. Introduction

In light heavy ion reactions, e.g. in the scattering of ^{12}C on ^{12}C, resonances in the cross sections have been observed which have been interpreted as caused by nuclear molecular configurations [1]. With the term "molecular configurations" we express the meaning that these configurations have simple cluster structures in contrast to compound nuclear configurations. The intention of this talk is to present "molecular" reaction theories, which describe the intrinsic structure of the nucleus-nucleus system in terms of molecular configurations.

We distinguish two different pictures which are used as basic concepts in reaction theories for light heavy ion collisions: the "atomic" or sudden picture and the "molecular" or adiabatic picture.

In the sudden picture it is assumed that the scattering process develops so fast, that the intrinsic structure of the nuclei does not appreciably change if the nuclei start to overlap. With such frozen configurations one has calculated potentials between nuclei in the framework of folding procedures and carried out coupled channel calculations for the inelastic excitation of the nuclei. Sudden potentials have typical soft cores at smaller distances which are due to the Pauli principle at lower relative velocities and due to repulsive forces (compression effects) at higher velocities.

The molecular or adiabatic description of the scattering process is microscopically based on the assumption that the nucleons move on molecular orbitals. Wether molecular orbitals are formed, depends on

---

[†] Supported by GSI, DFG and BMFT.

the ratio of two characteristic times, the collision time τ_c and the nuclear period or single-particle relaxation time τ_{sp}. In typical low-energy heavy ion collisions this ratio is of the order $\tau_c/\tau_{sp} \sim 2$. According to this ratio one expects that the valence nucleons have sufficient time to develop molecular orbitals during the reaction.

In the adiabatic picture the intrinsic configurations of the nucleus-nucleus system depend on the relative distance between the nuclei and have included static polarization effects. Molecular configurations are generally referred to an intrinsic coordinate system which rotates about an axis perpendicular to the line connecting the nuclear centers. The adiabatic configurations are perturbed by the rotational and radial relative motion of the nuclei. Only if these perturbations are small, it is meaningful to speak of intrinsic molecular configurations.

In principle, expansions of channel states in the states of the separated nuclei (sudden picture) or in the molecular states are completely equivalent. But it seems to be the most realistic approach to describe heavy ion collisions in the frame of molecular reaction theories. In that case the effects of the strong nuclear interaction between the nuclei are already incorporated in the channel functions, which are generally not much perturbed by the rotation of the intrinsic coordinate system (in complete analogy with the single-particle configurations in the strong-coupling model of Nilsson). Therefore, the representation of channel states in terms of a few molecular basis states is a good approximation for the intrinsic structure of nuclear molecules.

This review is concentrated on the various applications of molecular reaction theories to light heavy ion collisions. In Sect.2 we discuss collective molecular configurations. In Sect.3 we present the specific effects arising from molecular single-particle configurations. Finally in Sect.4, we derive a microscopical molecular reaction theory which makes use of a particle-hole formalism in connection with the two-center shell model.

2. Model for collective molecular states

2.1 Introduction: The energies of the resonances observed in $^{12}C+^{12}C$ have been explained by Cindro et al. [2] with the rotation-vibration model. This model describes the dynamics of a quadrupole-like deformed nucleus by the following degrees of freedom: the rotation of the system and β- and γ-vibrations around the staticly deformed nuclear shape. The energies of the eigenstates of the model

$$E_{IKn_2n_0} = (I(I+1)-K^2)\hbar^2/2\Theta + (\tfrac{1}{2}|K|+1+2n_2)E_\gamma + (n_0+\tfrac{1}{2})E_\beta \tag{1}$$

depend on the moment of inertia Θ, which has been fitted to the slope of the molecular band, and on the β- and γ-vibrational energies E_β and E_γ. The model predicts the energies of the resonances and the γ-transition probabilities between the states (Solem and Cindro [3]).

The rotation-vibration model gives a very physical description of the dynamics of the molecular configurations. But the model can not connect the states of the united system with the states of the separated nuclei, since it does not take the radial relative motion of the nuclei into account. Therefore, the rotation-vibration model has been generalized by a new model, which includes the radial relative motion and describes the continuous transition of the collective states of the separated nuclei into the molecular collective states of the united system. Such an extented model shall be discussed in the following (for details of the model see Ref. [4]).

2.2 The model in the laboratory system: Let us consider the scattering of two identical nuclei which have spin zero in the ground state. This system is described by the relative coordinate $\vec{r}=(r,\theta,\varphi)$ and quadrupole surface coordinates $\alpha_{2\mu}^{(1)}$, $\alpha_{2\mu}^{(2)}$ defining the nuclear shapes. For separated nuclei the surface is given by

$$R^{(i)} = R_0(1+\sum_\mu \alpha_{2\mu}^{(i)} Y_{2\mu}^*(\theta_i,\varphi_i)) \tag{2}$$

with $i = 1, 2$. The numbers 1 and 2 refer to nucleus 1 and 2. The asymptotic definition (2) has to be continued into the overlap zone of the nuclei as indicated in Fig. 1. For identical nuclei it is advantageous to introduce symmetric and antisymmetric quadrupole coordinates defined by

$$\alpha_\mu^\pm = \frac{1}{\sqrt{2}}(\alpha_{2\mu}^{(1)} \pm \alpha_{2\mu}^{(2)}). \tag{3}$$

In the overlap region the symmetric (antisymmetric) coordinates α_μ^+ (α_μ^-) describe quadrupole-type (octupole-type) excitations of the united system (see Fig. 1).

The Hamiltonian for the scattering of identical nuclei can be written as follows:

$$H = H_0(1) + H_0(2) + \frac{1}{2\mu}\vec{P}_r^2 + W(1,2,\vec{r}) \tag{4}$$

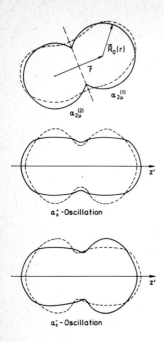

Fig. 1
Definition of the quadrupole coordinates; symmetric and antisymmetric oscillations of the nuclear shape.

The Hamiltonians $H_o(i)$ describe the collective low energy spectrum of the separated nuclei and can be expanded in the quadrupole coordinates $\alpha_\mu^{(i)} = \alpha_{2\mu}^{(i)}$ and their canonically conjugated momenta $\pi_\mu^{(i)}$ (model of Gneuss and Greiner [5]):

$$H_o(i) = \frac{1}{2B_2}\sqrt{5}[\pi^{(i)}\otimes\pi^{(i)}]^{[0]} + \frac{C_2}{2}\sqrt{5}[\alpha^{(i)}\otimes\alpha^{(i)}]^{[0]} + O(\alpha^{(i)3}) \tag{5}$$

In the relative kinetic energy in Eq.(4) we have inserted a constant reduced mass for simplicity. The interaction W between the nuclei can be generally constructed with the conditions of invariance under rotation, time reversal, parity and nucleus exchange. It yields:

$$\begin{aligned}W =\ & V(r)+iU(r)+I_2(r)\sum_\mu(\alpha_\mu^{(1)}+\alpha_\mu^{(2)})Y_{2\mu}^*(\theta,\varphi) \\ & +\sum_{L=0,2,4;M}\{J_L^\alpha(r)([\alpha^{(1)}\otimes\alpha^{(1)}]_M^{[L]}+[\alpha^{(2)}\otimes\alpha^{(2)}]_M^{[L]}) \\ & +K_L^\alpha(r)([\alpha^{(1)}\otimes\alpha^{(2)}]_M^{[L]}+[\alpha^{(2)}\otimes\alpha^{(1)}]_M^{[L]}) \\ & +J_L^\pi(r)([\pi^{(1)}\otimes\pi^{(1)}]_M^{[L]}+[\pi^{(2)}\otimes\pi^{(2)}]_M^{[L]}) \\ & +K_L^\pi(r)([\pi^{(1)}\otimes\pi^{(2)}]_M^{[L]}+[\pi^{(2)}\otimes\pi^{(1)}]_M^{[L]})\}Y_{LM}^*(\theta,\varphi)\end{aligned} \tag{6}$$

2.3 Transformation of the Hamiltonian to the rotating coordinate system - The collective two center model:

The Hamiltonian (4) can be transformed to a rotating coordinate system which has its z'-axis in the direction of the relative coordinate $\vec{r}$:

$$H = -\frac{\hbar^2}{2\mu}\frac{1}{r}\frac{\partial^2}{\partial r^2}r + \frac{1}{2\mu r^2}(\vec{I}-\vec{J}_{coll}')^2 + V(r) + iU(r) + H_{CTCM} \tag{7}$$

Here, $\vec{I}$ is the total angular momentum and $\vec{J}'_{coll}$ the angular momentum of the collective degrees of freedom with respect to the rotating coordinate system. With H_{CTCM} we denote the Hamiltonian of the Collective Two Center Model (CTCM). It is given by:

$$H_{CTCM} = \frac{1}{2} \sum_\mu \{ \frac{1}{B_\mu^+(r)} |\pi_\mu'^+|^2 + C_\mu^+(r) |\alpha_\mu'^+ - \beta(r)\delta_{\mu 0}|^2 \}$$

$$+ \frac{1}{2} \sum_\mu \{ \frac{1}{B_\mu^-(r)} |\pi_\mu'^-|^2 + C_\mu^-(r) |\alpha_\mu'^-|^2 \} - \frac{1}{2} C_0^+(r) \beta^2(r) \qquad (8)$$

$$+ O(\alpha_\mu'^3)$$

The collective two center model describes the intrinsic collective configurations of the nucleus-nucleus system as function of the internuclear distance. The static deformation parameter $\beta(r)$ measures the adiabatic polarization of the nuclear system and is proportional to the linear coupling potential $I_2(r)$ given in Eq. (6):

$$\beta(r) = -\sqrt{\frac{5}{2\pi}} I_2(r) / C_0^+(r) \qquad (9)$$

The mass parameters $B_\mu^\pm$ and stiffness constants $C_\mu^\pm$ can be analytically expressed in terms of the coefficients $J_L^{\alpha,\pi}$ and $K_L^{\alpha,\pi}$ defined in Eq. (6) (see Ref. [4]).

In the special case of identical nuclei which have vibrator spectra if they are separated, the Hamiltonian H_{CTCM} describes 10 decoupled harmonic oscillators with r-dependent frequencies:

$$\omega_\mu^\pm(r) = (C_\mu^\pm / B_\mu^\pm)^{1/2} \qquad (10)$$

The eigenvalues of H_{CTCM} depend on the internuclear distance and the absolute value $|\mu|$ of the projection of the collective angular momentum on the intrinsic z'-axis. For large separations they pass into the energies of the separated nuclei. The energies can be presented in a level diagram like the single-particle energies of the two-center shell model.

The states of H_{CTCM} have a physical meaning inside the reaction region, where they are strongly coupled to the rotating coordinate frame without large perturbations by the rotational and radial couplings induced by the relative motion. The condition of a strong coupling is not fulfilled in the touching range of the nuclei, where the Coriolis coupling in Eq. (7) predominates.

2.4 Application to the $^{12}C+^{12}C$ system: Fig. 2 presents the assumed energies of the 1-phonon states of the symmetric oscillations for $^{12}C+^{12}C$.

For large separations these energies approach the energy of the first 2^+ state of ^{12}C. In various estimates we found that the K=0 phonon $(K=|\mu|)$, describing a β-oscillation in the united system, has the lowest excitation energy inside the interaction region [4].

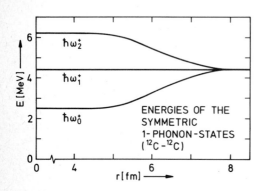

Fig.2 Energies of surface phonons

In the first calculations we restricted the phonon basis to the ground state and even 1-phonon states. This basis is sufficient to describe the elastic channel and the inelastic single excitation of the 2^+ state in the ^{12}C-^{12}C scattering. The basis states, which are diagonal in H_{CTCM}, are given as follows (I= even):

$$\phi^{IM}_{K=0,\lambda=0} = \phi^{IM}_{gs} = \sqrt{\frac{2I+1}{8\pi^2}} D^{I*}_{MO} |0\rangle \tag{11a}$$

$$\phi^{IM}_{K\geq 0,\lambda=1} = \sqrt{\frac{2I+1}{16\pi^2(1+\delta_{KO})}} (D^{I*}_{MK}\beta^{\dagger}_K + D^{I*}_{M-K}\beta^{\dagger}_{-K}) |0\rangle \tag{11b}$$

In order to study the influence of the rotational motion on the molecular states we diagonalized the kinetic energy of rotation in the basis (11):

$$\langle\psi^{IM}_{s'}|\frac{1}{2\mu r^2}(\vec{I}-\vec{J}_{coll})^2 + H_{CTCM} - 5\hbar\omega_\infty - E^I_s(r)|\psi^{IM}_s\rangle = 0, \tag{12}$$

where $\quad \psi^{IM}_s = \sum_{K\lambda} a^I_{K\lambda,s}(r) \phi^{IM}_{K\lambda} \tag{13}$

$s = 1, 2, 3, 4.$

For each even angular momentum I>0 we got four eigenvalues $E^I_s(r)$, ordered with increasing energy. In the upper part of Fig.3 we have drawn the corresponding probabilities $|a^I_{K\lambda,s}|^2$ as function of the internuclear distance. Two well-separated regions can be distinguished: the asymptotic (r>7fm) and molecular (r<6fm) zones. For large separations the states can be classified by the quantum number ℓ of the orbital angular momentum (see lower part of Fig.3). In the molecular zone the energetically lowest state (s=1) contains the unperturbed ground state and small admixtures of the 1-phonon states, whereas the

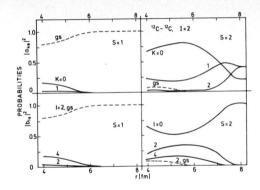

Fig.3

Upper part: probabilities of the basis states (11) in the states (13) for s=1 and 2 and I=2.

Lower part: probabilities for different basis states, which are classified by the quantum number ℓ of orbital angular momentum.

first excited state (s=2) shows a major excitation of the K=0-phonon. From this picture we conclude, that the coupling of the molecular states to the rotating coordinate frame is strong inside the interaction zone. Otherwise the molecular states would be completely mixed by the rotational coupling.

The energies $E_s^I(r)$ have to be added to the real potential $V(r)$ which we have taken from the work of Fink et al. [6]. In Fig.4 the resulting potentials are depicted for the two energetically lowest channels s=1, 2 with I=10. Assuming that the remaining radial coupling term does not influence the energies of the quasimolecular resonances, we have calculated the resonances in these potentials. As shown in Fig.5a,

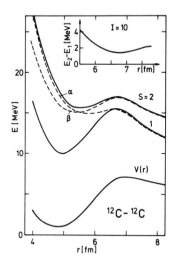
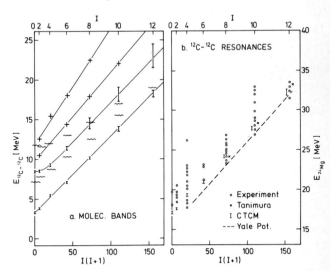

Fig.4
The potentials $V(r)$ and $V(r)+E_s^I(r)$ for s=1,2 and I=10.

Fig.5
Molecular resonances in ^{24}Mg, a) calculated bands, b) comparison of gs-band with experimental data and other calculations.

we found four resonances for each angular momentum I>0, which are grouped into a ground state band and K=0, 2, 1 bands according to the molecular basis state (11), which has the largest probability in each resonance. The K=0 and 2 bands can be interpreted as the β- and γ-bands in the notation of the rotation-vibration model. For the two lowest bands the heights of the Coulomb barriers and the widths of the states are shown in Fig.5a. In Fig.5b we compare the molecular resonances of the ground state band of the model with experimentally observed resonances and resonance energies calculated by Tanimura [7].

2.5 Enhancement of excitation of molecular channels at avoided level crossings:
After the diagonalization of the rotational coupling in Eq.(12) the coupling between the various molecular channels, described by the channel functions ψ_s^{IM}, is reduced to a single operator, namely the radial coupling term ($\sim\partial/\partial r$). The radial coupling between two molecular channels is enhanced at internuclear distances, where avoided level crossings occur. At these points the character of the molecular channel wave functions changes rapidly with the internuclear distance so that the radial coupling can strongly mix the two channels involved in the avoided level crossing. This effect is known as the Landau-Zener effect in molecular physics [8].

The enhancement of the excitation at avoided level crossings is closely related to the double resonance mechanism. This statement is clarified in Fig.4, where the diagonalized potentials are drawn by solid lines and the unperturbed potentials by dashed lines. The resonances in the unperturbed potentials fulfill the condition for the double resonance mechanism. After the interaction between these configurations is diagonalized, the potentials show an avoided level crossing (see inset in Fig.4) as the remainder of the former crossing of the unperturbed curves. The avoided crossing ensures that the energetically higher lying molecular channel gets strongly excited. Therefore, we conclude that the Landau-Zener effect in the adiabatic (molecular) or strong coupling picture is the analogue of the double resonance effect in the sudden or weak coupling picture.

3. Molecular single-particle configurations

In contrast to the case of collective molecular resonances, observed in various heavy-ion systems, molecular single-particle effects are not well established yet. Recently von Oertzen et al. [9] proved that the neutron orbits in the grazing collision of ^{12}C on ^{13}C are of molecular nature. Another example is the neutron transfer reaction

$^{13}C(^{13}C,^{12}C)^{14}C$ measured by Korotky et al. [10]. The excitation function shows a resonant-like behaviour, which can not be reproduced by finite-range, full-recoil DWBA calculations.

Molecular single-particle effects can be described by the Two-Center Shell Model (TCSM). Up to now this model has been used for the calculation of internuclear potentials in the framework of the Strutinsky formalism, for generating basis states in the generator coordinate method or Hartree-Fock formalism, and for the dynamical treatment of molecular single-particle effects in light heavy ion reactions. In this section we discuss the latter application of the two-center shell model.

3.1 Excitation mechanisms for molecular single-particle states: The basis for a dynamical treatment of molecular single-particle states are realistic two center level diagrams [11]. Their systematic study leads to an understanding of the possible molecular single-particle effects, before coupled channel calculations have been carried out. As an example we present the level diagram for the system $^{16}O+^{25}Mg$ in Fig.6 [12]. In this calculation the parameters of the asymmetric two-center shell model have been fitted to the experimental neutron single-particle levels near the Fermi surface of the separated nuclei ^{16}O and ^{25}Mg for $R\to\infty$ and of the united system ^{41}Ca at R=0. The ^{25}Mg nucleus is described with an intrinsic deformation built into the potential of the symmetric TCSM. Therefore, for large internuclear distances the single-particle spectrum of ^{25}Mg is in accordance with the spectrum obtained by the Nilsson model (see Fig.6).

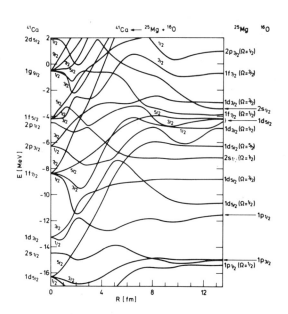

Fig.6
The neutron level diagram for $^{16}O+^{25}Mg\to^{41}Ca$.

Level diagrams as shown in Fig. 6 can be used for the prediction of possible reactions. There are two main excitation mechanisms between molecular single-particle states, namely the radial and rotational coupling.

The radial coupling, $<\lambda\Omega|\frac{\partial}{\partial R}|\lambda'\Omega>$, causes transitions near avoided crossings between levels with the same value of Ω, whereas the rotational or Coriolis coupling, $<\lambda\Omega|j_{\pm}|\lambda',\Omega\pm1>$, leads to transitions between levels with Ω and $\Omega\pm1$, which are allowed to cross in the level diagram of the TCSM. Here, Ω is the quantum number of the angular momentum component in the direction of the internuclear axis, and λ denotes the other quantum numbers. The promotion of nucleons at points of avoided level crossings is completely analogous to the promotion process of quasi-molecular electrons in atomic collisions according to the model of Fano and Lichten. The transition probability between the levels at avoided crossings can be estimated with the formula of Landau and Zener [8]. Since the transition probability depends on the relative velocity of the nuclei and, therefore, on the incident energy, we can expect signatures of avoided level crossings in the cross sections as function of the incident energy. Such signatures would be unique signs for the formation of molecular single-particle orbitals during heavy ion collisions.

Let us consider some possible consequences of level crossings for the $^{16}O+^{25}Mg$ reaction [12]. In Fig. 6 we recognize many pairs of levels between which the radial and rotational couplings can cause the excitation or transfer of nucleons. The possible inelastic excitation processes and neutron transfer reactions are schematically shown in Figs. 7 and 8, respectively. In these figures the numbers denote the individual ways of neutrons through the level diagram which get excited or transferred during the collision. The radial coupling at an avoided crossing is indicated by a straight arrow and the rotational coupling by a wavy arrow. For example, in Fig.7 we note that the $2s_{1/2}$ level of ^{25}Mg, which is occupied initially by the valence neutron, has an avoided crossing with the $1d_{3/2}$ ($\Omega=1/2$) level of ^{25}Mg near 5.3 fm. Here the neutron is excited to the $1d_{3/2}$ ($\Omega=1/2$) state by radial coupling (process No.1). Other inelastic and transfer processes can easily be traced in Figs. 7 and 8 by following the corresponding process numbers.

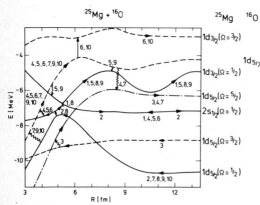

Fig. 7
Processes (No.1-10) leading to inelastic excitation in the reaction $^{16}O+^{25}Mg$.

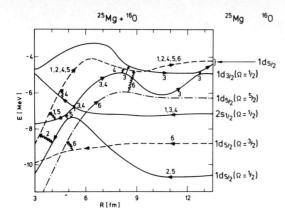

Fig. 8
Processes (No.1-6) leading to a neutron transfer in the reaction $^{16}O+^{25}Mg$

Such studies help one to select important inelastic and reaction channels and simplify coupled channel calculations.

3.2 Molecular particle-core model: In the following we review the molecular particle-core model which was applied by Terlecki et al. [13] for the calculation of the elastic and inelastic differential cross sections of the $^{13}C-^{13}C$ scattering. The extension of the model for neutron transfer reactions has been worked out in Ref.[14].

3.2.1 The particle-core model in the laboratory system: The model describes heavy ion reactions induced by loosely bound nucleons, which move in molecular orbitals during the scattering process. Therefore, the model treats the scattering of nuclei, which can be divided up into cores and loosely bound nucleons. The latter nucleons are denoted as extra nucleons in the following and described by molecular single-particle states in the framework of the two-center shell model. The mean field of the TCSM is generated by all nucleons and contains the adiabatic polarization effects between the nuclei.

The coordinates of the model are the coordinates $\vec{R}_{C1}$, $\vec{R}_{C2}$ of the centers of mass of the cores and N coordinate vectors $\vec{r}_1...\vec{r}_N$ of the extra nucleons (see Fig.9). In addition it is possible to treat the excitation of the cores by collective coordinates which are disregarded for simplicity in the following. Then the kinetic energy of the model is given by:

$$T = \frac{1}{2MC_1}\vec{P}_{C1}^2 + \frac{1}{2MC_2}\vec{P}_{C2}^2 + \sum_{i=1}^{N}\frac{\vec{p}_i^2}{2M} \qquad (14)$$

For the solution of the scattering problem we transform the coordinates to the center of mass coordinate $\vec{R}_{cm}$ of the total system, to the relative coordinate of the two nuclei and N particle coordinates (see Fig. 9). The particle coordinates are measured from the center of mass,

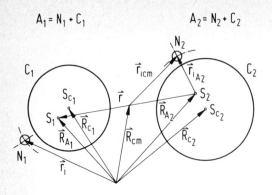

which is assumed to be the origin of the particle coordinates in the two-center shell model.

Fig.9
Definition of coordinates in the particle-core model.

$$\vec{R}_{cm} = \frac{1}{A}(C_1 \vec{R}_{C1} + C_2 \vec{R}_{C2} + \sum_{i=1}^{N} \vec{r}_i) \tag{15a}$$

$$\vec{r} = \frac{1}{A_1}(C_1 \vec{R}_{C1} + \sum_{i=1}^{N_1} \vec{r}_i) - \frac{1}{A_2}(C_2 \vec{R}_{C2} + \sum_{i=N_1+1}^{N} \vec{r}_i) \tag{15b}$$

$$\vec{r}_{icm} = \vec{r}_i - \vec{R}_{cm} \tag{15c}$$

The kinetic energy of the system given in Eq.(14) is transformed to the new coordinates as follows:

$$T = \frac{1}{2AM}\vec{P}_{cm}^2 + \frac{1}{2\mu}\vec{P}_r^2 + \frac{1}{2M}\sum_{i=1}^{N}\vec{P}_{icm}^2$$

$$+ \frac{1}{\mu}\vec{P}_r (\frac{A_2}{A}\sum_{i=1}^{N_1}\vec{P}_{icm} - \frac{A_1}{A}\sum_{i=N_1+1}^{N}\vec{P}_{icm}) - \frac{1}{2AM}(\sum_{i=1}^{N}\vec{P}_{icm})^2 \tag{16}$$

Here, μ denotes the reduced mass. The first term in (16) is the kinetic energy of the center of mass, the second one the kinetic energy of the relative motion of the nuclei and the third one the kinetic energy of the extra nucleons in the center of mass system. Whereas the last term in (16) can be neglected for $N/A \ll 1$, the fourth term describes the coupling of the relative motion with the single-particle motion and is important for large relative velocities $\vec{v}_r = \frac{1}{\mu}\vec{P}_r$. When the fourth term is neglected, one obtains unphysical excitations of the extra nucleons at large internuclear distances.

3.2.2. <u>The transformation of the Hamiltonian to the rotating coordinate system</u>: Since the two-center shell model and its wave functions are conveniently written in a coordinate system where the centers lie on the z-axis, it is advantageous to introduce a rotating coordinate system with the z'-axis along the direction of the relative coordinate.

The rotating coordinate system is fixed with respect to the laboratory system by the Euler angles φ, θ which are the spherical polar angles of the relative coordinate $\vec{r}$. The transformation of the Hamiltonian to the rotating coordinate system yields the following expression, after the kinetic energy of the center of mass motion has been subtracted:

$$H = -\frac{\hbar^2}{2\mu r}(\frac{\partial}{\partial r}+D)^2 r + \frac{(\vec{I}-\vec{J}_a')^2}{2\mu r^2} + W_{C_1 C_2}(r) + \sum_{i=1}^{N} h_{TCSM}(i)$$
$$- \frac{1}{2A_1 M}(\sum_{i=1}^{N_1} \vec{p}_{icm}')^2 - \frac{1}{2A_2 M}(\sum_{i=N_1+1}^{N} \vec{p}_{icm}')^2 \quad (17)$$

$$D = \frac{1}{A}(A_2 \sum_{i=1}^{N_1} \frac{\partial}{\partial z_{icm}'} - A_1 \sum_{i=N_1+1}^{N} \frac{\partial}{\partial z_{icm}'}) \quad (18)$$

$$\vec{J}_a' = \sum_{i=1}^{N_1}(\vec{r}_{icm}' - \frac{A_2}{A}\vec{r}) \times \vec{p}_{icm}' + \sum_{i=N_1+1}^{N}(\vec{r}_{icm}' + \frac{A_1}{A}\vec{r}) \times \vec{p}_{icm}' + \sum_{i=1}^{N} \vec{s}_i' \quad (19)$$

$$h_{TCSM}(i) = \frac{\vec{p}_{icm}'^2}{2M} + U_{TCSM}(\vec{r}_{icm}', \vec{p}_{icm}', \vec{s}_i', r) \quad (20)$$

Here, we denote the coordinates of the extra nucleons in the rotating coordinate system by $\vec{r}_{icm}'$ and their spin operators by $\vec{s}_i'$. The operator D and the special structure of the operator $\vec{J}_a'$ arise from the fourth term in Eq.(16). This form of the kinetic energy operator in (17) assures that the radial and rotational coupling matrix elements between the molecular single-particle states vanish for large internuclear distances. The interaction between the cores is contained in the complex potential $W_{C1C2}(r)$. The fourth term in (17) is the Hamiltonian of the two-center shell model for the extra nucleons. For the two-center potential U_{TCSM} in (20) we assume that the two-center distance can be replaced by the radial relative coordinate r. In principle one has to add a residual interaction between the extra nucleons in (17) which may be disregarded for simplicity.

3.2.3 The molecular wave functions: The wave functions solving the scattering problem can be written as follows:

$$\psi_{IM} = \mathcal{A}(1...N) S(C_1, C_2) \sum_{\alpha K} R_{\alpha K}^{I}(r) D_{MK}^{I*} \phi_{\alpha K}(\vec{r}_{icm}', r) \quad (21)$$

The radial wave function $R_{\alpha K}^{I}(r)$ depends on the total angular momentum I, the projection quantum number K of the angular momentum on the intrinsic z'-axis and a further number α which classifies the various molecular single-particle configurations described by $\phi_{\alpha K}$. The rotation of the coordinate system is described by the functions D^{I*}_{MK} depending on the Euler angles. The intrinsic wave functions are products of the eigenfunctions of h_{TCSM}. The wave functions ψ_{IM} are assumed to be antisymmetrized in the coordinates of the extra nucleons. They have to be symmetrized for the exchange of the cores if they are identical, as it is the case for the $^{13}C-^{13}C$ scattering discussed below.

Using the ansatz for the wave functions in (21) and the Hamiltonian (17) we can set up a system of coupled equations for the radial wave functions $R^I_{\alpha K}(r)$. This system has to be solved with proper boundary conditions for $R^I_{\alpha K}(r\to\infty)$, which are related to the scattering matrix (see Ref.[14]).

3.2.4 Application to the $^{13}C-^{13}C$ scattering:

The molecular particle-core model has been applied to the $^{13}C-^{13}C$ scattering by Terlecki et al. [13], who have calculated the elastic cross section and the inelastic ones for the single and mutual excitation of the $1/2^+$ (3.09MeV) state of ^{13}C which has a strong E1-transition to the $1/2^-$ ground state of ^{13}C. In this calculations the $^{13}C+^{13}C$ system was described by two ^{12}C cores and two extra neutrons. Neutron transfer and core excitation channels have been neglected. The various parameters and details of the calculations may be taken from Ref.[13].

Fig.10 shows the constituents of the radial coupling matrix element for a special transition. As already mentioned the radial coupling matrix elements vanish for large internuclear distances and are large at avoided level crossings.

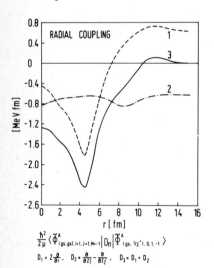

Fig.10

A special matrix element of the radial coupling

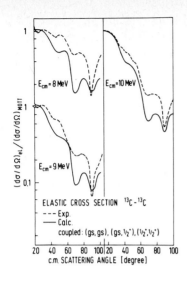

Fig. 11 Angular distributions for elastic $^{13}C-^{13}C$ scattering. The experimental data are measured by Helb et al. [15].

In Fig. 11 we compare the calculated elastic differential cross sections with the experimental data of Helb et al. [15]. The inelastic 90° cross sections for the single and mutual excitation of the $1/2^+$ (3.09MeV) state in ^{13}C are drawn in Fig. 12 (solid lines). The excitation functions reveal intermediate structures which arise when quasibound resonances are excited via the double resonance mechanism. There are yet no experimental data of inelastic cross sections available for comparison.

To investigate the effect of the radial and rotational coupling terms we have calculated the inelastic excitation functions separately with the radial coupling and with the rotational one. The results are shown in Fig. 12. Whereas the rotational coupling is of minor influence on the inelastic cross sections, the radial coupling plays the dominant

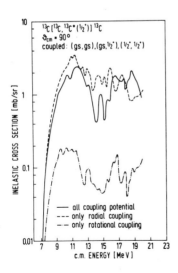
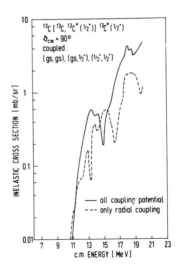

Fig. 12 The 90° differential cross sections for the single and mutual excitation of the $1/2^+$(3.09MeV) state in the $^{13}C-^{13}C$ scattering.

role in the excitation of the molecular single-particle states. The radial coupling leads to the excitation of nucleons when their orbitals, first centered at the individual nuclei, pass into the molecular states of the united system.

4. Microscopic model for a molecular treatment of nucleus-nucleus scattering.

Nuclear systems, showing resonances of molecular nature in their cross sections, are mostly built up by even-even nuclei. As an example we mention the $^{12}C+^{12}C$ system. These systems can not be described with the molecular particle-core model formulated in Sect.3. Also the particle-core model does not include the excitation of the core nucleons and the antisymmetrization between the cores and extra nucleons. However, compared with the complex apparatus developed for the resonating group and generator coordinate methods, the molecular reaction theory of Sect.3 and the two-center shell model are quite simple and versatile tools for describing the various reaction channels in heavy ion scattering. Therefore, there is a demand for a reaction theory on the basis of the TCSM which allows to describe microscopically the excitation and transfer of nucleons in a particle-hole formalism based on molecular single-particle states. It is the aim of this section to present the principle constituents of such a reaction theory which will be applicable for a wide range of scattering problems.

4.1 Hamiltonian in coordinate space:
Let us start with the Hamiltonian of a system of A nucleons given by

$$H = T+V = \sum_{i=1}^{A} \frac{\vec{p}_i^2}{2M} + \sum_{i<j} V(|\vec{r}_i-\vec{r}_j|) \tag{22}$$

This Hamiltonian should describe the scattering of two nuclei with A_1 and A_2 nucleons where $A_1+A_2=A$. For that purpose we transform the A particle coordinates $\vec{r}_i$ and two redundant coordinate vectors $\vec{S}_1$ and $\vec{S}_2$ into the center of mass coordinate $\vec{R}_{cm}$, the relative coordinate $\vec{r}$ and A particle coordinates $\vec{r}_{icm}$, referred to the center of mass:

$$\vec{r} = \frac{1}{A_1} \sum_{i=1}^{A_1} \vec{r}_i - \frac{1}{A_2} \sum_{i=A_1+1}^{A} \vec{r}_i \tag{23a}$$

$$i = 1,\ldots A_1: \quad \vec{r}_{icm} = \vec{r}_i - \frac{1}{A} \sum_{i=1}^{A} \vec{r}_i + \vec{S}_1 \tag{23b}$$

$$i = A_1+1,\ldots A: \quad \vec{r}_{icm} = \vec{r}_i - \frac{1}{A} \sum_{i=1}^{A} \vec{r}_i + \vec{S}_2 \tag{23c}$$

A further transformation of these coordinates onto a rotating coordinate system with the z'-axis along the direction of the relative coordinate $\vec{r}$ yields the following expression for the kinetic energy:

$$T = \frac{1}{2AM} \vec{P}_{cm}^2 - \frac{\hbar^2}{2\mu}(\frac{1}{r}\frac{\partial}{\partial r}r + D)^2 + \frac{1}{2\mu r^2}(\vec{I}-\vec{J}_a')^2$$
$$+ \frac{1}{2M}\sum_{i=1}^{A}\vec{P}_{icm}'^2 - \frac{1}{2A_1 M}(\sum_{i=1}^{A_1}\vec{P}_{icm}')^2 - \frac{1}{2A_2 M}(\sum_{i=A_1+1}^{A}\vec{P}_{icm}')^2 \quad (24)$$

The operators D and $\vec{J}_a'$ have the same structure as those in Eqs.(18) and (19); only the particle numbers N_1 and N_2 have to be replaced by A_1 and A_2, respectively. After introducing the relative coordinate by Eq. (23a), the operator of the kinetic energy is no more symmetrical in the coordinates of the nucleons. Therefore, the transition from the coordinate space of the nucleons to the formalism of second quantization can not be carried out straightforwardly.

4.2 The wave functions in coordinate space: We take the following ansatz for the wave functions describing the scattering of the nuclei:

$$\Psi_{IM,\mathcal{A}} = \mathcal{A}\Psi_{IM} \quad (25)$$

$$\Psi_{IM} = \sum_{\alpha,K} R_{\alpha K}^I(r) D_{MK}^{I*} \phi_{\alpha K}(1...A,r) \quad (26)$$

$$\phi_{\alpha K} = \frac{1}{\sqrt{A_1!}} \begin{vmatrix} (1) \cdots \\ \vdots \\ (A_1) \end{vmatrix} \cdot \frac{1}{\sqrt{A_2!}} \begin{vmatrix} (A_1+1) \cdots \\ \vdots \\ (A) \end{vmatrix} \quad (27)$$

$$\mathcal{A} = 1 - \sum P_{1-nucl.exch.} + \sum P_{2-nucl.exch.} - \cdots \quad (28)$$

The intrinsic channel wave functions $\phi_{\alpha K}$ are antisymmetrized in the nucleons with numbers $1...A_1$ and $A_1+1....A$ separately. The operator $\mathcal{A}$ antisymmetrizes the nucleons between these two groups. Since the direct reactions take place in the touching zone of the nuclei, only the first terms of $\mathcal{A}$ and the terms, where identical nuclei are exchanged ($A_1=A_2$), play the dominant role in the overlap of Ψ_{IM} with $\Psi_{IM,\mathcal{A}}$.

4.3 Transition to Fock space: The form of the intrinsic channel functions suggests a possible definition of creation and annihilation operators. We define two groups of Fermi operators, separately for the nucleons $1...A_1$ and $A_1+1....A$. Let us refer these operators to a complete set of orthonormal single-particle states of the two-center shell model, where the quantum numbers of the wave functions are abbreviated by α:

$$h_{TCSM}(\vec{r}'_{cm},r)\varphi_\alpha(\vec{r}'_{cm},r) = \varepsilon_\alpha(r)\varphi_\alpha(\vec{r}'_{cm},r) \qquad (29)$$

As in Eq. (20) we assume that the two-center distance is equal to the relative coordinate. Then we define the following creation and annihilation operators of Fermi-type, namely

i) for particles $1...A_1$: $a_\alpha^\dagger$, a_α, (30)

obeying the anticommutation relations

$$[a_\alpha^\dagger, a_\beta]_+ = \delta_{\alpha\beta}, \quad [a_\alpha, a_\beta]_+ = [a_\alpha^\dagger, a_\beta^\dagger]_+ = 0$$

ii) for particles $A_1+1...A$: $b_\alpha^\dagger$, b_α, (31)

obeying the anticommutation relations

$$[b_\alpha^\dagger, b_\beta]_+ = \delta_{\alpha\beta}, \quad [b_\alpha, b_\beta]_+ = [b_\alpha^\dagger, b_\beta^\dagger]_+ = 0$$

Since the two sets of operators $(a_\alpha^\dagger, a_\alpha)$ and $(b_\alpha^\dagger, b_\alpha)$ operate in different Fock spaces, they commute, e.g.

$$[a_\alpha^\dagger, b_\beta]_- = [a_\alpha, b_\beta]_- = 0 \qquad (32)$$

With these operators we can transform the intrinsic channel wave function (27) into Fock space:

$$\phi_{\alpha K} = a_{\alpha_1}^\dagger a_{\alpha_2}^\dagger \ldots a_{\alpha_{A_1}}^\dagger |0\rangle \cdot b_{\alpha_{A_1+1}}^\dagger \ldots b_{\alpha_A}^\dagger |0\rangle \qquad (33)$$

The operator T of kinetic energy in Eq.(24) is symmetric in the two sets of coordinates $\vec{r}'_{1cm}....\vec{r}'_{A_1cm}$ and $\vec{r}'_{A_1+1cm}....\vec{r}'_{Acm}$. Therefore, it is straightforward to rewrite T and then H in terms of the creation and annihilation operators defined above. In a further step we have fixed the Fermi levels for the two nuclei and defined particle and hole operators in terms of the particle operators. After this step it is possible to characterize the excited states in terms of particle-hole configurations with respect to the ground state configuration.

4.4 Expansion of the antisymmetrization operator: The relative coordinate $\vec{r}$ is changed in the antisymmetrization procedure, which leads to nonlocal potentials in an exact treatment. Because we want to have a local reaction theory, we expand the differences with respect to $\vec{r}$ into Taylor series. Then we can write the antisymmetrization operator $\mathcal{A}$, defined in Eq. (28), in lowest order using the creation and annihilation operators:

$$\mathcal{A} = 1 - \sum_{\alpha\beta} a_\alpha^\dagger a_\beta b_\beta^\dagger b_\alpha$$

$$- \frac{A}{A_1 A_2} \{ \sum_{\alpha\beta\gamma} <\alpha|\vec{r}'_{cm}|\beta> (a_\gamma^\dagger a_\beta b_\alpha^\dagger b_\gamma - a_\alpha^\dagger a_\gamma b_\gamma^\dagger b_\beta) \} \quad (34)$$

$$\cdot \{ \vec{\nabla}_r + \sum_{\delta\varepsilon} <\delta|\vec{\nabla}_r|\varepsilon> (a_\delta^\dagger a_\varepsilon + b_\delta^\dagger b_\varepsilon) \} + \ldots$$

The first term inside the last curly parentheses operates on the relative wave function, the second term arises from the r-dependence of the single-particle wave functions. Since the full antisymmetrization operator is complicated, we suggest to take $\mathcal{A}$ in the form (34), which is a good approximation if the reactions develop in the touching region of the nuclei.

Whereas the full antisymmetrization operator $\mathcal{A}$ is Hermitian, $\mathcal{A} = \mathcal{A}^\dagger$, and commutes with H, the expression (34) does not fulfill these relations. In order that the coupled channel equations become an Hermitian system of equations (beside absorptive potentials), we suggest to derive the coupled channel equations for the radial wave functions from the following Schrödinger equation:

$$\{ \frac{1}{2}(H\mathcal{A} + \mathcal{A}^\dagger H^\dagger) - \frac{1}{2}(\mathcal{A} + \mathcal{A}^\dagger) E \} \psi_{IM} = 0 \quad (35)$$

Here, $\mathcal{A}^\dagger$ and $H^\dagger$ are the Hermitian adjoints of $\mathcal{A}$ and H and ψ_{IM} is the partly antisymmetrized state vector defined in Eq.(26). Coupled channel calculations on the basis of Eq.(35) are in progress.

<u>4.5 Applications</u>: The reaction theory, developed in this section, can be used for a microscopic treatment of inelastic excitation and nucleon transfer in the framework of the TCSM. The theory does not require a separation of the system into cores and extra nucleons. The antisymmetrization is included in a consistent manner which complicates the simple picture of the TCSM considerably, but can not be avoided with a simpler procedure.

Within this method molecular boson operators can be constructed from pairs of Fermi operators in order to describe collective molecular excitations. Such a procedure finally leads to a microscopic foundation of the collective two center model introduced in Sect.2. It would allow a microscopical calculation of the collective excitation energies, transition potentials and mass parameters.

This theory makes no a priori assumptions about the suddeness or adiabaticity of the interaction potentials between the nuclei. Therefore, it can be applied for studying the effects of the intrinsic excitation on the relative motion, especially the effects arising from the diabatic occupations of the levels at avoided level crossings. Also this reaction theory can be used to derive microscopical absorptive potentials on the basis of molecular single-particle states, as discussed in Ref.[16].

5. Summary and conclusions

In this review we have presented molecular reaction theories for the scattering of light heavy ions. The molecular theories describe the intrinsic structure of the nucleus-nucleus system in a coordinate system rotating about an axis perpendicular to the direction of the internuclear relative coordinate. The intrinsic molecular states are adiabatic states; i.e. they describe the structure of the system for fixed nuclear centers. We distinguish molecular collective, single-particle and many-particle states. The type of the molecular basis states, which one actually selects for a satisfactory description of a scattering problem, depends on the special aspects one is interested in.

The intrinsic molecular states are perturbed by the rotational and radial couplings. Both types of couplings are induced by the relative motion of the nuclei. Only if these couplings do not completely mix the molecular states inside the reaction zone, it is justified to claim the physical existence of molecular intrinsic configurations. In heavy ion scattering at incident energies not too high above the Coulomb barrier, the internuclear interaction is sufficiently strong enough to couple the configurations strongly to the rotating coordinate system. Therefore a molecular description of these reactions seems to be the most physical one.

Like in molecular physics the crossing of molecular levels plays an important role for the excitation mechanism. In the cases we have investigated, we observed that the radial coupling at avoided level crossings is more effective than the rotational coupling. It can be stated that the enhanced excitation at avoided level crossings (Landau-Zener effect) is the analogue of the double resonance mechanism in the non-adiabatic scattering theory.

Acknowledgement: The work reported here has been carried out in a friendly collaboration with Prof. Walter Greiner and R. Koennecke of the University of Frankfurt.

References:

[1] D.A. Bromley, invited talk at the International Workshop on Resonances in Heavy Ion Collisions, Bad Honnef 1981, in these proceedings
[2] N. Cindro et al., Phys. Rev. Lett. 39 (1977) 1135; F. Cocu et al., J.de Phys. Lett. 38 (1977) 421; N. Cindro, J.Phys.G. 4 (1978) L23
[3] J.C. Solem and N. Cindro, Fizika 13 (1981) Suppl. 1, 19
[4] D. Hahn, G. Terlecki and W. Scheid, Nucl.Phys. A325 (1979) 283
[5] G. Gneuss and W. Greiner, Nucl.Phys. A171 (1971) 449
[6] H.J. Fink , W. Scheid and W. Greiner, Nucl.Phys. A188 (1972) 259
[7] O. Tanimura, Nucl.Phys. A309 (1978) 233
[8] L. Landau, Phys.Z.Sow. 2 (1932) 46; C. Zener, Proc.Roy.Phys.Soc. A137 (1932) 696
[9] W. von Oertzen, B. Imanishi, H.G. Bohlen, W. Treu and H. Voit, Phys.Lett. 93B (1980) 21
[10] S. Korotky et al., Bull.Am.Phys.Soc. 24 (1979) 13; D. A. Bromley in Ref. [1]
[11] J. Park, W. Greiner and W. Scheid, Phys.Rev. C21 (1980) 958
[12] J. Park, W. Scheid and W. Greiner, preprint 1981, University of Giessen
[13] G. Terlecki, W. Scheid, H.J. Fink and W. Greiner, Phys.Rev. C18 (1978) 265
[14] J. Park, W. Scheid and W. Greiner, Phys. Rev. C20 (1979) 188
[15] H.D. Helb et al., Nucl.Phys. A206 (1973) 385
[16] G. Terlecki, D. Hahn et al., Lecture Notes in Physics 89 (1979) 410

STRUCTURE AND FORMATION OF MOLECULES[*]

Ulrich Mosel
Institut für Theoretische Physik, Universität Giessen
6300 Giessen, West Germany

1. INTRODUCTION

In recent years a number of suggestions have appeared in the literature that the observed molecular resonances, for example in $^{12}C + ^{12}C$, may be connected with the population of secondary, shape-isomeric minima in the potential energy surfaces of the compound system. For a discussion I refer particularly to the talk given by E. Cosman at this meeting[1].

In the first part of my talk I plan, therefore, to discuss some details of a calculation that Chandra and myself did some years ago and that predicted the existence of special shape-isomeric states all the way out to the nucleus - nucleus configuration. I will then briefly discuss the dynamical conditions for formation of such a state. The strong interactions needed have consequences for angular momentum - energy mismatched channels that will be discussed next. Finally follows a discussion of the influence of strongly coupled direct excitations of the heavy ion optical potential.

---

[*]Work supported by BMFT and GSI Darmstadt.

2. ISOMERIC MOLECULAR CONFIGURATIONS

About 10 years ago, when theoretical calculations were successful in reproducing the properties of shape isomeric states observed in the fission of heavy elements, the hope was that similar calculations might produce a second minimum also in the interaction barrier of light systems like e. g. $^{12}C + ^{12}C$. This second minimum could then support the molecular state.

The calculations done then were all unsuccessful to generate such a structure (see e.g. ref. 2). In 1977, however, Chandra and myself did a new calculation in which a standard constraint, that of axial symmetry of the nucleus-nucleus system, was relaxed[3]. This had a drastic effect on the ion-ion potentials as indicated in figure 1.

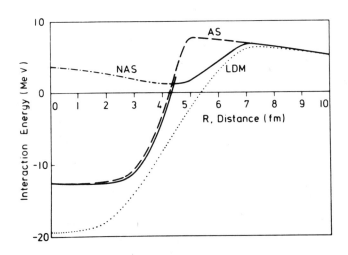

Fig. 1

For details see text.

The curve AS is the potential for a configuration in which the two oblate ^{12}C nuclei form an axially symmetric system, i.e. in which they have their symmetry axes parallel to the line connecting their centers. The curve labeled NAS gives the potential for a non-axially symmetric configuration. In it the two symmetry axes are parallel to each other and vertical to the separation line. As can be seen NAS is clearly lower than AS between r~4.5 fm and 7 fm due to a stronger nuclear attraction in this configuration.

The interesting result was that the configuration NAS was stable against small changes of the deformation all the way in to center distance R=0 fm. At R=0 fm, where the two center model merges into the Nilsson-model, the configuration NAS was found to be exactly that obtained by Leander and Larsson in a one-center Nilsson calculation for the potential energy surface of ^{24}Mg at $\varepsilon = 1.26$, $\gamma = 42°$ (ref. 4).

As can be seen from fig. 1 the two potential curves cross at $r \approx 4.5$ fm. Thus here a transition, for example from the NAS to the lower energy AS configuration, could occur. However, because of the quite different geometrical distribution of the wave functions in the two configurations the interaction between them is expected to be very small and thus transitions from one to the other will be strongly hindered. The configuration NAS, even though being higher in energy, may thus be nevertheless quite stable. In other words: if initially at large R the two nuclei are in NAS then they will stay in there with a high probability.

What actually happens in terms of the intrinsic structure at the crossing can be seen in fig. 2.

Fig. 2

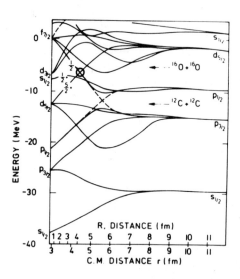

This figure shows that the two Carbon nuclei that occupy all single particle states at R → ∞ up to the $p_{3/2}$ shell stay fairly intact into about 5 fm. There an avoided level crossing (as indicated by dashed lines) happens: the formerly unoccupied $p_{1/2}$ state dives down through the Fermi-surface. Keeping the original occupation fixed, then at R ~ 3.5 fm another crossing happens, this time with a state coming down from the $d_{5/2}$ shell in the individual nuclei. The state finally reached at R = 0 fm (i.e. in ^{24}Mg) is that of 4 particles in the $f_{7/2}$ shell with 4 holes in the $p_{1/2}$ shell. Similarly, in ^{16}O + ^{16}O where analogous situations prevail[3], again 4 particles are in the $f_{7/2}$ shell but the 4 holes are now in the $d_{5/2}$ shell. The excited states formed in ^{24}Mg and ^{32}S, respectively, are thus simple excited quartet states. Their structure is very reminiscent of that of the famous 0_2^+ state in ^{16}O, in which 4 particles are promoted into the next higher ($d_{5/2}$) shell, that has captured the interest of many physicists in the late sixties.

One must note, however, that these (4p-4h)-quartets are <u>not</u> simply groundstate + α-particle configurations:

$$^{24}Mg^*(R = 0) \neq {}^{20}Ne_{gs} + \alpha$$

$$^{32}S^*(R = 0) \neq {}^{28}Si_{gs} + \alpha$$

and are thus not expected to decay strongly into the corresponding groundstates. Instead, due to the presence of holes in the $p_{1/2}$ or $d_{5/2}$ shell, respectively, the quartet states are expected to α-decay selectively to excited states in their daughter nuclei.

As the excited quartett states are structually quite different from the ground state configurations in the compound nuclei their decay into states of the CN is hindered. One thus expects small spreading widths of these states. This expectation is indeed borne out in a calculation by Deubler and Fliessbach[5]. These authors find that the only configuration in the CN that couples to a molecular configuration at large R with any noticeable strength is just the quartet state discussed here.

That the single particle level crossings discussed here are not dependent on the specific model used but are general consequences of the Pauli-principle was pointed out by M. Harvey in 1975 (ref. 6). Simplified correlation diagrams for harmonic oscillator potentials are shown for ^{12}C + ^{12}C (for the configuration NAS) in fig. 3 and for

$^{16}O + {}^{16}O$ in fig. 4.

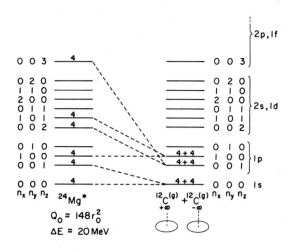

Fig. 3

(from ref. 6)

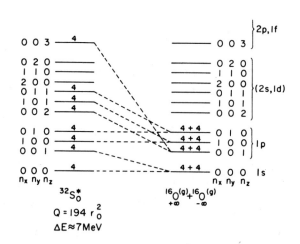

Fig. 4

(from ref. 6)

In these figures the z-axis is taken to be parallel to the line connecting the centers of the two nuclei. Thus, when moving the two nuclei over each other, the x- and y-degrees of freedom are not affected, their oscillator quanta remain constant. The number of quanta in z direction, however, has to change due to the Pauli principle (for details see ref. 6). One sees quite clearly the excitation of the N = 3 shell (i.e. the $f_{7/2}$ state) in the combined nucleus.

It is also interesting to note that in the inverse process,

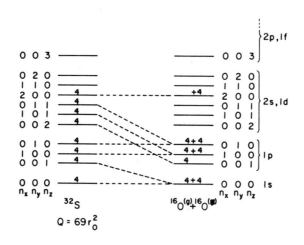

Fig. 5

(from ref. 6)

i.e. the "cold" fission of ^{32}S, one of the two ^{16}O fragments will be in an excited 0^+ (4p-4h) state with the particles in N = 2 shell. Thus, measuring the relative yield of the 0_2^+ state relative to the ground state in ^{16}O in fission of ^{32}S might give interesting complimentary information on the structure of these nuclei.

3. CONNECTION WITH REACTION THEORY

All these results and considerations are based on a static model of the nucleus-nucleus interaction. Fortunately, some support for these results comes also from reaction theory.

In the Perturbed Stationary State method, for example, as developed Imanishi and Tanimura[7] and later on used extensively by Tanimura and Tazawa[8] one can calculate the "directivity" of the nuclear symmetry axes, i.e. the distribution of angles that the individual axes form with the line connecting the centers of the two nuclei. The result of such a calculation, as obtained by Tanimura[9], is given in fig. 6 that shows the probability distribution of the angles just mentioned.

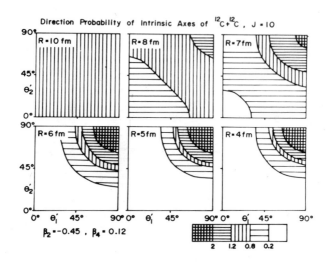

Fig. 6
(from ref. 9)

One sees clearly how at small distances in a $^{12}C + ^{12}C$ collision both nuclei direct their symmetry axes vertically to their centerline. Thus exactly the same configuration as the one labeled NAS in section 2 prevails at smaller distances.

The two different potentials have also been obtained by Cugnon et al.[10] who have incorporated the two different configurations into a reaction theory. The interesting result of their study is that additive combinations of the two potentials form the diagonal channel potentials whereas the difference between the two curves determines the coupling matrix elements to the lowest rotational states in ^{12}C.

All the considerations in section 2 on the possible population of the shape isomeric states contain the tacit assumption that a single nucleus has been formed. Thus in a rotational state the combined nucleus (for example, ^{24}Mg) is assumed to rotate "rigidly" as one body so that a part of the total, angular momentum is in internal rotation of the two ^{12}C nuclei. This is in contrast to a shape-resonant state where all the angular momentum is in relative motion.

One can now ask oneself the question what the conditions are for the formation of such a rigidly rotating system. There exists a quite general necessary condition for the transition from two nuclei without intrinsic rotation to a single rigidly rotating nucleus: the points where the two nuclei touch must be at rest relative to each other. This condition defines the "sticking limit" for the angular momentum transfer ΔL from the incoming orbital angular momentum L_i into intrinsic rotation:

$$\Delta L = L_i \frac{J_1 + J_2}{\mu R^2 + J_1 + J_2}$$

Here J_1 and J_2 are the intrinsic moments of inertia and R is the distance at which the nuclei touch. Assuming spherical nuclei and rigid rotor values for J_i for simplicity and a touching distance $R = R_1 + R_2$ gives the well known result (for symmetrical systems):

$$\Delta L = \frac{2}{7} L_i$$

This implies that, for example, in a ^{12}C + ^{12}C collision at $E_{c.m.} \approx$ 25 MeV, where the reaction is dominated by the L = 14 partial wave, $\Delta L = 4$. Thus, for the "sticking limit" both ^{12}C nuclei have to be excited to their first rotational 2^+ states (using experimental moments of inertia does not change this result).

The answer to the question asked above is thus that sticking can occur only in a mutual $(2^+, 2^+)$ excitation. If sticking should also be valid at smaller distances then higher rotational states would have to be included.

In the most important region at large R only the fully aligned mutual $(2^+, 2^+)$ channel in the ^{12}C + ^{12}C reaction can directly, i.e. without further (virtual) transitions, lead to a rigidly rotating molecule. This channel is thus the most selective one for the population of the shape isomer. This enhanced selectivity may have consequences for the search for such states in scattering[1] or γ-decay experiments[11,12].

4. STRONG COUPLING APPLIED TO MISMATCHED CHANNELS

The sticking condition discussed in section 3 is only a necessary condition. If the transition to one rigidly rotating system actually occurs is a much more complicated question that cannot be answered at present. However, it is clear that such a transition requires a strong, fairly long ranged interaction between the two nuclei so that nearly all nucleons can be locked into the common rotation.

Such a strong interaction is provided by the folding model. The coupling form factors are considerably stronger, shifted to smaller nucleus-nucleus distances and are broader in R (for a review see Tanimura and Tazawa, ref. 8) than in the gradient-coupling models.

Tanimura and myself[13] have recently applied this model to the description of the inelastic scattering to the mismatched 0_2^+ state in $^{16}O + ^{16}O$. The total cross section for this inelastic transition as obtained by Freeman et al.[14] shows clearly gross structure that is correlated with that in the well matched channel of the single 3^- excitation. This correlation is unexpected and cannot be obtained either in Austern-Blair[15] or conventional DWBA calculations that work quite successfully for the well matched channels[16]. The results of a new calculation are shown in fig. 7.

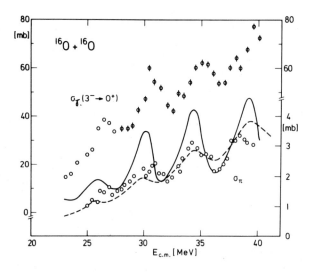

Fig. 7

Both the absolute height and the correlations are now well described. (The experimental 3⁻ cross section is the result of an inclusive measurement; it contains both the single and twice the mutual cross section. The calculated curve contains only the single excitation cross section).

Considerable physical insight can be obtained if the potentials are inspected that generate exactly the coupled channel wave functions. These potentials (obtained in the slightly different earlier calculation[13]) are shown in figure 8.

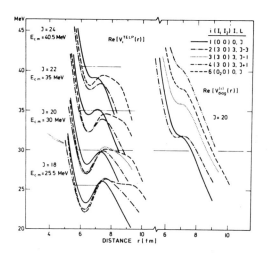

Fig. 8

On the right hand side the diagonal (input) potentials are shown for $J = 20$ for the different channels. On the left hand side the final potentials are plotted that contain all the channel coupling effects in them. One sees that the potentials are dramatically changed by the strong channel-channel couplings. The curves for the different channels now nearly coincide and all have a shape resonance at about the same energy (for a more detailed discussion see ref. 13).

Thus the mechanism that generates the gross structure is exactly the same as that responsible for bumps in the inelastic excitation[16] or the transfer[17] channels. What is different here, however, is that not the "bare" diagonal but only the "dressed" potentials, that contain already the effects of channel coupling, exhibit these shape resonances. Studies of mismatched channels thus offer an opportunity to learn considerably more about the reaction mechanism than those of matched ones that are dominated by simple dynamical matching conditions without any effect of the particular structure of the nuclei involved.

5. INFLUENCE OF STRONG DIRECT EXCITATIONS ON HEAVY ION POTENTIALS

J. Schiffer in his talk at this meeting[18] has pointed out some interesting connections between the absorptive potential and the shell structure of the ions involved. It is clear in this context that not the single particle states but collective states will be most strongly excited in inelastic heavy ion collisions. These in turn will be lowest in open shell nuclei.

A detailed theoretical study of the effects of these states on the heavy ion potential is often made difficult by the presence of strong transfer and break-up channels. However, a "clean" case is offered by the $^{12}C + ^{12}C$ system for which it is experimentally known that up to intermediate energies a few inelastic excitations exhaust the unitarity limit for the grazing partial wave and that all other L's contribute to fusion[19].

We have, therefore, started a program to treat all inelastic channels explicitly in a coupled channel calculation. Here I show first results of a calculation in which only the single 2^+ and the mutual $(2^+, 2^+)$ excitations in $^{12}C + ^{12}C$ are included[20]. Using Feshbach's formalism one can project on the elastic channel and thus calculate the polarization potential that contains the changes of the diagonal optical model potential due to the coupling to the other channels. The propagator needed for the projected potentials can here be calculated exactly in the subspace of explicitly treated states.

The polarization potentials obtained in this way are nonlocal and J- and E-dependent:

$$\Delta V^J(r,r') = \sum_{j\ell} U^J_{0j}(r) G^J_{j\ell}(r,r') U^J_{\ell 0}(r')$$

Here the U^J_{ik} are the coupling potentials and $G^J_{j\ell}$ is the (exact) propagator. For simpler interpretation these potentials can again be converted into a trivially equivalent local potential.

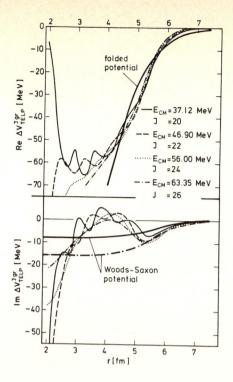

Fig. 9

In figure 9 the real and the imaginary parts of the polarization potentials are shown for different energies and angular momenta J (chosen to be the grazing ones at these energies). These turn out to be very strong for both the real and the imaginary part. The imaginary part is seen to be strongly surface peaked (all curves inside $r \lesssim 4$ fm are irrelevant since here the wave functions are vanishingly small due to absorption). The weakly positive parts of Im(ΔV) at $r \approx 4$ fm are due to a feedback from the excited into the elastic channel.

In fig. 10 finally the J dependence of the trivially equivalent local V at a fixed r = 5.5 fm is shown. It is seen that the real parts assume maximum values just at the grazing partial wave which couples most strongly. The absorptive parts are maximum for considerably lower J's but fall off more steeply towards larger J.

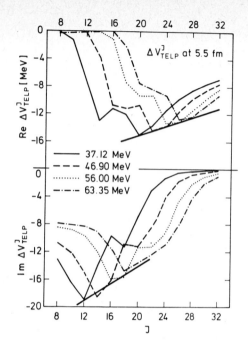

Fig. 10

6. SUMMARY

In this talk I have tried to show that the formation of a molecule depends on two conditions: a) the existence of a special configuration in the combined system that is directly coupled to the entrance and exit channels and b) a large structural difference between this configuration and that of compound nucleus so that the spreading width can remain small. As was shown the existence of such configurations is predicted in the two center model.

The "sticking limit" conditions are expected to enhance the selectivity for population of those molecular configuration. For the $^{12}C + ^{12}C$ case in the relevant energy regions the mutual (2_1^+, 2_1^+) channel is the one that corresponds to the sticking limit.

For "sticking" to occur not only the dynamical conditions must be right. Also strong interactions between the different channels are needed. It was shown that such a strong interaction model can explain the observed correlations between gross structures in well matched and mismatched channels in contrast to weak coupling models. The general mechanism for gross structure is here the same as elsewhere: shape resonances in the ion-ion potential are responsible for it.

Finally, some effects of the strongly coupled direct channels on the heavy ion potential were shown. It was pointed out that these couplings introduce strong energy - and angular momentum dependences.

REFERENCES

1) E. Cosman, invited talk at this meeting
2) T. D. Thomas and P. W. Riesenfeldt, Proc. Symp. Heavy Ion Scattering, Argonne 1971, Argonne National Laboratory, Report ANL-7837, p. 53
3) H. Chandra and U. Mosel, Nucl. Phys. A298 (1978) 151
4) G. Leander and S. E. Larsson, Nucl. Phys. A239 (1975) 93
5) H. Deubler and T. Fliessbach, Nucl. Phys. A238 (1975) 409
6) M. Harvey, Proc. Second Int. Conf. on Clustering Phenomena in Nuclei, College Park 1975, USERDA report ORO-4856-26, p. 549
7) B. Imanishi, in: Nuclear Molecular Phenomena, N. Cindro, ed., North Holland Publ. Co (Amsterdam 1978), p. 379
   O. Tanimura and B. Imanishi, Phys. Lett. 80B (1979) 340
8) O. Tanimura and T. Tazawa, Phys. Rep. 61 (1980) 253, and references therein
9) O. Tanimura, Nucl. Phys. A309 (1978) 233
10) J. Cugnon et al., Nucl. Phys. A331 (1979) 213
11) V. Metag et al., contribution to this meeting, and to be published
12) R. C. McGrath et al., contribution to this meeting, and Phys. Rev. C (1981), in press
13) O. Tanimura and U. Mosel, Phys. Rev. C24 (1981) 321
14) W. S. Freeman et al., Phys. Rev. Lett. 45 (1980) 1479
15) R. L. Philipps et al., Phys. Rev. Lett. 42 (1979) 566
16) O. Tanimura and U. Mosel, to be published
17) O. Tanimura, invited talk at this meeting
18) J. P. Schiffer, invited talk at this meeting
19) B. R. Fulton et al., Phys. Rev. C21 (1980) 198
20) R. Wolf et al., to be published

WEAK COUPLING MODEL APPROACH TO HEAVY ION MOLECULAR RESONANCE*

O. Tanimura

Institut für Theoretische Physik, Universität
Giessen, 6300 Giessen, West Germany

Prepared for presentation at the International Workshop on Resonances in Heavy Ion Collisions, Bad Honnef, October 1981

* Work supported by GSI, Darmstadt and BMFT

Recent experimental studies using ^{14}C nucleus exhibit prominent gross structures not only in the inelastic channels such as ^{14}C (^{12}C, ^{12}C) ^{14}C (3^-:6.73 MeV) and ^{14}C (^{16}O, ^{16}O)$^{14}C(3^-)$ but also in the transfer reaction channels such as ^{14}C (^{12}C, ^{13}C) ^{13}C ($5/2^+$: 3.854 MeV) and ^{14}C (^{16}O, ^{12}C) ^{18}O (2^+ : 1.982 MeV) in the $^{14}C+^{12}C$[1] and $^{14}C+^{16}O$[2], respectively. More importantly, peaks observed in the inelastic and transfer channels in the respective scatterings are apparently correlated with each other. These experiments confirm the suggestion given by Brussel group[3] that the reaction including ^{14}C such as $^{12}C + ^{14}C$, $^{14}C + ^{14}C$ and $^{14}C + ^{16}O$ are good candidates for exhibiting molecular resonance. Furthermore, these occurence of molecular resonance in these channels is consistent with the recent numerical prediction based on the numbers of open channels by Haas and Abe[4].

On the other hand, though there have been proposed models or interpretations to explain the inelastic resonances such as BCM[5], Austern-Blair model[6], barrier-top-resonance model[7], DWBA[8], weakcoupling model[9] and coupled channel model[10,17] there are no realistic models to explain and reproduce molecular resonance in the rearrangement channel except for a recent DWBA calculation given by Kondo and Tamura.[11] This may be due to the fact that DWBA is considered to produce no oscillatory behavior with energy and that coupled channel model is not yet practicable to treat rearrangement channel.

In the previous paper[9] we have presented the weak coupling model in order to explain the correlations of the molecular resonance arising in single and mutual excitation channels in $^{12}C + ^{12}C$, $^{14}C + ^{14}C$ and $^{16}O + ^{16}O$. From this analysis we have shown that the molecular resonance is caused by the overlap between independent enhancements of resonance in the entrance and exit channels. It seems important to study whether this interpretation is valid to the molecular resonance in the transfer channels.

In this article we present a consistent explanation for the molecular resonance both in the inelastic and transfer channels in terms of weak coupling model: i.e. DWBA with relevant resonance condition. Next we make the comparison between DWBA and Austern-Blair model and discuss about the origin of the gross structure of the excitation function.

In the system composed of (core + valence particle) and core, the parity dependence of the interaction plays an important role in the elastic and inelastic scatterings. The even-odd difference comes from the fact that system is symmetric with respect to the valence particle transfer. Therefore, it is reasonable to add the parity dependent potential to the distorting potential in our problem. Since we don't know precise form of the parity dependent potential, we take the same form as the distorting potential for simplicity.

It is reasonable to assume that inelastic excitation takes place in the same potential as the elastic one. Parameters used here are listed in Table 1.

Table 1. Distorting Potentials of Woods-Saxon Form.[a]

	V_0 (MeV)	V_π[b] (MeV)	r_I (fm)	nomalization
$^{12}C + ^{14}C$, el., inel.	-16.5	0.5	1.25	$S_3 = 0.148$
$^{13}C + ^{13}C(5/2^+)$	-16.5	0.5	1.25	$C = 0.94$[c]
$^{14}C + ^{16}O$, el., inel.	-15.5	1.0	1.3	$S_3 = 0.17$
$^{12}C + ^{18}O(2^+)$	-12.5	1.0	1.3	$C = 1.16$

a) the other parameters are the same:
   $W = -0.4 - 0.1(E_{c.m.} + Q_i)$, $r_0 = r = 1.35$, $a_0 = a_\pi = a_I = 0.35$.
b) parity depth is defined by $\pi(A_1)\pi(A_2) \cdot (-1)^L \cdot V_\pi$.
c) C is the product of spectoscopic amplitudes between initial and final states of transfer particl.

Fig. 1 shows the calculated results for
^{14}C (^{12}C, ^{12}C) ^{14}C (3^-) and ^{14}C (^{12}C, ^{13}C ($5/2^+$)) ^{13}C
together with the J-decomposition of the calculation
denoted by solid line. The calculations of inelastic
cross section with and without parity generate gross
structures, but that with parity reproduces the experiment
better than that without parity. Here we should notice
that the both experiments reveal the even-odd difference
in themselves if the sequential interger is assumed for the
observed peaks: The separation between peaks at 18.6 and
20.8 MeV is by 1 MeV smaller than that between peaks at 20.8

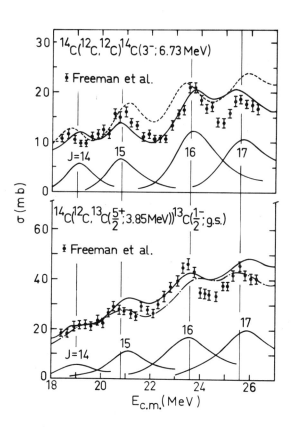

Fig.1
Integrated cross sections in $^{12}C+^{14}C$ scatterings. Solid and dashed curves in upper figure denote results with and without the parity dependence. Solid and dot-dashed curves in lower figure denote results without and with the parity in ^{13}C-^{13}C channel, but both curves include the parity in ^{12}C-^{14}C channel.

and 23.8 MeV and almose equal to that between peaks at 23.5 and 25.6 MeV. This consideration leads to the 0.5 MeV even-odd difference, which coincide with our choice of parity dependent potential. In the DWBA calculation for this transfer channel we used the same distroting potential as the elastic one.

Form factor was calculated such that the transferred neutron is initially bound by ^{13}C in a state of (n, l, j) = (0, 1, 1/2) and is finally trapped by ^{12}C in a state of (n, l, j) = (0, 2, 5/2). Radius and diffuseness of both bound states are 1.2 fm and 0.65 fm, respectively. The product of the spectroscopic amplitudes in initial and final state of the neutron was taken to be 0.94.

We don't know the parity dependence for ^{13}C (1/2)$^+$ ^{13}C (5/2$^+$), because elastic and inelastic scattering of ^{13}C + ^{13}C does not tell us about the odd-parity state. So we try to calculate the cross section with and without parity dependence for this channel. Calculation without parity generate a good agreement with the experimental oscillatory pattern. From the J-decomposition of this cross section we can suggest that observed peaks are due mainly to a single J-component. Introduction of the parity dependence in ^{13}C+^{13}C does not improve the cross section very much, as seen in dot-dashed curve. We only remark that the same sign of V_π as the elastic channel may be favourable in comparison with the experiment. More important is that the model reproduces the observed correlation between inelastic and transfer channels with the assignement of the same spins. Unfortunately, the spins have not been assigned yet experimentally.

Fig. 2 shows the calculated results for $^{14}C\,(^{16}O,\,^{16}O)\,^{14}C\,(3^-)$ and $^{14}C\,(^{16}O,\,^{18}O\,(2^+))\,^{12}C$ together with their J-decompositions. Calculation with $V_\pi = 1$ MeV reproduces well both the observed inelastic cross sections and the transfer reaction cross sections. Here we have assumed dineutron transfer from ^{12}C to ^{16}O. Dineutron is initially bound by ^{12}C in a state of $(n, l, j) = (1, 0, 0)$ and is bound by ^{16}O in a final state of $(n, l, j) = (0, 2, 2)$.

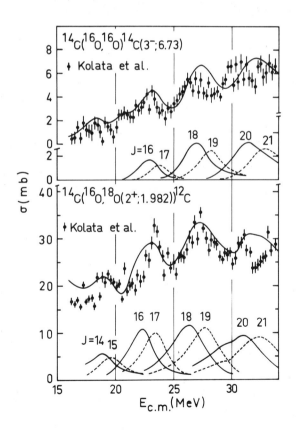

Fig. 2
Integrated cross section for the inelastic excitation (upper figure) and for transfer reaction (lower figure) in $^{14}C + ^{16}O$.

Radius and diffuseness parameters of both bound state are 1.45 fm and o.65 fm. The product of sepctroscopic amplitudes in initial and final state was taken to be 1.16. We have taken a relatively large radius parameter in order to take into account the fact that transfer process of valence particle is enhanced in the contact region as suggest by Seglie et al.[12] and von Oertzen et al.[13]

In the inelastic channels the observed separation between peaks is about 4 MeV, which is about twice the separation between resonances predicted for sequential integer values. Therefore, it seems reasonable to assigne them in terms of even or odd integer band. Barenhardt et al.[14] have observed resonances at 19.5, 23 and 27.5 MeV in the elastic $^{14}C + ^{16}O$ scattering. Kolata et al.[2] have suggested that $J = 16^+$ assignment for 23.5 MeV peak give an angular distribution in good agreement with the data of Bernhardt et al.[14] This may confirm our calculation, in which peaks are due to even integer value with small odd component.

In the $^{12}C + ^{18}O$ channel, Webb et al.[15] have observed resonances at $E_{c.m.}(^{12}C + ^{18}O) = 19.0$ and 21.8 with spin assignement of 15^- and 17^-, respectively, which corresponds to 19.9 and 22.7 MeV in $E_{c.m.}$ of $^{14}C + ^{16}O$ system, respectively. Kolata et al.[2] have suggested that peaks at 23.5 MeV may favor $J = 16^+$ assignement from the consideration of the limitation of the fusion yield. Thus we cannot deduce a definite conclusion for the spin assignement from the experiment at present. From the lower part of this figure, we can see that calculation produces a good description of the observed oscillatory structure which are formed by (even-odd) pair contribution of J. These pair J are correlated with the inelastic channel resonances. Moreover our model roughly reproduced the inelastic $^{12}C + {}^{18}O$ (2^+) cross section observed by Freeman et al.[16] Thus, it should be noticed that our model gives a consistent description of both inelastic and transfer reactions with the use of the fixed optical potential parameters.

In order to clarify the correlation, it is very interesting to investigate energy behavior of wave function as discussed in ref. 9.

Fig. 3 shows energy dependence of $|u_i^{(+)}(R_0)|^2$, where $u_i^{(+)}(R_0)$ denotes a wave function of the aligned coupling channel. Upper and lower panels correspond to $^{14}C + {}^{16}O$ and $^{12}C + {}^{14}C$, respectively. The DWBA amplitude is defined by the integration of the product of the wave functions of entrance and exit channels and the form factor. If the interaction is localized enough in the surface region, which is the case of our problem, then the cross section is roughly proportional to $|u_{el.}^{(+)}(R_0)|^2 |u_i^{(+)}(R_0)|^2$ where R_0 is the interaction radius. Obviously, the wave functions have resonance-like behaviors which are caused by the respective potential pockets. We can see that the behaviors of $|u_{el.}^{(+)}(R_0)|^2, |u_{inel.}^{(+)}(R_0)|^2$ and $|u_{tr.}^{(+)}(R_0)|^2$ are all correlated with each other.

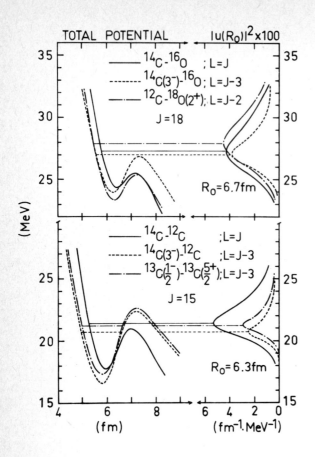

Fig. 3
Total potential (left part) and the energy dependence of the squared wave functions (right part). Upper and lower figures correspond to $^{14}C+^{16}O$ ans $^{12}C+^{14}C$ scatterings, respectively.

Therefore, we can find that the inelastic cross section proportional roughly to the overlap of $|u^{(+)}_{el.}|^2$ and $|u^{(+)}_{inel.}|^2$ is correlated with the transfer cross section proportional to the overlap of $|u^{(+)}_{el.}|^2$ and $|u^{(+)}_{tr.}|^2$, since the product of the functions are illustrated by the overlap of the two functions. In other words, these three channels are all well matched one another in this angular momentum. This interpretation is closely related to that of Band Crossing model[5] and double resonance model[10].

It is very interesting to compare between Austern-Blair model and DWBA. First, let's discuss about the conclusion from Austern-Blair model. Erb[16] argued that since this model with $\delta_{max} = 0.5$ produces oscillatory structure with energy clearly, the observed structure is not due to resonance but due to diffraction. However, a detailed investigation may alter this conclusion.

Fig. 4 shows $^{16}O + ^{16}O$ (3^-) cross sections by means of Austern-Blair model for 3 sets of parameters of δ_{max} and βR. For comparison, we have taken $\delta_{max} = 0.0$, 0.5 and $\pi/2$. βR is regarded as normalization constant of the cross section. Obviously, these three sets of calculations produces little difference one another as far as structure is concerned. In another word,

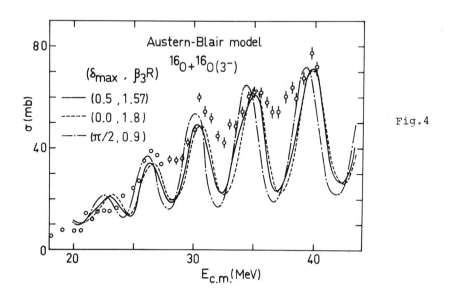

Fig.4

the oscillatory structure does not depend on the phase shift δ of this model. Therefore, we cannot deduce a conclusion whether structure comes from resonance or non-resonance diffraction.

We should notice that Austern-Blair model make use of only the elastic phase shift. Moreover, the phase shift should reproduce the excitation function of elastic scattering. Now let's look at the elastic excitation function. Erb et al.[16] calculated this function for $^{16}O + ^{16}O$ by using $\delta_{max} = 0.5$. The resulting cross section agrees fairly well with the data at about $90°$, but it is not in good agreement at the forward angles. On the contrary, Gobbi potential gives us better description of the data at whole angles. However, if we insert the Gobbi potential phase shift into the Austern-Blair model, the resultant inelastic cross section does not produce clear structure.

Fig. 5 shows the comparison between DWBA and Austern-Blair model for $^{16}O + ^{16}O(3^-)$ excitation function with the use of Gobbi potential. Data points are taken from the measurement by Kolata et al.[18] Obviously, these two curves are quite different and DWBA produces a relevant structure. This means that Austern-Blair model is not good approximation to DWBA in this problem and that the model may not be able to produce the relevant structure because Gobbi phase shift is more realistic to than the Austern-Blair parametrization.

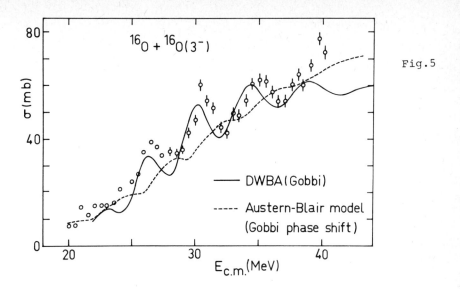

Fig.5

Fig. 6 shows the comparison between Gobbi potential S-matrix and the strong absorption S-matrix. From this figure we may say that the strong absorption parametrization is not relevant in two points. One is tail behavior of $|S_L|$ and the other is rising behavior of phase shift. This different behavior of phase shift generates a large difference in $|dS/dL|$.

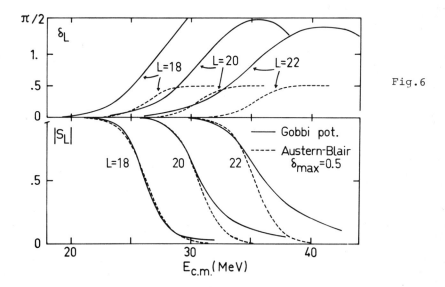

Fig.6

Fig. 7 shows the energy dependence of $|dS/dL|$ and $|dS/dL'|$ in the $^{16}O + ^{16}O$ scattering. Clearly we can see that $|dS/dL|$ of Gobbi potential S-matrix generates much broader structure than that in the strong absorption model.

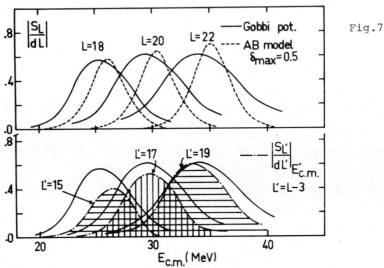

Fig.7

Therefore, the overlap between $|dS/dL|$ and $|dS/dL'|$, which are illustrated by shaded area, has no prominent structure. This is the reason why Austern-Blair model with Gobbi phase shift does not generate clear structure.

Fig. 8 shows DWBA calculations with Gobbi potential with modifications. Solid curve shows the cross section without modification, dashed curve shows the calculation with reduction of the real potential depth by a half

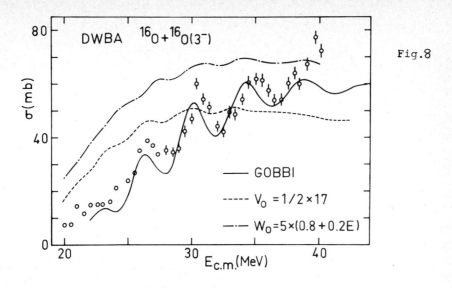

Fig. 8

and dot-dashed curve shows the calculation with 5 times deeper imaginary depth than the original value. Here we notice that Gobbi potential phase shift generates a resonance or nearly resonance at the grazing angular momenta. From this figure we can deduce the fact that shallowering the real part gives actually no structure, since the potential has no clear pocket at these angular momenta and that deepening the imaginary part smear out the structure, since the resonance enhancement are smeared out by the absorption. Moreover, we obtain the same illustration as in Fig. 3 for Gobbi potential. Namely, enhancements are due to the overlap between initial and final resonant wave function.

In view of these discussions, we may summarize as follows:

1) Austern-Blair model say nothing as to resonance or non-resonance diffraction.

2) Austern-Blair model may not be relevant in the problem of molecular resonance.

3) DWBA with relevant potential pocket and surface transparency gives us good and consistent description of the observed structure in both inelastic and transfer channels.

4) Prominent structure is caused by the overlap of the resonance or resonance-like enhancement between initial and final channel.

The author would like to thank U. Mosel, P. Braun-Münzinger and P. Paul for the valuable discussions.

References

1) R. M. Freeman et al., Phys. Lett. $\underline{90B}$, 229 (1980)

2) J. J. Kolata et al., Phys. Rev. $\underline{C23}$, 1056 (1981)

3) P. H. Heenen and D. Baye, Phys. Lett. $\underline{81B}$, 295 (1979)

4) F. Haas and Y. Abe, Phys. Rev. Lett. $\underline{46}$, 1667 (1981)

5) T. Matsuse, Y. Kondo and Y. Abe, Prog. Theor. Phys. $\underline{59}$, 1009 (1978)

6) R. L. Phillips et al., Phys. Rev. Lett. $\underline{42}$, 566 (1979)

7) W. A. Freedman et al., Phys. Lett. $\underline{87B}$, 179 (1979)

8) L. E. Cannell et al., Phys. Rev. Lett. $\underline{43}$, 837 (1979)

9) O. Tanimura and U. Mosel, to be published in Phys. Lett. $\underline{B}$

10) W. Scheid, W. Greiner and R. Lemmer, Phys. Rev. Lett. $\underline{25}$, 176 (1970)

11) Y. Kondo and T. Tamura, preprint of University of Texas

12) R. J. Ascuitto, J. F. Petersen and E. A. Seglie, Phys. Rev. Lett. $\underline{41}$, 1159 (1978)

13) W. von Oertzen et al., Phys. Lett. $\underline{93B}$, 21 (1980)

14) K. G. Bernhardt et al., Proceedings of international conference on Resonances in Heavy Ion Reaction, Hvar, Yugoslavia, 1977, edited by N. Cindro, P 367

15) M. P. Webb et al., Phys. Rev. Lett. $\underline{36}$, 779 (1976)

16) K. A. Erb, Report at international conference on Resonant Behavior of Heavy Ion Systems, Aegean Sea, Greece, June 1980

17) O. Tanimura and T. Tazawa, Phys. Rep. $\underline{61}$, 253 (1980)
    O. Tanimura and U. Mosel, Phys. Rev. $\underline{C24}$, 321 (1981)

18) J. J. Kolata et al., Phys. Rev. $\underline{C19}$, 2237 (1979)

MULTISTEP TRANSFER OF NUCLEONS AND THE FORMATION OF MOLECULAR ORBITALS

W. von Oertzen[*], B. Imanishi[+]

Hahn-Meitner-Institut

für Kernforschung, Berlin, Germany

The question of multistep transfer is studied in the framework of coupled reactions channel analysis. Neutron transfer and direct excitation of single particle states in the system $^{12}C + ^{13}C$ are calculated in a coupled channel formalism. Angular distributions and excitation functions for excited states are discussed. It is found that the coupling between ground states and excited states is weaker than the coupling between excited states. Inelastic excitation tends to increase the transfer coupling via polarisation (hybridization effect) of the orbitals. Thus in states of the total system with positive parity the coupling between ground and excited states is strong, while it is weak in negative parity states. The formation of molecular orbitals of the neutron in the $^{12}C + ^{13}C$ system is illustrated and their significance for the understanding of the different reaction channels are discussed.

[*] Also Fachbereich Physik, Freie Universität Berlin

[+] on leave from Institute for Nuclear Studies
University of Tokyo, Japan.

I. Introductory remarks

A long standing problem in the interaction of two nuclei is the description of molecular phenomena connected to valence nucleons. The formation of molecular orbitals for nucleons in a collision has been discussed since the early work of Breit et al (1). Molecular wave functions for the active nucleons represent an alternative basis for the description of nucleon transfer between nuclei (2,3,4). It is, however, obvious that the usefullness of such an approach is connected to the validity of the adiabatic approximations. If the interaction between two nuclei, due to transfer or direct excitation (or a combination of both) is sufficiently strong and the collision time long, multiple interactions will occur.

In such situations a description with eigenstates of the separated nuclei needs a complete coupled reaction channel (CRC, ref. 5) treatment; however, some physical phenomena may not become evident if the asymptotic states are used as a basis. The use of two center eigenstates (molecular wave functions) may bring advantages in the description and a clearer understanding of the physics underlying the process (e. g. parity dependent Majorana potentials in the case of elastic transfer (2)).

Therefore the assesment of higher order processes in the transfer of nucleons between nuclei during a collision is the primary step in the discussion of the physics of molecular orbitals. This can be done using a CRC calculation with assympotic nuclear basis states. Two step contributions have been observed in single nucleon and two nucleon transfer reactions where strong inelastic transitions are involved (6,7). Two step transfer processes have been discussed in the context of two nucleon transfer (8,9) and in inelastic transfer (10) and inelastic scattering generally (6,11,12).

In the following we will discuss:

> i) conditions for systems where two steps are favourable and look into the experimental evidence for two step transitions.
>
> ii) a CRC analysis of the $^{12}C + ^{13}C$ system
>
> iii) relation between the CRC results and the description with molecular orbitals.

II. Two step transitions

In order to make many steps one has to do one and two steps first. A typical diagramm for a two step transfer is shown in fig. 1; sequential transfer, or two nucleon transfer or sequential transfer of the same nucleon leading back to the original (elastic) channel or an excited state. The two step transfer of the same nucleon can best be studied in cases where it contributes to inelastic scattering (6,11,12). A further possibility is the combination of a transfer process with direct excitation.

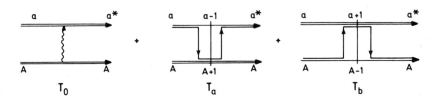

Fig. 1: Schemes for two step transitions for reactions between nuclei leading to excitation of nucleus a.

The condition for large two step contributions are identical with the conditions for multistep transitions (for each step):

  a) large spectroscopic amplitudes
  b) small binding energies
  c) small effective Q-values in the intermediate states.

Conditions a) and b) taken together give a large total width for the valence particle and large transfer matrix elements. Condition c) emphasises the importance of dynamical matching of the reaction (11).

The importance of the Q-value of the second step has been demonstrated rather convincingly in cases where the first step is an inelastic transition, and the second step one nucleon transfer (6,13). Thus for one proton transfer reactions with ^{12}C and ^{16}O on ^{62}Ni we obtain the following conditions if projectile excitation is considered: ^{62}Ni ($^{12}C,^{11}B$) ^{63}Cu, $Q_o = -9,8$ MeV, if the transfer occurs via $^{12}C_2^+$ (4.43 MeV) we obtain $Q_{4.43} = -5.4$ MeV / $Q_{opt} = -5.5$ MeV.

Thus transitions via the 2^+ state of ^{12}C contribute strongly because of the good dynamical matching and the smaller binding energy in the excited state. For ^{62}Ni ($^{16}O, ^{15}N$) ^{63}Cu we have, $Q_o = -6.00$ MeV and for transfer via $^{16}O_3^+$ (6.13 MeV), $Q_{6.13} = +0.128$ MeV / $Q_{opt} = -4.2$ MeV.

In this two step reaction a dynamical mismatch is produced by the inelastic excitation, and the reaction is dominated by the single step.

Similar results for the importance of the Q-value of the intermediate step were obtained in calculations for two step transfer contributions to inelastic scattering of $^{18}O_0+ \rightarrow {}^{18}O_2^{*}+$ (11).

As a further example we show the result for the excitation of $1/2^+$ state in ^{13}C in the scattering of ^{13}C on ^{16}O (12). The ground state Q-values for the first step are very favourable ($^{12}C + {}^{17}O$, $Q_0 = -0.805$ MeV) and large cross sections for the single neutron transfer are observed. The second step leading back to the excited state of ^{13}C ($1/2^+$, 3.09 MeV) is slightly mismatched (Q-values -2.29 MeV and -1.42 MeV) starting with the ground state of ^{17}O or with the $S_{1/2}$ state respectively. The coupling scheme for the two step transfer via the $^{17}O + {}^{12}C$ channels is shown in fig.2. In a coupled reactions channel calculation we find that more than 70 % of the cross section observed is due to two step transfer. Higher order contributions were found to be negligible in this system and the transitions considered.

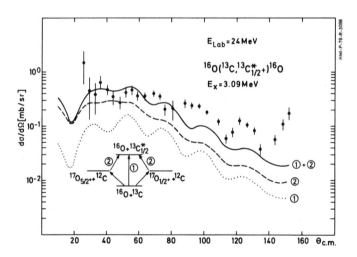

Fig. 2: Two step transfer transitions contributing to the inelastic excitation of $^{13}C_{1/2}+$(3.09 MeV). The full curve shows the full calculation including direct (1) and two step (2) transitions.

We can summarise this section with two conclusions: i) For transitions between ground states and low lying single particle states two step transfer can be observed if the single particle binding energies are 5 MeV or smaller and the spectroscopic factors $>$ 0.5.

ii) inelastic excitations can enhance transfer transitions apreciable by inducing better dynamical conditions and by reducing the effective binding energy.

Concerning the last point a further comment is necessary. The combination of an inelastic transition and subsequent transfer can also be viewed as a polarization of the ground state wave function (14,15). Polarization effects of this type are usually weak in the tails of the wave functions. As noted in ref. 15 this polarization effect can be expected to give effects for weakly bound states, which have strong transitions to continuum states. A complete static mixing calculation for the $f_{7/2}$ state in the $^{40}Ca(^{16}O, ^{15}N)^{41}Sc$ reaction, however, gives an overestimate of a factor 50 (ref. 15). In our calculations we find that the mixing becomes important only for cases where states of different parity are involved (sect. iV and ref. 18). A different approach, where polarizations of single particle states due to molecular orbit phenomena occur as suggested in ref. 14, are expected to be weak for ground state transitions, because multi step transfers between ground states are generally weak (see sect. III). The suggestions of ref. 14 are thus in contradiction with conclusions drawn here in sect. III. and IV.

III. A coupled reaction channel analysis of the $^{12}C + ^{13}C$ system

The $^{12}C + ^{13}C$ system was one of the first systems where large elastic transfer contributions were observed (16,17). Arguments for multiple transfer in the elastic channel were discussed extensively for this system (2) and it was found that with a strong dependence on the imaginary potential, contributions of second order could be ascertained. A complete understanding of the system has to incorporate the low lying single particle states. A reanalysis of the data (18,19) and further measurements of angular distributions of the excited states of ^{13}C (20) give a full view of the higher order effects. A thorough analysis of the $^{12}C + ^{13}C$ system in CRC analysis was therefore undertaken (21).

The scheme of the relevant states and transitions is shown in fig. 3. The excited states of ^{13}C have a large single particle strength and low binding energy (1.86 MeV and 1.1 MeV for the $s_{1/2}$ and $d_{5/2}$ states respectively). These circumstances make ^{13}C more favourable for multistep transfer than ^{17}O, where the binding energies of the $s_{1/2}$ and $d_{5/2}$ states are over 3.2 MeV (see fig. 4).

The coupled equations for the system of states in fig. 3 with a Hamiltonian which contains all relevant interactions have been solved with the neglection of recoil terms (21) in the dynamic variables. In the transition matrix elements the recoil terms partially drop out because they have been symmetrized. Nonorthogonality terms have been included in first order.

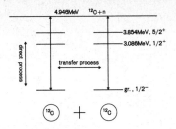

Fig. 3: Scheme of single particle states and transitions in the $^{12}C + ^{13}C$ system.

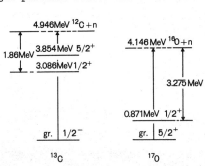

Fig. 4: Scheme of single particle states for the $^{13}C + ^{16}O \rightarrow ^{12}C + ^{17}O$ system.

Using the data for the ground state and the excited states (20) a parameter search was performed in order to fix the coefficients of fractional parantage CFP of the ^{13}C states ($(CFP)^2 = S$ - spectroscopic factor in the usual definition) and the potential parameters for the $^{12}C + ^{12}C$ core interaction. The potential parameters were in addition determined by a fit to the 90° excitation function of the ground state (22). The standard values used here are for $E_{CM} \geq 10$ MeV

$V = -(11 + 0.45\ E_{CM}, \text{MeV})$; $r_o = 1.35$ fm; $a_o = 0.2$ fm
$W = -(1.5 + 0.03\ E_{CM}, \text{MeV})$; $r_i = 1.35$ fm; $a_i = 0.3$ fm.

Below 10 MeV the same (constant) values as for $E_{CM} = 10$ MeV are used. Fig. 5 shows a comparison with data: at higher energies a stronger increase in absorption is needed and a better fit is obtained with another choice of, $W = (+1.2 - 0.3\ E_{CM})$ MeV. The imaginary potential is weak compared to earlier choices and reflects the inclusion of states in the coupled channel analysis. In the standard set of CFP (coefficients of fractional parantage) values, for the fit of the data at 7.8 MeV the value for $s_{1/2}$ state, $CFP_2 = 0.55$ had to be choosen rather small as compared to the expected values of $CFP_2 = 0.8 - 0.9$ (ref. 23). It is possible to use larger values of CFP_2, with stronger absorptive potentials. The latter bring the cross section down to the experimental values, however, the shape is not any more correctly described. From experience with calculations using channel coupling between 1 and 2 alone we found that the inclusion of a further channel, the $^{12}C_{2+}$ state, would

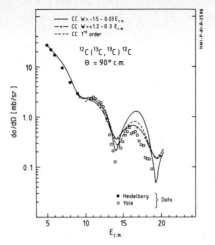

Fig. 5: Excitation function for the elastic scattering of ^{12}C on ^{13}C at $90°$; the curves are explained in the text.

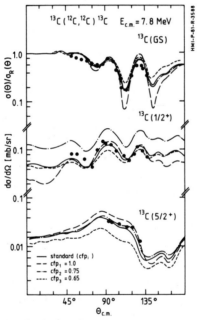

Fig. 6: Effects of the variations of CFP values of the single particle states of ^{13}C in the coupled channel calculation for angular distributions of the ground state and excited states of ^{13}C. The "standard" CFP values in this analysis are: $CFP_1 = 0.8$, $CPP_2 = 0.55$, $CFP_3 = 0.85$.

allow the use of CFP_2 values of 0.8 - 0.9. The "standard" CFP values for the present analysis are given in fig. 6, where the calculations are compared with the data for angular distributions.

An increase of the CFP_2 value to 0.75 increases the $s_{1/2}$ - cross section (at $90°$ CM) by a factor 2, (see fig. 6). This is a little more than the ratio $(0.75/0.55)^2 = 1.85$ which is expected in a first order calculation. A corresponding increase of CFP_1 from 0.8 to 1.0 $(1.0/0.8)^2 = 1.56$ gives for the $s_{1/2}$ state only an increase of 10 % of the cross section at $90°$ CM, similarly the cross sections for the $d_{5/2}$ state respond to changes of CFP_1 and CFP_3 in a non linear way, although less pronounced as the $s_{1/2}$ state.

The strength of the coupling interactions was tested in several steps. First for all 3 channels and the full interaction the coupled equations were solved by iteration of the nondiagonal parts of the direct and transfer interaction. Fig. 7 shows the comparison of the full coupled channel calculations with iterations of $1, 2^{nd}$ and 3rd order. We find that the iterations do not converge, the divergence is even faster at the higher energy $E_{CM} = 9.88$ MeV.

In fig. 8 we show, how the contributions of the transfer interaction and the direct excitation contribute (they are shown separately.). In this case at $E_{CM} = 9.88$ MeV the agreement with the data becomes worse. We assume that this is mainly due to the missing coupling to the $^{12}C_2+$ state, which would reduce the cross sections of channels 2 and 3 and probably also shift the structure in the angular distributions.

The question now arises, which transitions lead to the divergence of the iterations? We performed several calculations with one or two channels switched off and with interactions (direct or transfer) switched off. The result of these calculations can be summarized as follows.

a) channel 1 alone (elastic transfer)
   The difference in cross section between 1st and 2nd iteration is 10 %. This result is consistent with earlier findings that the elastic transfer process in $^{12}C + ^{13}C$ involves effects of second order.

b) channel 1 and 2 alone
   Convergence is acchieved after 3 iterations with $CFP_2 = 0.55$ at $E_{CM} = 7.8$ MeV (for $CFP_2 = 0.9$ convergence is accieved after 9 iterations).
   The differences in inelastic cross sections between 1st and 2nd iteration are 5 %, between 2nd and 3rd are 60 %. The difference between the 2nd and 3rd iteration reduces to 8 % if the direct interaction is switched off. This fact illustrates

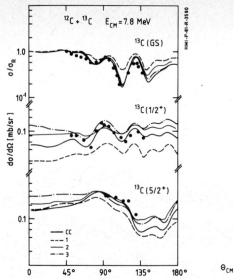

Fig. 7: Calculations as in Fig. 6 using "standard" parameters by iterating the coupled channels. The calculation starts to diverge at the 7th order.

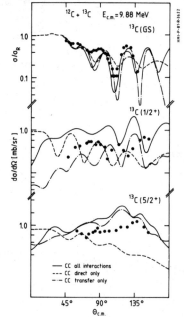

Fig. 8: Calculations using coupled channels for the full interaction and with direct and transfer interactions alone for the $^{12}C + ^{13}C$ reactions at E_{CM} = 9.88 MeV.

the importance of the mixing of $p_{1/2}$ and $s_{1/2}$ state - the s-p hybridization discussed in ref. 18.

c) channels 2 and 3

Calculations using either direct or transfer interaction show divergence in the iterations!

From the results cited in points a) - c), we conclude that the transitions between low lying states with higher order transfer contributions are not very likely to be found in heavy ion reactions. Second order contributions are observed in favourable cases. Very strong transitions are, however, observed between higher lying states and the iteration procedure diverges if such transitions are included. This picture reminds of the situation in deep inelastic collisions (24), where after a few transitions, which are of first order, the systems is locked via strong coupling into a process where all kinetic energy is dissipated into internal excitation. Fig. 9 summarizes this general point; for an illustration it is shown where in nucleus-nucleus collisions the strong coupling transitions can be expected. Eventually only distorted two center states exist for single particle states with positive binding energy.

Another important point, which emerges from the coupled reaction calculation is the spin polarization of the ^{13}C in its $p_{1/2}$ and $s_{1/2}$ states. In Fig. 10 we show the result at two energies; the fully coupled calculations as well as the calculations for 1st and 2nd order are shown. The remarkable fact which emerges from this result is that the ground state ($p_{1/2}$) shows spin polarization due to the coupling (the spin polarization vanishes for the ground state in first order without L-S force, whereas for the $s_{1/2}$ state, dynamical spin polarization is obtained due to the Q-value (25)). Actually a decompositon of the total diagonalised potential in terms of L and S shows a splitting which could be described by an L.S-term.

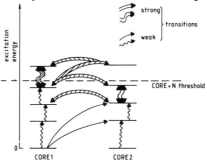

Fig. 9: A general coupling scheme for the interaction of two nuclei. Strong transitions (transfer and direct) occur between states close to or at the nucleon binding threshold. The effective Q-values must be small.

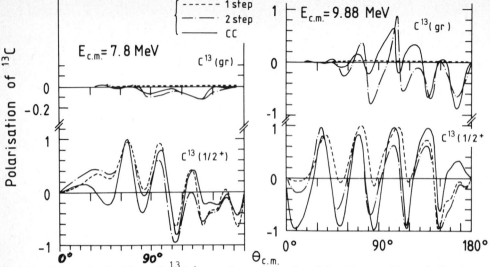

Fig. 10: Spin polarization of ^{13}C (gs and $s_{1/2}$) produced by the coupling of the inelastic and transfer channels.

IV. Intrinsic molecular orbits for the ^{12}C + ^{13}C system

After having convinced ourselves that transfer transitions with two steps and for higher excited states with <u>many steps</u> are a reality it seems worth while to turn to the adiabatic description (26) of the reaction process. It is hoped that insight into the physics of molecular phenomena can be gained in that way.

As an example the inclusion of inelastic direct excitation was found to produce sizeable effects, which can be understood in terms of hybridization (27,18). The case of ^{13}C is in that respect unique in nuclear physics that low lying states of the same spin but opposite parity can be mixed (the s-p mixing in molecular physics occurs at the same atomic configurations for Carbon). This mixing of sd-orbits with p-orbits introduces a pronounced asymmetric extension of the neutron orbit (fig. 11). Mixing effects of this type can play an important role in other systems for the coupling between excited states, where single particle levels from two major shells are in close vicinity.

In order to discuss molecular orbitals we choose a set of basis functions $\phi_{nljk}^{IM\pi}(\vec{r},\vec{R},\sigma)$ with total angular momentum I, its Z component M, parity π, and k the projection of I and j on the molecular axis given by the distance vector $\vec{r}$. These wave functions are obtained using the single particle wave functions $\varphi_{n\ell j}(\vec{R},\sigma)$

$$\phi_{n\ell jk}^{IM\pi}(\vec{r},\vec{R},\sigma) = \frac{\sqrt{2I+1}}{\sqrt{2(1+\delta_{k0})}\sqrt{4\pi}} \left\{ D_{Mk}^{I*}\varphi_{n\ell jk}(\vec{R},\sigma) + (-)^{\pi+I+\ell-j} D_{M-k}^{I*}\varphi_{n\ell j-k}(\vec{R},\sigma) \right\}$$

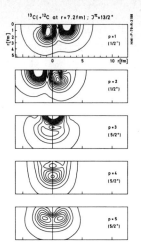

Fig. 11: Neutron orbits of ^{13}C polarized due to the presence of another ^{12}C core. The asymmetric distortions for the p = 1 ($p_{1/2}$) p = 2 ($s_{1/2}$) orbits are due to parity mixing (s-p hybridization).

If we want to look into the instrinsic molecular states, we choose a set of basis functions $\Phi_p^{IM\pi}$ which are obtained by diagonalizing the wave function $\phi_{n\ell jk}^{IM\pi}(\hat{r},\vec{R},\sigma)$ with all interactions in the total interaction except for the radial kinetic energy terms (adiabatic approximation).

$$\Phi_p^{IM\pi}(\hat{r},\vec{R},\sigma) = \sum_{n\ell j} \sum_k (1+P_{12}) \phi_{n\ell jk}^{IM\pi}(\hat{r},\vec{R},\sigma) B_{n\ell jk}^{I\pi}(r)$$

The mixing of the states is described by the amplitudes $B_{n\ell jk}^{I\pi}(r)$. In this diagonalisation the mixing due to the transfer and direct interactions as well as due to the Coriolis interaction can be seen. The new states will correspond to the instrinsic states of the rotating molecular state (RMO, see also ref. 19 and 28).

As discussed in ref. 19 the direct and transfer interactions induce the mixing between states of different parity and l but the same k-quantum number. The coriolis interaction mixes the states with different k-quantum numbers. In order to illustrate the properties of the system we show in figs. 12 and 13 density distributions (18,29)

$$\rho_p^{I\pi}(\vec{R},r) = \sum_\sigma \int d\hat{r} |\Phi_p^{IM\pi}(\hat{r},\vec{R},\sigma)|^2$$

of such molecular neutron orbits as function of distance between the two cores. Several remarkable features can be learned by inspection of these graphs:

1) In the <u>positive parity state</u> ($I^\pi = 9/2^+$) the p and s-state retain their identity up to about 8 fm.

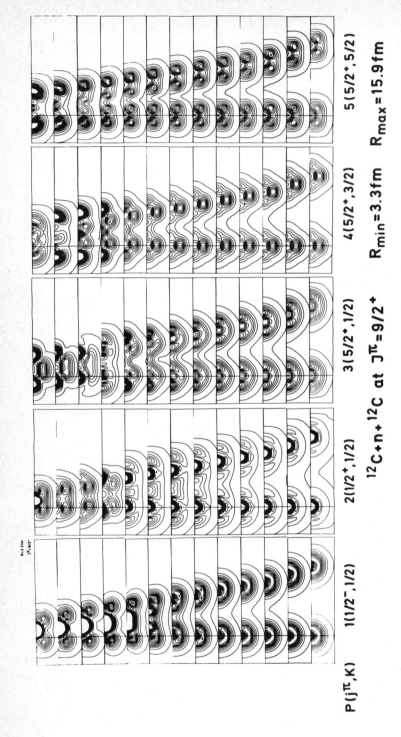

Fig. 12: Density distributions of the neutron in the instrinsic rotating molecular frame. At large distances the density distribution of the asymptotic states $p_{1/2}$, $s_{1/2}$ and $d_{5/2}$ (k = 1/2, 3/2, 5/2) is shown (in principle the neutron has probabilities at core 1 or core 2 which depend on the interaction; 0. and 1.0 in the incident channel). From these plots many physical phenomena can be deduced (see text), s - p hybridization, large overlap for positive parity states, pseudo-crossing of states p = 2 and p = 3 for r = 12.0 fm. The change of distances is in steps of 0.9 fm (with a few exceptions at larger distances).

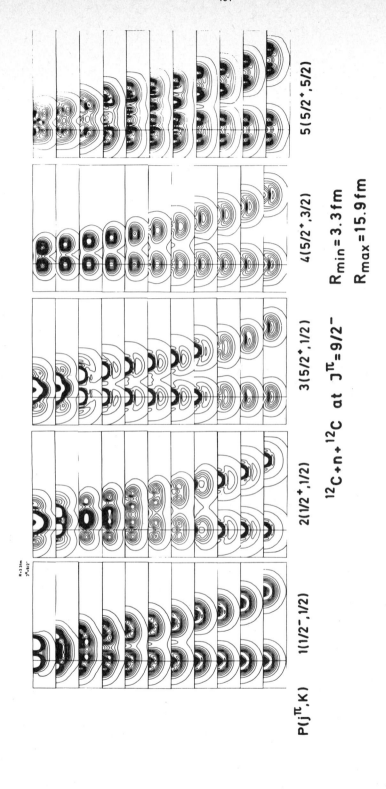

Fig. 13: as figure 12 for negative parity

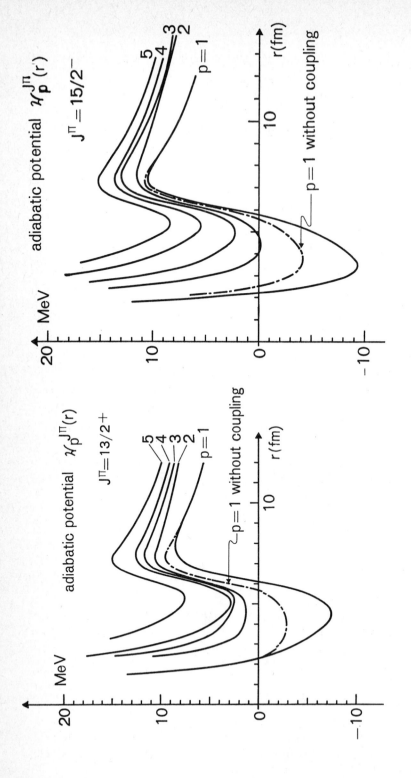

Figs. 14, 15: Adiabatic potentials of the molecular orbitals in the body fixed frame. Transitions between these states are obtained by radial coupling. Note the crossing of states p = 2 and 3 in $J^\pi = 15/2^-$ at ca. r = 12 fm.

2) The overlap between the $s_{1/2}$ states is seen to occur already at distances of 15.9 fm$_o$ for positive parity, for negative π it occurs much later.
3) In the 8th figure from the bottom ($I^\pi = 9/2^+$) and the following figures we observe the effect of p-sd hybridization by a strong shift of the densities to the center.
4) In the negative parity states ($9/2^-$) the overlap between states is seen only at small distances because of the node at the center.
5) States p = 2 and p = 3 interchange their role at a distance of approximately r = 12 fm. Such an effect is understood in terms of a level pseudo-crossing which is well seen in the adiabatic potential curves for the states p = 2 and p = 3, (fig. 13).

The adiabatic potential curves $V_p^{I\pi}(r)$ are obtained by diagonalization of the original total interaction in the radial coupled equation $V_{tot}^{I\pi}(r)$

$$V_p^{I\pi}(r) = \left\{ (B^{I\pi}(r))^{-1} V_{tot}^{I\pi}(r) B^{I\pi}(r) \right\}_{pp}$$

In the adiabatic approximation transitions between states are now induced by the radial coupling terms $\frac{d}{dr}$ and $\frac{d^2}{dr^2}$ in the new basis. Strong coupling occurs in this representation if the adiabatic potential energies come close or have pseudo-crossings. These adiabatic potentials are shown in figs. 14 and 15. We notice that strong coupling is expected between channels p = 2,3 and 4. This result confirms the findings in the coupled channel calculation that strong coupling occurs between $s_{1/2}$ and $d_{5/2}$ states leading to divergence if the solution is calculated using iterations.

As a conclusion, the coupled channel calculations and the adiabatic approach, support the concept of formation of molecular orbitals of nucleons. This aspect becomes dominant for states at the nucleon binding threshold and evokes processes which finally merge into deep inelastic reactions, where nucleons are exchanged freely between the two cores.

References

1) G. Breit, Proceed. of third Conf. on Reactions Between Complex Nuclei, Asilomar (1963) p. 97. and references cited there.
2) W. von Oertzen and H. G. Bohlen, Phys. Rep. 19C (1975) 1 and references therein.
3) F. Becker, S. Joffily, C. Becchria and G. Baron, Nucl. Phys. A221 (1974) 475
4) J. Y. Park, W. Scheid and W. Greiner, Phys. Rev. C6, 1565 (1972), Phys. Rev. C20, 188 (1979)
5) W. R. Coker, T. Udagawa, H. H. Wolter, Phys. Rev. C7 (1973) 1154
6) K. S. Low, Proceed. European Conf. on nuclear Physics with Heavy Ions Caen, Jour de. Phys. C5 (1976) 15 and Int. Conf. on Nuclear Structure, Tokyo; J. Phys. Soc. Japan 44 (1978)

7) N. Glendenning, Proceed. Int. Conf. on reactions between complex nuclei, Nashville, 1974. Vol. 2, p. 137 (North Holland Publ. Co.)
8) T. Kammuri, Nucl. Phys. A 259 (1976) 343
9) D. H. Feng, T. Udagawa and T. Tamura, Nucl. Phys. A274 (1976) 262
10) G. Baur and H. Wolter, Phys. Lett. 51B (1974) 205
    B. Imanishi, H. Ohmishi, O. Tanimura, Phys. Lett 57B (1975) 309
11) Q. K. K. Liu, W. von Oertzen, H. Wolter, Proc. Conf. on Clustering Aspects of Nuclear Structure and Reactions, Manitoba (1978) F27
12) W. Bohne, K. Grabisch, I. Hergesell, Q. K. K. Liu, H. Morgenstern, W. von Oertzen, W. Galster, W. Treu and H. Wolter, Nucl. Phys. A332 (1979) 501
13) J. C. Peng, M. C. Mermaz, A. Greiner, N. Lisbona and K. S. Low, Phys. Rev. C15 (1977) 1331
14) E. A. Seglie, J. F. Petersen, R. J. Ascuitto, Phys. Rev. Lett. 42 (1979) 956
15) G. Delic, K. Pruess, L. A. Charlton and N. Glendenning, Phys. Lett. 69B (1977) 20, see also Int. Conf. on Nuclear Structure, Tokyo; J. Phys., Soc. Japan 44 (1978) 272
16) A. Gobbi, U. Matter, J. L. Perrenoud, P. Marnier Nucl. Phys. A112 (1968) 537
17) H. G. Bohlen, W. von Oertzen, Phys. Lett. 37B (1971) 451
18) B. Imanishi, W. von Oertzen, Phys. Lett. 87B (1979) 188
19) W. von Oertzen, B. Imanishi, H. G. Bohlen, W. Treu, H. Voit, Phys. Letters 93B (1980) 21
20) H. Voit, H. Fröhlich et al to be published
21) B. Imanishi, W. von Oertzen to be published
22) L. T. Chua, Thesis Yale University; U. Weiss, D. Fick, K. D. Hildenbrand, W. Weis, G. R. Plattner and I. Sick, Nucl. Phys. A274 (1976) 253
23) S. Cohen and D. Kurath Nucl. Phys. A101 (1967) 1; L. Grunbaum and M. Tomaselli Nucl. Phys. A160 (1971) 437
24) A. Gobbi and W. Nörenberg in Heavy Ion collisions Vol.2 p. 127 (ed. R. Bock, North Holland 1980)
25) P. Wust, W. von Oertzen, H. Ossenbrink, H. Lettau, H. G. Bohlen, W. Saathoff and C. A. Wiedner Z. f. Physik A291 (1979) 151
26) J. S. Briggs, Report on Progress in Phys. 39 (1976) 217
27) Linus Pauling. The nature of chemical Bond (Cornell U. P., 1960)
28) B. Imanishi and W. von Oertzen, contribution to this conference.
29) B. Imanishi, Genshikaku Kenkyu 23 (1979) 133, in Japanese.

VALIDITY OF THE ADIABATIC MOLECULAR ORBITAL CONCEPT
IN THE INTERACTION OF HEAVY IONS

B. Imanishi[*,**] and W. von Oertzen[**]

[*] Institute for Nuclear Study, University of Tokyo, Tokyo 188, Japan
[**] Hahn-Meitner-Institut für Kernforschung, Berlin, Germany

Molecular orbital formation of nucleons in the interaction of heavy ions is of interest because it defines the intrinsic states constructed from the system of colliding particles and also it furnishes us with new aspects on heavy ion reaction mechanism[1]. Up to now, however, there is little evidence for the formation of molecular orbitals. Here, we consider, in connection with CRC theory, in what situation the molecular orbitals are formed and how the concept of molecular orbital is useful in understanding the scattering, $^{13}C(^{12}C,^{12}C)^{13}C^*$(gr.,1/2$^-$; 3.086MeV,1/2$^+$; 3.854MeV; 5/2$^+$) (ref. 2,3)
We assume that "rotating" molecular orbitals (RMO) are formed with the basis functions defined in refs. 1-3. In the RMO representation transitions between the states of Φ_p and Φ_q are induced by the coupling interaction containing the radial kinetic energy operator T(r) (radial coupling) $(A^{-1}[T,A])_{pq} = \Delta\mathcal{V}^{(1)}_{pq}(r)d/dr + \Delta\mathcal{V}^{(2)}_{pq}(r)$.

If the radial coupling is a small perturbation, Φ_p is a good basis function describing the scattering process (adiabatic approximation). Fig. 1 shows the S-matrix elements between $\alpha=1$ and 2 and $\alpha=1$ and 3 ($\alpha=1,2,3 \Rightarrow (n\ell jK) = (1p1/2,K=1/2), (2s1/2,K=1/2), (1d5/2,K=1/2)$; calculations of CRC (CC), 1st-order perturbation in the CRC representation (PWR 1-step) and 1st-order perturbation in the RMO representation (ADB 1-step) are shown. At low incident energies the ADB 1-step calculation is a much better approximation of the exact calculation (CC) than the PWR 1-step calculation. In addition, elastic scattering was well described at low incident energies only with the adiabatic potential. In fig. 2 it is shown that the coupling interactions $\Delta\mathcal{V}^{(1)}_{pq=1}$ (i = 1 and 2) for (p,q) = (2,1) and (3,1) reach peak values at two points of r, $\sim$7.5 and $\sim$2 fm. Calculations show that the outside peak at $r\sim$7.5 fm is responsible for the transitions at low incident energies due to direct and transfer interaction. The sharp peak inside is due to radial coupling and it becomes rapidly appreciable with the increase of the energy and destroys the adiabaticity of the scattering process. Then, the ADB 1-step approximation becomes bad.

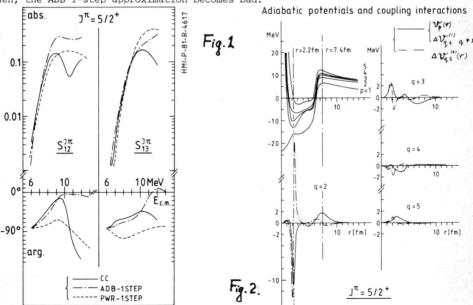

Fig. 1

Fig. 2.

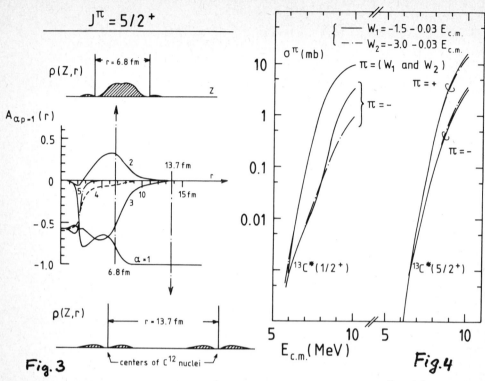

Fig. 3

Fig. 4

In fig. 3 each α-component of a RMO wave function with p = 1 and $J^{\pi}=5/2^+$ is drawn as a function of the relative distance r. At distances from 9fm to 2fm a state with the components of α=1, 2 and 3, each of which belongs to the same K-quantum number of 1/2, is formed and each α-component changes rather smoothly with the change of the distance, while in a narrow region around a distance of $\sim$2fm the components change rapidly with additional mixing of K-quantum numbers. In the case of the minus parity of the total system the pure state (1p1/2, K = 1/2) retains its identity up to small distance of $\sim$4.5fm, around this distances strong K-quantum number mixing occurs. Generaly, such strong mixing of K-quantum numbers occurs with big changes of α-components of the RMO state and induces strong diabatic radial coupling interactions between different RMO states.

For <u>plus parity</u> states we find that for distances larger 2fm pure K = 1/2 rotating molecular orbitals can be defined. The rotational motion is thus separated from the intrinsic motion. The corresponding density distributions $\rho_p^{I\pi}(x,z,r) = \Sigma \int dr |\Phi_p^{IM\pi}|^2$ are also shown in fig. 3. A strong concentration of the neutron is observed suggesting a ^{12}C-n-^{12}C configuration.

The fact that in the <u>minus parity</u> states the transition between p = 1 and p = 2 is induced mainly at distances smaller 4.5fm suggests that these transitions are in competition with the absorption process due to the imaginary potential. This has been tested by increasing the imaginary part of ^{12}C-^{12}C potential from W =-1.5 - 0.03 E_{CM} to W = -3.0 -0.03 E_{CM}. As is shown in fig. 4 the minus parity cross sections are strongly affected whereas the plus parity cross sections are not (the latter occuring at r >6fm).

1) W. von Oertzen and B. Imanishi, invited paper in this conference.
2) B. Imanishi, W. von Oertzen, Phys. Lett. <u>87B</u>(1979)188.
3) W. von Oertzen, B. Imanishi, H. G. Bohlen, W. Treu, H. Voit, Phys. Lett. <u>93B</u> (1980)21.

Resonances in $^{16}O+^{16}O$ and the Systematic Occurrence of $J^\pi=8^+$ Resonances in Heavy Ion Resonant Systems*

M. Gai, E.C. Schloemer, J.E. Freedman, A.C. Hayes, S.K. Korotky[†],
J.M. Manoyan, B. Shivakumar, S.M. Sterbenz, H. Voit[††], S.J. Willett,
and D.A. Bromley

Wright Nuclear Structure Laboratory, Yale University, New Haven, Ct. 06511, U.S.A.

We have studied[1] the resonance-like structure of the $^{16}O+^{16}O$ system at $E_{cm} \approx 16$ MeV. Excitation functions for the elastic channel and α_0 and α_1 groups all show narrow structures at that energy. The angle integrated ($17° \leq \theta_{cm} \leq 90°$) yield of the α_0 group, measured in steps of 50 keV (cm), showed a weak structure at $E_{cm} \approx 15.8$ MeV and pronounced structures at $E_{cm} = 15.9$ and 16.1 MeV. Fifteen angular distributions were measured in the range $15.50 \leq E_{cm} \leq 16.4$ MeV.

The $^{16}O(^{16}O, \alpha_0)^{28}Si$ angular distributions were analyzed via a partial wave decomposition[1] and the extracted S matrix elements are shown in Fig. 1. The decomposition was restricted as follows: (1) We require a good fit ($X^2 \leq 3$) for each angular distribution. (2) The extracted S_ℓ are required to reproduce the total cross section. (3) Since these angular distributions arise from only a few ℓ-waves within a sharp window around the $\ell=10$ grazing partial wave[1] ($8 \leq \ell \leq 12$), we were able to choose measuring angles (θ_0) at which the observed cross section reflects a single partial wave.[1] The extracted S_ℓ are required to reproduce the cross section measured at θ_0. These three requirements indeed produce unambiguous extraction of the S matrices, as shown in Fig. 1.

We find a weak $J^\pi = 10^+$ resonance at $E_{cm} \cong 15.8$ MeV and $J^\pi = 8^+$ resonances at $E_{cm}=15.9$ and 16.1 MeV. It is also clear that in the $\ell=10$ grazing partial wave we obtain a background (non-resonant) gross energy dependence. That gross structure appears to be related to the narrow ℓ-window around the $\ell=10$ grazing partial wave.[1] That background amplitude dominates the cross section measured at $\theta_{cm}=90°$.[1]

The $J^\pi=8^+$ resonances located here appear to be related to $J^\pi=8^+$ resonances found in many other heavy ion systems which are currently under study at Yale. These systematics are shown in Fig. 2. The prominent structures in most cases have been shown to be resonances but they do not appear in inelastic scattering channels, even when these channels are well matched. In all cases except $^{12}C+^{12}C$ the $\ell=8$ is very different from the grazing partial wave. The 8^+ resonances appear at $E_{cm} = 4 \times 2.8 + N \times 2.4$ MeV, where N is an integer, and they appear to be different from the "diatomic" barrier resonances of $^{12}C+^{12}C$. The new resonances appear to be of "polyatomic" character[2], namely, they correspond to excitation of substructures within the interacting nuclei. The available evidence suggests that alpha particle clusters play a dominant role.

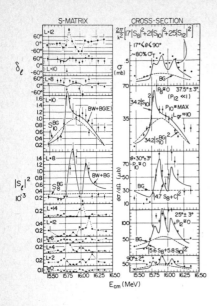 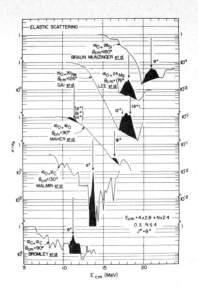

Fig. 1. Extracted S matrix elements and measured cross section for $^{16}O(^{16}O, \alpha_0)^{28}Si$. At $\theta_{cm} \cong 37.5°$ the cross section is entirely given by the $\ell=10$ grazing partial wave that shows a narrow resonance superimposed on an energy dependent background.

Fig. 2. Systematic occurrence of $J^\pi = 8^+$ resonances in heavy ion resonant systems

---

*Supported in part by USDOE Contract No. DE-ACO2-76ERO3074.
†Current address: Bell Laboratories, Holmdel, New Jersey 07733.
††Permanent address: Physikalisches Institut der Universität Erlangen-Nurnberg, West Germany.

1. M. Gai, E. C. Schloemer, J. E. Freedman, A. C. Hayes, S. K. Korotky, J. M. Manoyan, B. Shivakumar, S. M. Sterbenz, H. Voit, S. J. Willett, and D. A. Bromley, Yale Preprint 3074-676.
2. M. Gai, Bull. Amer. Phys. Soc. 24 (1979) 843; ibid., 25 (1980) 592; and Proc. Int. Conf. Nucl. Phys. Berkeley, 1980, LBL-11118, Univ. of California, pp. 424-25; and M. Gai and D. A. Bromley, to be published.

Schematic models of resonances: predictions and comparison

N. Cindro and D. Počanić

Rudjer Bošković Institute, 41001 Zagreb, Croatia, Yugoslavia

Four schematic models of resonances have been proposed recently[1-4]. The aim of these models is to explain the bulk properties of resonances on the basis of certain simple assumptions about their origin and nature. Explicit or implicit in all the four models is the quasimolecular concept developed twenty years ago by Bromley et al.[5], where resonances are seen as originating from long-lived configurations consisting of two orbiting nuclei.

The orbiting-cluster model[1] (OCM) combines the quasimolecular picture with the resonance-window concept of Greiner and Scheid[6]. The observability of resonances, connected with their width, is related to the spreading of discrete resonant states into the complex compound-nuclear states. Resonance observation is possible if the width Γ is smaller than the resonance spacing. In the model, Γ is parametrized essentially by a single parameter, the level density, and resonances are expected in systems which display a zone of relatively low compound-nucleus level density in the E_x-J plane.

The model of level densities at grazing angular momenta[2] (MLDGAM) is based on similar physical assumptions. Here the resonances are seen as generated by isolated high spin states formed by heavy-ion entrance channels which populate regions of low level densities in the compound system. Consequently, the main criteria for resonance observation in the MLDGAM are the same as in the OCM. However, significant differences are found in calculating the level densities; also, the MLDGAM neglects considering the other ingredient of $\Gamma^\downarrow$, the matrix element $<cn|V|el>$ which the OCM takes into account in an approximate way. Thus, the predictions of the two models differ considerably.

The model of weak absorption into direct channels[3] (MWADC) is based on the assumption that the resonant behaviour of heavy-ion systems arises from the weak absorption due to the small number of open direct-reaction channels. Hence the occurrence of resonances would be related to a surface transparent interaction rather than to the formation of a compound nucleus in a region of low level densities. This is a major difference from the two previously described models.

The effective-barrier model[4] (EBM) is based on physical grounds similar to the MWADC: the occurrence of resonances is associated with the weakening of absorption for the near-grazing partial waves, i.e. again surface transparency. The origin of this transparency is in the fact that conservation of angular momentum and parity reduces the number of channels available to carry out the grazing angular momenta in some systems. Unlike the MWADC, the EBM does not calculate the number of open channels explicitly, relying, instead, on a simple barrier rule; nevertheless, the predictions of the two systems are very similar.

The comparison of predictions of the four schematic models[1-4] for lighter systems is given in the table.

Table

Composite system	Entrance channel	Predictions[a]				Experimental evidence of resonances[e]
		OCM	MLDGAM[b]	MWADC[c]	EBM[d]	
^{21}Ne	^{9}Be+^{12}C	-	YES	-	-	indications
^{24}Mg	^{12}C+^{12}C	YES	YES	YES	(YES)	observed
	^{10}B+^{14}N	NO	NO	NO	-	not observed
^{25}Mg	^{12}C+^{13}C	NO	YES	YES	-	controversial
^{26}Mg	^{12}C+^{14}C	NO	-	YES	(YES)	indications
	^{13}C+^{13}C	NO	-	YES	-	not observed
^{28}Mg	^{14}C+^{14}C	NO	-	YES	(YES)	indications
^{26}Al	^{12}C+^{14}N	NO	YES	-	-	indications
^{28}Si	^{12}C+^{16}O	YES	YES	YES	(YES)	observed
^{30}Si	^{12}C+^{18}O	YES	-	NO	-	observed
	^{13}C+^{17}O	NO	-	YES	-	not observed
	^{14}C+^{16}O	YES	-	YES	(YES)	observed
^{32}S	^{16}O+^{16}O	YES	-	YES	(YES)	indications
	^{12}C+^{20}Ne	YES	-	NO	(NO)	observed
^{36}Ar	^{12}C+^{24}Mg	YES	-	-	YES	observed
^{40}Ca	^{12}C+^{28}Si	YES	-	-	YES	observed
	^{16}O+^{24}Mg	YES	-	-	YES	observed
^{44}Ti	^{16}O+^{28}Si	YES	-	-	YES	observed
^{56}Ni	^{28}Si+^{28}Si	YES	-	-	YES	observed

[a] Systems for which a given model predicts resonance observation are labelled with "YES". "NO" - resonance observation not expected.

[b] The MLDGAM was applied only to the five composite systems shown. It is evident, however, from the formulation of the model that it is inadequate to predict resonances in heavy systems.

[c] The MWADC calculations were performed only for C+C, C+O and O+O systems.

[d] The EBM was applied only to α-like systems with $A \geq 36$ [4]. Predictions for lighter systems, listed in parenthesis in the table, were obtained by applying the effective barrier rule as described in Ref. 4.

[e] For a detailed list of resonance data up to Jan 1st 1981, see Ref. 7.

References

1. N. Cindro and D. Počanić, J. Phys. G: Nucl. Phys. **6** (1980) 359
2. S.T. Thornton, L.C. Dennis and K.R. Cordell, Phys. Lett. **91B** (1980) 196
3. F. Haas and Y. Abe, CRB/PN Report No 80-22, CRN Strasbourg 1980
4. D. Baye, Phys. Lett. **97B** (1980) 17
5. D.A. Bromley, J.A. Kuehner and E. Almquist, Phys. Rev. Lett. **4** (1960) 365 and Phys. Rev. **123** (1961) 878
6. W. Greiner and W. Scheid, J. Phys. **C6** (1970) 91
7. N. Cindro, La Rivista del N. Cimento, in press

ON THE STRUCTURAL SIMILARITY OF NUCLEAR MOLECULES

Niels Marquardt

(Institut f. Exp.-Physik III, Ruhr-Universität Bochum, Germany)

Several experimental facts will be discussed which indicate that there exists a common configuration inherent to all quasimolecular resonances so far observed in heavy-ion reactions. A comparison of the moments of inertia of the nuclear rotational bands seen as series of resonances in different systems of composite mass $A \leq 44$ (Fig. 1) and of partial and total widths of the corresponding intermediate structures reveals evidence for a structural similarity of these resonance phenomena. With very few exceptions, resonances have been observed only in reactions where ^{16}O is one of the interacting particles or where the resonant ^{16}O reduced partial width is large.

This concept is further supported by Fig. 2, where the moments of inertia given in Fig. 1 are plotted versus the corresponding compound-nucleus mass number A (besides some other structural information of 4 N nuclei). The striking feature is that the moments of inertia of all resonances fall on a straight line which crosses the A axis at $A = 16$. This linear relation, represented by the thick line of Fig. 2, and the disappearing $2\Theta/\hbar^2$ values for $A \leq 16$ can hardly be reproduced by the usual assumption of dumb-bell configurations. Better agreement is obtained by placing A-16 valence nucleons in orbitals in the equatorial plane around a nearly spherical ^{16}O core. Such a torus-like configuration of oblate

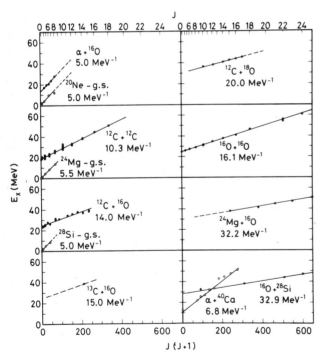

Fig. 1: Survey of known rotational bands in different sd-shell nuclei; moments of inertia $2\Theta/\hbar^2$ are given.

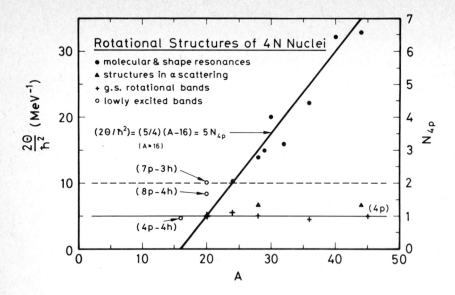

Fig. 2: Moments of inertia of rotational bands as a function of mass number A of compound nucleus.

deformation allows for the largest angular momenta. The rather "cold" ^{16}O core acts as a spectator which itself does not contribute appreciably to the total moment of inertia. The valence nucleons surrounding the core might be in a more or less equilibrated compound state or in the form of α clusters. A classical estimate leads to a reasonable radius parameter of r_o = 1.2 fm describing the distance between the center of the core and the torus.

An extrapolation of the rotational bands exhibited in Fig. 1 towards the crossing with the E_x axis gives the energies of the band heads with J = 0. The variation of these band-head energies with the mass number A of the compound nucleus follows closely the variation with A of the ^{16}O binding energies of sd-shell nuclei and further supports the proposed model.

In summarizing, it is suggested that the resonant configurations in different sd-shell nuclei are similar and consist basically of an inert ^{16}O core surrounded by nucleons or nucleon clusters.

IX. SUMMARY OF THE CONFERENCE

SUMMARY OF THE CONFERENCE

Walter Greiner
Institut für Theoretische Physik
Johann Wolfgang Goethe-Universität
Frankfurt am Main, Germany

I have been looking forward to this meeting for some time. Many old friends with common interests are gathering at Bad Honnef. When I drove along the autobahn on Sunday evening, joined by my friend D.A. *Bromley*, the air was clear, the sun setting and a colorful sky presented itself together with the green picturesque countryside so rich and impressive, that I was transformed into the right mood: full of expectations. These expectations have been fulfilled to a considerable extent as I shall now outline.

The conference started with the excellent opening address of D.A. *Bromley*. An opening talk is a summary at the beginning. Hence my speech will be the second summary you have to listen to. Let me also say from the beginning that I am not willing to discuss and mention every detail I learned at this meeting. There were many interesting points raised and data presented which I have no time to go into again. Instead I would like to take up a few key questions and key informations which were discussed here, and incorporate them into my view of the status of our field: The search for and study of nuclear quasimolecules.

1. What is a nuclear quasimolecule?

A molecule, as in atomic physics, consists of two (or more) centers which are bound together by valence particles, i.e. electrons in the atomic case and nucleons in the nuclear case. An atomic (or electronic) quasimolecule is one of those short-living intermediate systems formed in heavy ion scattering, in which electrons feel - during the collision - the binding to both atomic nuclei. Similarly a nuclear (or nucleonic) quasimolecule is produced in close collisions of two nuclei and here the outer nucleons are feeling both force centers during the collisions, orbiting more or less around both and binding them together for times longer than the collision time. Of course, depending on the ratio of this molecular time and the collision time, $\frac{\tau_{mol}}{\tau_{coll}}$, a whole variety of more or less pronounced intermediate phenomena do occur. And this is what we do observe: Molecules of virtual and quasibound type, longer living intermediate (nearly compound) systems (Fig. 1). Even the shape (fission) isomers might be considered to be nuclear molecules in this respect. I believe, it is essential for such systems that the potential between the two centers has a certain "pocket", i.e. one or more minima, so that the two nuclei are captured, delayed, resonating during their relative motion. The complete analogy to the atomic case should be stressed. There the potential minimum between two atoms comes due to certain two-center-orbitals with lower energy, which expresses the

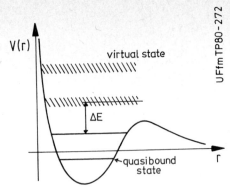

Fig. 1. Virtual and quasibound states in a typical nucleus-nucleus potential. ΔE is the energy transferred to the intrinsic structure of one or both nuclei.

binding effect due to the valence electrons. I remind of the works of *London*, *Heitler*, *Teller*, *Hylleraas* and others. In nuclear physics the two-center shell structure of the valence nucleons causes the additional binding, as has been described to us by W. *Scheid*, U. *Mosel*, *Imanishi* and von *Oertzen* . The nuclear molecular systems are, however, much more complex than the atomic ones: First, because of the strong interaction many more nucleons participate in the molecular interaction, generally depending on the overlap and fragmentation of the two subsystems. This leads quickly to a wash-out of the two-center-structure and thus of the molecular structure. A kind of self-consistent rearrangement (modification) of the individual centers themselves takes place. Secondly there are many more closed and open channels - in particular there are also collective degrees of freedom - which are impossible to treat and to consider altogether in any actual problem. At higher energies compression effects are becoming more important than at lower energies. Therefore a number of approximations have to be introduced:

2. Limited number of channels treated - Molecular channels

Substructure is introduced through the double-resonance mechanism (Fig. 1) and coupling to other dominant collective states and cluster states. Experimental evidence for that came through the reports and works of *Bromley*, *Erb*, *Cormier*, *Cossmann*, *Voit*, *Schiffer*, *Braun-Munzinger*, *Paul*, *Freeman*, *Marquardt* and others, who saw new substructure in elastic and inelastic excitation functions of various systems. A number of eleastic scattering cross sections have nicely been summarized by *Konnerth*, *Eberhard* et al. (Fig. 2). Typical inelastic excitation functions for the ^{16}O-^{16}O system have been demonstrated by *Peter Paul* (Fig. 3). He correctly mentioned that all the various excitation functions have to be considered and compared together with a particular theoretical nuclear model. This has been my stand-point for many years. I just remind you on the first calculations of this kind by *Fink*, *Scheid* and myself by showing here the most recent calculations on the ^{12}C-^{12}C-system

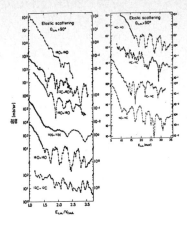

Fig. 2. Elastic scattering cross sections for various surface transparent systems (data assembled by Konnerth, Eberhard et al.).

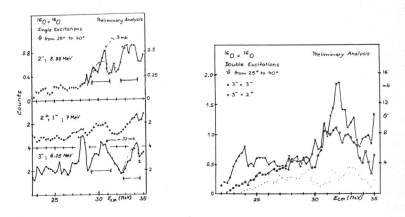

Fig. 3. Various inelastic excitation functions for the ^{16}O-^{16}O-system as presented by P. Paul.

by *Koennecke, Guirguis and Scheid* (Fig. 4). The novelty of these investigations is that the coupling between the relative motion of the ions and the collective coordinates $\alpha_{2\mu}$ has been taken into account up to second order in $\alpha_{2\mu}$. Also the importance of the 4^+-state of ^{12}C at 14.08MeV has been demonstrated. In general, a good qualitative and semi-quantitative agreement of the theory with experiments can be noticed. There is no need that the calculated structure agrees in every detail with the experiments. The general over-all semi-quantitative agreement is satisfying to me. It demonstrates that what we observe, namely the structure,

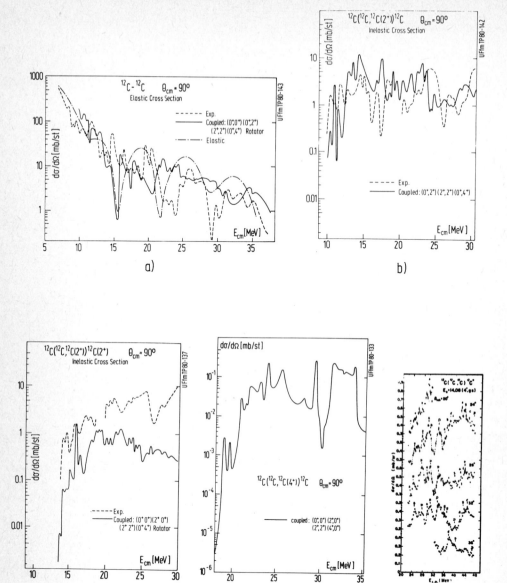

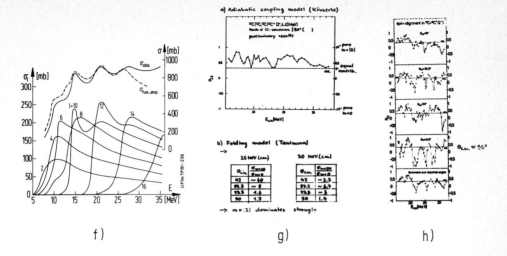

f) g) h)

Fig. 4. Elastic and inelastic excitation functions for the $^{12}C-^{12}C$ system according to a coupled channel calculation by Koennecke, Guirguis and Scheid (Journal of Physics G 1981). This is the improved model of the early work of Fink and Scheid (Nucl. Phys. A188(1972)259) by including linear and quadratic coupling terms in the collective variable α. As figs. 4a) to 4d) show, the agreement between theory and experiment is overall quite good. Particularly also the fusion cross section 4f) looks nice. Fig. 4d) shows the inelastic cross section to the 4^+-state in ^{12}C at 14.08 MeV. Data for that excitation have been presented by the Notre-Dame-group (Kolata et al.) and are displayed in Fig. 4e). They are, however for the more forward angles and not as the theory (Fig. 4d) for 90°. Figs. 4g and 4h show the theoretical calculation of the alignment and the experiments of the Munich-group respectively.

resonances and fluctuations alike are due to the coupling of molecular channels. Hence the molecular picture is proven. I draw also attention to the alignment calculation (4g) and the extensive measurements of alignment by the Munich group (4h). Two points ought to be made, namely:1) Perhaps the inelastic 2^+-cross section should be divided out of the alignment measurement, as *Bromley* did it. This will probably result in a more smooth excitation function for it. 2) Nevertheless the very sharp structures seen in the alignment can only be reproduced by the calculations, if the imaginary potential in the inelastic molecular channels is practically zero. Up to now elastic and inelastic channels were treated with the same imaginary potential. This need not be so. Hence, it seems, that we might learn from these measurements how the imaginary potential in the inelastic channels behaves. Also Fig. 4c suggests that the imaginary potential should be smaller (if not zero) in the 2^+-channel. Let me make a comment on *Paul's* and *Mosel's* so-called "mismatched" channels. This is a new name for the long-known fact, that the interaction can shift the unperturbed position of resonances. As is - by the way - the band-crossing model of *Kondo* and of others of the "Japanese family", which is nothing else than another formulation of the double-resonance mechanism explicitly assuming rotational structure for the molecular states. The latter has its advantages of being rather schematic and therefore easier to handle. Cross- and intermediate structure of the $^{12}C-^{12}C$-system is also observed in the classical radiative capture reaction $^{12}C(^{12}C, \gamma_{23})^{24}Mg$ which has been reported in a fine experiment by *Sandorfi* (Fig. 5).

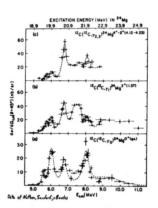

Fig. 5. $^{12}C+^{12}C\rightarrow^{24}Mg+\gamma$ radiative capture measurements by Sandorfi. The following conclusions can be drawn from these results:
(1) Previously observed quasimolecular $^{12}C+^{12}C$ resonances appear in radiative capture - but only at a relatively low level consistent with statistical γ-ray emission.
(2) A new class of resonances appears in $^{12}C(^{12}C, \gamma)$ - and for these both Γ_c and Γ_γ are much larger than statistical predictions.

Previously observed quasimolecular $^{12}C-^{12}C$ resonances do appear in this reaction, but only at relatively low level consistent with statistical γ-ray emission. Calculations of this process do not yet exist, but are called for! The virtual resonances are the molecular doorway-states in the sense as *Ingrid Roetter* discussed it so nicely at the meeting. The coupling to excited nuclear states introduces intermediate structure. The coupling to even more involved structure or simply the coupling of very many molecular channels can lead to fluctuations, which is visually obvious from Fig. 4. The intermediate structure via the double-resonance-coupling for the $^{28}Si-^{28}Si$ system was discussed by *Langanke*. A fluctuation on top of an intermediate structure has been carefully investigated for the 14^+-state at 19.8 MeV in the $^{16}O+^{12}C$-reaction by *Braun-Munzinger* (Fig. 6). *Voit* continued his precise work on total reaction cross section measurements close to the Coulomb-barrier. I remind you with Fig. 7 once more to his results for the $^{16}O+^{12}C$-system. The large number of resonances he observed stimulates the interest of a theoretician to construct a proper model for their explanation. Obviously a number of molecular

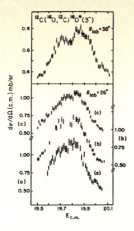

Fig. 6. Braun-Munzinger's high resolution measurement of the 14^+-structure at 19.8 MeV in the $^{16}O+^{12}C$-reaction reveals a substructure, which is likely a molecular fluctuation.

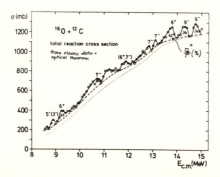

Fig. 7. The total cross section measurements for the $^{16}O+^{12}C$-reaction by Voit et al. reveal a number of resonances in the vicinity of the Coulomb-barrier.

channels do couple also in this case and generate the many structures. The coupling of many molecular channels has also nicely been demonstrated by *M. Freeman*, who reported a number of transfer-excitation functions for the $^{14}C-^{14}C$-system, which I display again in Fig. 8. It should be mentioned that the large structure introduced by coupling various molecular channels (see Figs. 4) is certainly coming out of a direct reaction model and, nevertheless, shows fluctuation-properties - as has been shown many years ago by *Scheid and Jansen* -, but can nearly be seen with the bare eye. I make this point again to comment the long discussions here on what can be called a resonance and what not. My point of view is simple: In so many resonating strongly coupled channel systems all features from "sure" resonances to

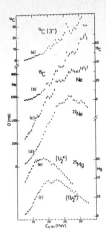

Fig. 8. Freeman's measurements of various final channels of the $^{14}C+^{14}C$ system demonstrate again the coupling of very many nuclear molecular channels.

statistical fluctuations can occur; and all within the direct interaction model of molecular channels. Hence fluctuations do not disprove the molecular model. There appear molecular fluctuations. The proof of the molecular model comes from the overall consistency in so many different aspects. These observations are independent of the particular theoretical models used. The localization of the molecular band, particularly the spin-asignment is not always easy. An especially nice example has been presented at this meeting by the *Münster-group*, who measured the angular distribution in very fine energy steps for the $^{16}O(^{16}O, \alpha)Si^{28}$ g.s. reaction. The results of *Gaul et al.* are displayed in Fig. 9, showing that the angular momentum

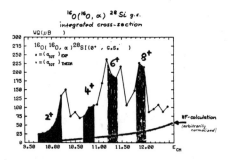

Fig. 9. The angular momentum strength as measured by the Münster-group for $^{16}O(^{16}O, \alpha)^{28}Si$ g.s. The method works here particularly well because the α-reaction diminishes the background.

strength is not always at the peak of the resonances. This is an experience known from the (coupled channel) calculations for quite some time. *Dr. Evers* presented again his information on the analogous reactions $^{12}C(^{12}C, n)$ ^{23}Mg and $^{12}C(^{12}C, p)^{23}Na$ (see Fig. 10). He sees some differences between the proton and neutron channels. These measurements are known to me since the Agaeen Sea conference.

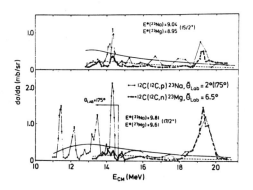

Fig. 10. Excitation function of $^{12}C(^{12}C, p)$ and $^{12}C(^{12}C, n)$ for two strongly populated analogue states in the mass 23 nuclei with J = 15/2+ and 17/2+ respectively.

I was at first a little surprised by these results, but after recognizing that the neutron emission threshold is around 12 MeV, I am convinced that this is not an isospin-effect, but a threshold effect. Before making conclusions about isospin-nonconservation the **barrier-effects**, different for protons and neutrons, should be first taken out of the data.

3. The Untreated Channels - Imaginary Potential

The large amount of channels arising from semi-molecular and compound structure of the nearly amalgamated system cannot be treated explicitly. Here the introduction of the imaginary potential is helpful. As mentioned by *Scheid*, the imaginary potential is given through the decay of the molecular channel into these very many untreated states

$$W(r) = -i\pi |<\psi_{comp}|V|\psi_{Mol}>|^2 \rho(E,I) ,$$

where $\rho(E, I)$ is the precompound level density as a function of energy E and spin I. It is believed that the energy and spin dependence of the imaginary potential is given simply by $\rho(E, I)$. This need not to be true everywhere, but is perhaps a good starting point. We are thus lead to the idea of the molecular window: In the vicinity of the Yrast-band there exists a "window" of low level density, so that the molecular channels lying therein are not damped out; they couple more or less only under themselves (Fig. 11a). This guarantees that the molecular states coupled to the elastic channel through the double-resonance mechanism are not damped to such extend that the structure would be lost. *Koennecke, Guirguis and Scheid* at Frankfurt and Gießen have determined the strength and the moment of inertia parameter entering the level density in the imaginary potential such that elastic excitation functions were reproduced satisfactorily (see Figs. 4). The resulting imaginary potential in the E-ℓ-plane is shown in Fig. 11b. It can be seen that the molecular

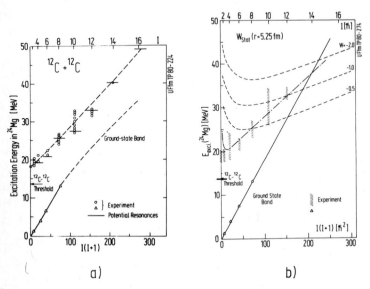

Fig. 11a: The Yrast band and the molecular band in ^{24}Mg. The circles indicate observed molecular resonances. Their very existence proves the idea of Fink et al. (Nucl. Phys. A188 (1972) 259) that there must exist a molecular window along the Yrast band, in which the imaginary potential (proportional to the level density) is small.

Fig. 11b: The imaginary potential in the E-I-plane of the form:

$$W(r,E,I) = \alpha\, N(r)\, \rho(E,I)$$

$$\rho(E,I) \sim \frac{(2I+1)}{\sigma^3}\, e^{2\sqrt{a(E-V)} - \frac{(I+1/2)^2}{2\sigma^2}}, \quad \sigma^2 = \frac{\theta}{\hbar^2}\sqrt{\frac{E-V}{\beta}}$$

with $\alpha = .3$ MeV and $\beta = .2$ MeV^{-1} as obtained by Koennecke, Guirguis and Scheid by fitting the elastic and inelastic excitation functions. The location of the observed molecular resonances is indicated by the dashes. It might be necessary that for higher E and I the above formula has to be modified. In particular there are indications (see Fig. 4c and the discussion of the spin alignment) that there is a different (much smaller) imaginary potential for the inelastic channels at work. This indicates an additional E-I-dependence and probably also an individual structure of the transition matrix element depending on the special molecular configuration.

window exists according to this model up to spins 14 or beyond, depending on the behaviour of the Yrast band. The molecular band structure for various light-ion-systems has been assembled by Marquardt (Fig. 12a). Further evidence for a molecular band in the $^{12}C+^{24}Mg$-reaction (^{30}Ar-system) was reported by Čaplar, Vorvopoulos et al. (Fig. 12b). All this adds to our knowledge of nuclear molecules in a variety of systems.

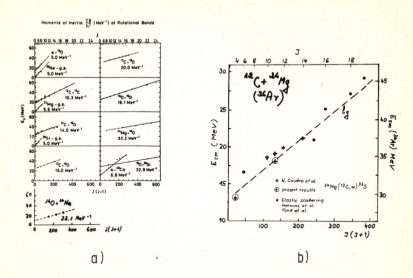

Fig. 12. a) Molecular band structure of various systems assembled by Marquardt et al. b) Nuclear molecular band observed in $^{24}Mg(^{12}C, \alpha)^{32}S$-reaction. These results were presented by R. Čaplar, G. Vourvopoulos, X. Aslanoglou and D. Počanić.

4. Further Developments

The new accelerators with variable energy and high resolution coming to work have shed already some light on this conference. The Argonne group of *Walter Henning, John Schiffer et al.* showed us the resonance structure appearing in $^{24}Mg(^{16}O, ^{12}C)^{28}Si$ (g.s.) reaction, which seems to indicate an equidistant spacing (Fig. 13). Some of his data were obtained with the new Argonne machine. In other words, a kind of oscillator real potential must exist (Fig. 14). Then, because of the well-known degeneracy of various oscillator states with different angular momenta, those resonances should contain various angular momenta. I dare to make such a prediction and look forward to learn what future angular distributions analysis will tell us. The first three states seem to have a peculiar angular momentum sequence, which *J. Schiffer* tried to explain as a grazing window effect cutting through various (involved) bands (Fig. 15). One has to postulate, that three quasi-molecular bands exist, that their moments of inertia is nearly equal and that they have different, appropriate band-head energies. Then the molecular window may cut out the spin sequence 20, 23, 26 as illustrated in Fig. 15. One wonders what the spins of the higher resonances are and one also wonders whether the spin of the lower

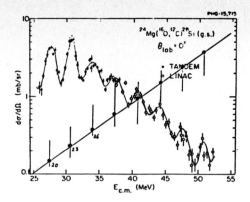

Fig. 13. Excitation function for ^{24}Mg(^{16}O, ^{12}C)^{28}Si$_{gs}$ transition at $\theta_{lab}=\theta_{cm}=0$ measured at Argonne tandem (Phys. Rev. C21 (1980) 1810) and at the Argonne superconducting lineac by Henning and his colleagues. The solid line is drawn to guide the eye. The dashed dots (and the line through them) indicate the relative positions of peak cross sections. The angular momenta for the lower 3 structures are predominantly J=20, 23, 26. The higher ones are currently being measured. These data were first reported at the Agaeen Sea Conference 1 1/2 years ago.

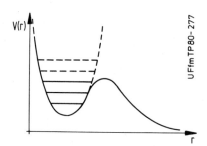

Fig. 14. Oscillator structure of the quasimolecular potential minimum as inferred from Henning's data. At equidistant spacings appear a number of degenerate resonating partial waves (virtual quasimolecular states).

resonances is really the one reported. This deserves to be checked! If the above suggested oscillator picture is correct, each of the observed bumps should consist of a cluster of various angular momentum states. A next higher bump should have a higher component than the one before. I am pretty sure that here also significant fine structure will show up (due to the double resonance mechanism) if looked with higher resolution and in other channels. Other very stimulating results came from *Russel Betts*, who measured cross sections for elastic and inelastic scattering of ^{28}Si+^{28}Si

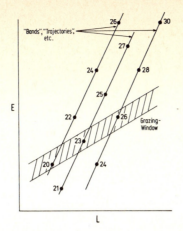

Fig. 15. Illustration of a grazing window and three involved molecular bands as a possible model for understanding the peculiar angular momentum sequences observed for the first three resonances of Fig. 13.

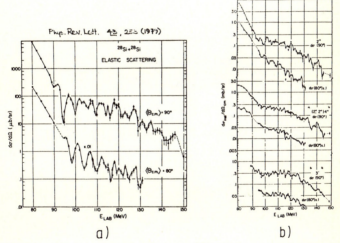

Fig. 16. a) Elastic excitation functions for $^{28}Si+^{28}Si$ at two angles (Fig. 16c).
b) Various inelastic excitation functions for the $^{28}Si+^{28}Si$-system at two angles.

(Figs. 16a-c). The inelastic excitation functions have now been well established. The spin assignments came from the observation of angular distributions. Clearly these data are *extremely stimulating*. They not only yield the presently highest spin observed in a nuclear system (I=42), but also indicate that nuclear quasi-molecular structures are found in heavier systems. Fine structure is also seen, as all the excitation functions are in their basic features very similar to those of $^{12}C+^{12}C$ and other already well-studied systems. The fact, that the $^{16}O+^{40}Ca \rightarrow {}^{28}Si+^{28}Si$

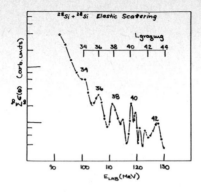

Fig. 16. c) The angle-integrated (between 60° and 90°) elastic scattering cross section for $^{28}Si+^{28}Si$ shows definite cross structure whose angular momenta are identified.

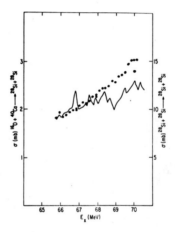

Fig. 17. The cross section $^{16}O(^{40}Ca, ^{28}Si)Si^{28}$ (dotted points) in comparison to the $^{28}Si+^{28}Si \rightarrow {}^{28}Si+^{28}Si$ cross section.

cross section is of equal magnitude as $^{28}Si+^{28}Si \rightarrow {}^{28}Si+^{28}Si$ (see Fig. 17) indicates the strong coupling of these particular molecular channels. Since ^{16}O and ^{40}Ca are magic clusters, the potential energy surface of ^{56}Ni should have a pronounced molecular minimum for this particular fragmentation. I am sure that if one looks closer, one will find a number of isomeric resonances, similar as what we have learned from the resonance study of *K. Erb* and *D.A. Bromley* for ^{24}Mg (see later). I mention, that *Langanke* gave already some quantitative hints this morning about the fine structure, which is probably quite well accounted for by the double resonance mechanism. Another

feature passed by nearly unnoticed this morning. It is the mass fragmentation of
$^{16}O+^{40}Ca$ measured also by Betts (Fig. 18). Namely, *Werner Scheid* in his excellent

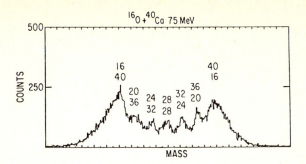

Fig. 18. The mass-fragmentation of the ^{56}Ni-system produced in the $^{16}O+^{40}Ca$ molecular channel at 75 MeV shows typical enhancements for α-cluster transfer.

theoretical talk this morning had no time to mention the molecular fragmentation theory, which I illustrate now: The mass- and charge-transfer and in particular also the α-particle transfer can most lucidly be described in the fragmentation theory, which has been developed in Frankfurt over the last 8 years. - H.J. Fink, W. Greiner, R.K. Gupta, S. Liran, J.A. Maruhn, W. Scheid, and O. Zohni, Proc. of the Int. Conf. on Reactions between Complex Nuclei, Nashville (1974) 21, North-Holland, Amsterdam. K.H. Ziegenhain, H.-J. Lustig, J. Hahn, J.A. Maruhn, W. Greiner, and W. Scheid, Fizika 9 (1977) 559, Supplement 4. J.A. Maruhn, W. Greiner, and W. Scheid, in "Heavy-Ion-Collisions", ed. by R. Bock, vol. II (North-Holland, Amsterdam, 1980) chapter 6. J. Hahn, J.A. Maruhn, A. Sandulescu, and W. Greiner, J. Physics G 7 (1981) 785. -
It is a quantum-mechanical theory in contrast to the rather ad-hoc diffusion models used in deeply inelastic heavy ion collisions. Its basic elements are these: One divides the nucleus into two parts (see Fig. 19) with the mass-fragmentation coordinate $\eta = \frac{A_1-A_2}{A_1+A_2}$ and the relative distance $\vec{r}$ between the two fragments with nucleon

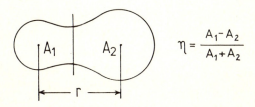

Fig. 19. Illustration of the basic coordinates in fragmentation theory.

number A_1 and A_2 respectively. More coordinates may easily be introduced, if necessary. The Hamiltonian consists then of a kinetic energy of relative motion, $-\frac{\hbar^2}{2\mu(\eta)}\Delta_r$, a kinetic energy in the fragmentation degree of freedom, $-\frac{\hbar^2}{2B_{\eta\eta}}\frac{\partial^2}{\partial\eta^2}$, and a potential $V(r,\eta)$. The latter can be calculated in the two-center-shell-model. The Schrödinger equation then reads

$$\left[-\frac{\hbar^2}{2\mu(\eta)}\Delta_r - \frac{\hbar^2}{2B_{\eta\eta}}\frac{\partial^2}{\partial\eta^2} + V(r,\eta) \right] \psi(r,\eta)$$

It can be solved in various ways, which can all be found in the literature. The main point I want to make is that the potential $V(r,\eta)$ is a rather periodic function of η. It is illustrated in Fig. 20: If the relative motion (r-degree of freedom) is relatively adiabatic, as e.g. in a fission process, the Schrödinger equation

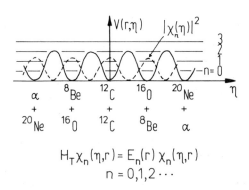

$$H_T \chi_n(\eta,r) = E_n(r) \chi_n(\eta,r)$$
$$n = 0, 1, 2 \cdots$$

Fig. 20. For fixed relative position the fragmentation potential is a periodic function of η, revealing the α-cluster structure (particular strong binding) of the system. Here, as a schematic example, the $^{12}C+^{12}C$ system is displayed. The collective energy levels for the stationary fragmentation Schrödinger equation are indicated by horizontal lines. A solution $\chi(\eta, r_{fixed})$ of such a state is indicated. It also reveals periodic structure.

for fixed r then reads

$$\left[-\frac{\hbar^2}{2B_{\eta\eta}}\frac{\partial^2}{\partial\eta^2} + V(r_{fixed}, \eta) \right] \chi_n(\eta, r_{fixed}) = E_n \chi(\eta, r_{fixed})$$
$$n = 0, 1, 2 \ldots$$

Then $\chi_n(\eta,r)$ clearly also reveals periodic structure and $n = 0,1,2$ numerates the energy levels (which might be quasi-energy bands as in a crystal). The probability for mass fragmentation

$$P(\eta) \, d\eta \sim \left| \chi(\eta, r_{fixed}) \right|^2$$

then also shows periodic structure. This illustrates the strong correlations in the molecular channel wavefunctions

$$Q_c = \chi_n(\eta, r) \, Y_{IM}(\hat{r})$$

with the α-, 2α-, 3α .... etc. transfer. We are dealing here with <u>molecular-mass-transfer-resonances.</u> Such resonating mass transfer has been predicted by us on many occasions (in Proceedings of the International Symposium on Future Directions in Studies of Nuclei far from Stability, Nashville 1979; in Proceedings of the Int. Conf. on Reactions between Complex Nuclei, Nashville (1974),21). We have here an outstanding example for it. Essential is, that the fragmentation potential reveals the α-cluster structure. Even in the ^{40}Ca-potential (Fig. 21) this feature is still very obvious. One expects therefore collective resonance - fragmentation (transfer) and this seems to be precisely

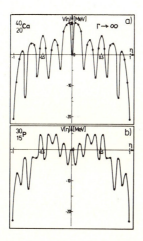

Fig. 21. The fragmentation potential for ^{40}Ca and for ^{30}P also reveals the periodic structure. In the case of ^{40}Ca the structure is due α-clustering. The same is true for the ^{56}Ni-fragmentation potential.

seen by *Betts*. It would be most interesting to measure the mass distributions in small energy steps (excitation function) to look for the resonance transfer we predicted many years ago. The magnitude of the transfer cross section should show maxima and minima as a function of energy.

Theory predicts that molecular resonances can be found up to Nd-Nd collisions (Morovic et al. Z. Naturforsch. 31a (1976) 327) and it may even play a role in deep inelastic reactions. I show some preliminary calculations by *Koennecke* on Si-Si (Fig. 23) and Ca-Ca (Fig. 24) scattering. Clearly the imaginary potential (molecular window) has to be investigated much more carefully before quantitative comparison can be made. One should take these preliminary results, however, as a stimulus to try out even heavier systems (e.g. Ni-Ni). Here I rely on *Karl Erb* and would strongly like to urge him to investigate this system, particular its elastic, inelastic excitation functions and its mass transfer. I know that this takes time, but the Oak Ridge accelerator is ideally suited for such **studies** and it would be a most

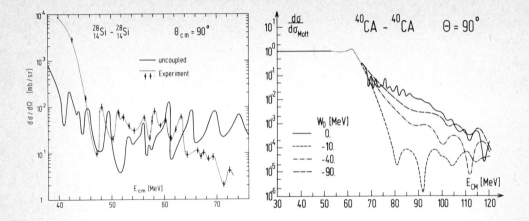

Fig. 22. A very preliminary calculation of the Si - Si elastic excitation function evaluated by Koennecke and Guirguis. The calculation is based on a real potential calculated from an extended liquid drop model and an imaginary potential $W(E,I)$ described earlier.

Fig. 23. Schematic studies of the $^{40}Ca-^{40}Ca$ elastic excitation function in its dependence on the imaginary potential. The calculation is based on a real potential calculated from the two center shell model and a Woods-Saxon type imaginary potential, where the depth parameter was varied causing different structures in the cross section at higher energies.

significant contribution from this laboratory to the field of molecular nuclear physics. Please, Karl, go to work!

Before I turn to theory let me make some remarks on the "possible dynamical symmetries in nuclear molecular spectra" reported by *Allan Bromley* and *Karl Erb*. What they basically do is the very successful fitting (see Fig. 24) of a number of resonances in the $^{12}C-^{12}C$-system with a formula - based on a Morse-potential -

$$E(\nu,L) = -D + a(\nu+1/2)^2 + cL(L+1)$$

The group-theoretical music composed by Iachello for that is unnecessary.
Essential is only that there are certain vibrational states on which rotational bands are built. The work of *Cindro* has to be seen in the same spirit. *Cindro* had used the rotation-vibration-model for which the analogous formula reads

$$E(n_o,n_2,I) = E_o + E_\beta(n_o+\frac{1}{2}) + E_\gamma(n_2+\frac{|K|}{2}+1) + \frac{\varepsilon}{2}I(I+1)$$

Here, the various constants have a physical meaning: E_o is the shift of the molecular potential minimum above the ground state, E_β and E_γ are the β- and γ-vibrational energies in such a potential well and ε is the reciprocal moment of inertia of the nuclear molecule. A slight anharmonicity in the nuclear potential of the

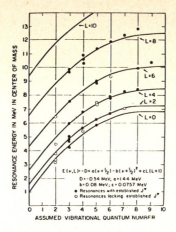

Fig. 24. The fit of the quasimolecular resonances close to the Coulomb barrier by a Morse-type-potential according to K. Erb and D.A. Bromley.

form

$$V(R) = \frac{C_2}{2}(R-R_o)^2 - C_4(R-R_o)^4$$

will immediately introduce square terms proportional to $-C_4'(n_o+\frac{1}{2})^2$. This demonstrates the closeness of Cindro's formula to that of Erb and Bromley. What I want to say is this: Stay at the physical side. The group-theoretical language of Iachello is just fog which covers the beautiful physics. This - by the way - is also true for the IBA-model. I just mention here that *Moshinsky* has recently shown (Nucl. Phys. A338 (1980) 156) that the IBA-model is mathematically completely equivalent to the potential energy approach we have developed in Frankfurt during the last ten years. The latter has the additional advantage that it is physical: One sees from a potential energy surface where a nucleus has its minima, maxima, soft degrees of freedom, instabilities etc. From the fitted parameters, needed for constructing the Hamiltonian from the Casimir operators of a certain Lie-group and its subgroups, one cannot learn very much about the features interesting to a physicist. After this side-remark let me come back to the most interesting observation of *Erb* and *Bromley*. If from their fitted constants a and b the Morse-type potential is reconstructed, it seems to give long-range interaction between the ^{12}C-nuclei (up to 10-15fm see Fig. 25). This can certainly not happen in the entrance channel. If this turns out to be true (Cindro should improve the rotation vibration-model as I suggested and supplement it by a Coulomb-barrier to see how he can fit all the data - it simply might be that the Morse-potentials are a too stringent class of potential functions), it could only mean that there must exist isomeric minima in ^{24}Mg of very elongated type: Perhaps of six α-particles in a chain ⓐⓐⓐⓐⓐⓐ, which yield the longest ranging interactions of ^{12}C-nuclei.

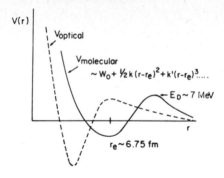

Fig. 25. The Morse-type potential deduced by Bromley and Erb from their fits of the molecular energy levels displayed in Fig. 24. The dashed curve shows the well-known molecular potential of the entrance channel. The comparison of both shows that the near- and sub-Coulomb resonances fitted by Bromley and Erb must correspond to an extremely stretched molecular configuration of ^{24}Mg (six α-particles in a chain).

Mosel and Chandra get also isomeric structures within the two center shell model (they used β- and γ-deformed centers!) Perhaps their result has something to do with what we have just discussed. The strong coupling *Tanimura and Mosel* introduced to explain the pair production cross section for the decay of the second 0^+-state in ^{16}O-^{16}O collision, as measured by *P. Paul* (Fig. 26), also indicates that there is a peculiar role of deformed individual configurations in the molecular reactions. The

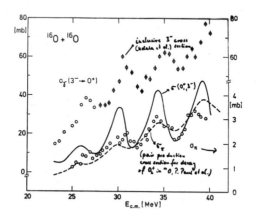

Fig. 26. The Tanimura-Mosel calculations, using very strong coupling potentials, fit the new measurements of Peter Paul of the conversion of the molecularly excited second 0^+-state in ^{16}O rather well. This state can not be directly excited, except if it has some compression component in its wavefunction; but it can be excited in two or more steps.

fact, that more - center configurations create isomerism in light nuclei, I know for years from the diploma theses of *Lustig* and *Hahn* and the doctoral thesis of *Degheidy* at Frankfurt. I demonstrate a few of these aspects in the following figures 27. One notices that such isomerism arises in quite different nuclei. Perhaps it is a rather general phenomenon. In fact, *Bromley and Erb* plotted also the resonances $^{16}O-^{12}C$ in the now familiar pattern (Fig. 28). The very existence of this "molecular order" suggests that there exists such isomerism also in ^{28}Si). I wonder how the $^{16}O-^{16}O$, i.e. the ^{32}S resonance pattern and that of other systems will look like.

We see here the importance of investigations as those of the Erlangen group. If the resonances they observe below and in vicinity of the barrier can be ordered in a "molecular order" of the *Bromley-Erb-* or *Cindro* type it probably means the discovery of molecular isomerism. The results of *Strayer and Cusson*, reported here by *Bromley* and displaced again in Fig. 29 point into the same direction. According to their TDHF-calculation the excited α-particle-chain-type 2^+-excited ^{12}C bounces forth and back through the spherical ^{12}C. This indicates that we have to expect a strong coupling of shape-isomeric molecular configurations to the standard molecular entrance channels.

To round-off my arguments, I remind you also to the coordinate-dependend masses as a cause for isomerism and more minima in the molecular potential. This was first demonstrated by *Fink, Scheid and Greiner* (Journal of Physics, G1 (1975) 685). I shall come back to this point in the next section.

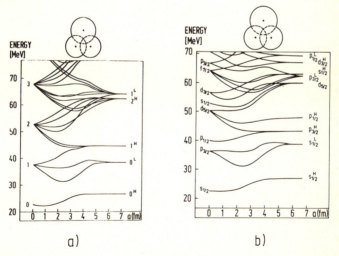

Fig. 27. a) Three center break-up of a harmonic oscillator into one big and two equal smaller fragments (oscillators). The distance between the centers is denoted by a.
b) The same as in Fig. 27a, but now with spin-orbit-force included. Here the realistic nuclear shells are reproduced in each limit.

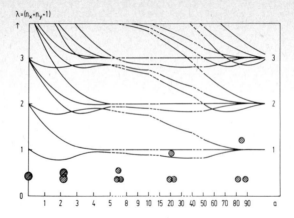

Fig. 27. c) Break up of an harmonic oscillator into three equal fragments. The specific sequence of the break-up is illustrated in the lower part of the figure.

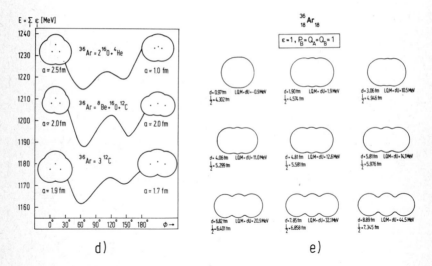

Fig. 27. d) The potential energy structure for various three-center break-ups of ^{36}Ar. One notices that a whole variety of molecular-type of isomerism can exist.
e) The three-center break up of ^{36}Ar into a chain of equal ^{12}C-nuclei.

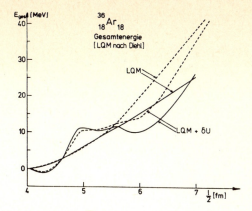

Fig. 27. f) The potential energy for the three-center break-up of ^{36}Ar shows two isomeric structures. The break-up is according to the geometry displaced in figure 27e. Liquid-drop model plus shell corrections are indicated.

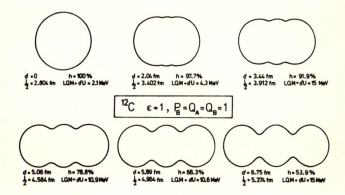

Fig. 27. g) Geometrical shapes of the chain-like break-up of ^{12}C into three α-particles.

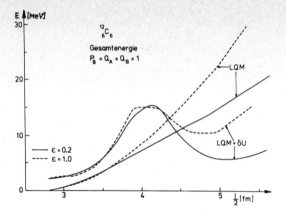

Fig. 27. h) Potential for the chain-like break-up of ^{12}C into three α-particles (geometry is shown in Fig. 27g) shows an isomeric molecular-type structure, which can be identified with the second 0^+-state of ^{12}C.

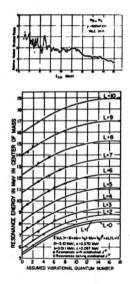

Fig. 28. Molecular ordering of levels below and close to the Coulomb barrier for the $^{16}O+^{12}C$-system (from D.A. Bromley's report).

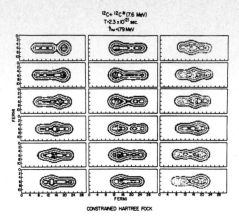

Fig. 29. TDHF-Calculations of M. Strayer and Ron Cusson for the central collision of two ^{12}C; one of them being in the excited 2^+-state. The bouncing forth and back of the α-particle chain through the spherical ^{12}C is impressive. This calculations again demonstrates that we have to expect exotic isomeric structures coupling to the molecular channels of entrance.

5. The importance of the two center shell model degrees of freedom.

The two-center degree of freedom is for the binary molecule (and this we always produce in the entrance channel of a nucleus-nucleus collision) the most important one. The two-center shell model (TCSM) describes the underlying shell structure. The question arises, where in nuclear molecular physics the two center shell structure is important. We know, of course, that it determines the potential for the nucleus-nucleus collision, but are there perhaps specific features, which can be observed? *Mosel* and *Chandra* pointed out that level crossing effects can effectively lead to two types of potential, namely for some channels more adiabatic ones, for others more sudden type potentials. This is connected with the coexistence of sudden and adiabatic configurations, which I suggested some years ago (see e.g. Scheid's talk at the Hvar-meeting 1981, where this has also been discussed — North-Holland-Pub. Comp., Amsterdam, editors Cindro, Ricci and Greiner). Mosel incorporated this idea in his strong coupling calculations with Tanimura, which I mentioned earlier. A specific level-crossing situation of a so-called diabatic level with an adiabatic one is shown in Fig. 30.

There is also another indication that the two center-structure works. Namely, according to the report by *Konnerth*, the Munich group of *Eberhardt* has found that in the $^{12}C-^{14}C \rightarrow ^{12}C$ (2+, 4.43)$-^{14}C$ process the excitation of the 2+state in ^{12}C is very depressed compared to its excitation in e.g. $^{12}C-^{12}C$. (see Fig. 31). One would have expected it naively to be similar in both cases. However, due to the two neutrons in ^{14}C the $p_{1/2}$-level is partially blocked for the 2^+ excitation, as I indicated in Fig. 30. Hence we understand the depletion of the excitation in this particular

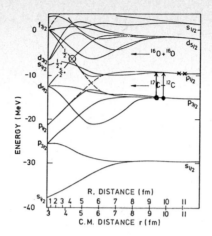

Fig. 30. The two center level diagram for light systems. The $^{12}C-^{12}C$ entrance and the $^{16}O-^{16}O$ entrance are indicated. Also the crossing of a diabatic with an adiabatic level is shown. The arrows (↑↑) indicates the excitations needed for the 2^+-state of ^{12}C and the crosses (xx) indicate the two neutrons of ^{14}C, which block these excitations.

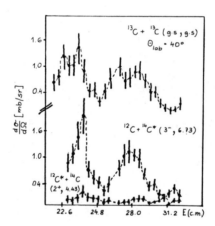

Fig. 31. Excitation functions for various excit channels having $^{12}C+^{14}C$ as entrance channel. The depletion of the 2^+-excitation in ^{12}C is obvious from the data shown at the bottom of this figure.

reaction! *Park* and *Scheid* talk for some time now about the possibility of resonance-transfer of nucleons due to crossing of two-center-shell-model-levels of nucleons. The idea is illustrated in Fig. 32: The odd neutron of ^{25}Mg will end up in the collision with ^{16}O in an ^{17}O-orbit. The possible points of crossing-transfer are indicated by wavy lines.

Let me again draw attention to the work of *Fink* and *Scheid* of about 8 years ago, in which they argue that due to the crossing or pseudo crossing of two-center-orbitals the mass $B_{RR}(R)$ for the relative motion of the two deviates from the reduced mass. In fact, it can become several times larger than the reduced mass, reflecting the rigidity of the nuclei to be attached to their original orbits. This structure in the mass $B_{RR}(R)$ is equivalent to second (or more) minima in the effective ion-ion potential (see Fig. 33). Hence we should keep in mind that the coordinate-dependent mass of two nuclei can create resonance-type structures. Up to now this aspect of nuclear molecular physics has not been demonstrated in detail theoretically. I call for more work in this direction.

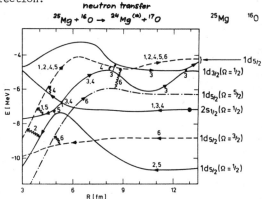

Fig. 32. Resonance-neutron transfer due to crossing of two-center-orbitals. At the crossing points the transfer of a nucleon is particularly favored (according to Park and Scheid).

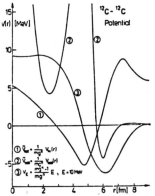

Fig. 33. Interaction potentials calculated from coordinate dependent masses. Curve 2 shows the additional potential term which enters because of the coordinate dependence.

6. Recent developments

At Yale-University D.A. *Bromley* and his collaborators M. *Gai* and *Rex Keddy* have bombarded ^{14}C- with α-particles and excited various states of ^{18}O. Especially interesting is the very strong E1-transition between the 1^--state at 4.45 MeV and the 0^+-state at 3.63 MeV. Normally E1-transitions are of the order of 10^{-5} Weiskopf-units (see Fig. 34). This transition is by factor 100-1000 larger, however. It indicates in my mind that we are dealing in case of the 1^--state with a nuclear molecule with strong dipole moment. It must have the structure of the form

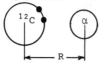

where the two neutrons keep sticking to the ^{12}C-core. Possibly even some neutron probability of the α-particle is dragged over to the ^{12}C-core because of two-center-shell effects. The unequal distribution of neutrons relative to the protons causes the dipole moment and hence the <u>polar molecule</u>, similar as in the NH_4-molecule. If this picture checks also with the moment of inertia and a few other E2-transitions, we will have here the first polar nuclear molecule.

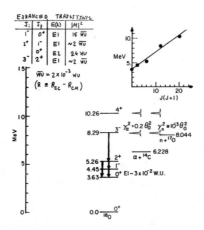

Fig. 34. The level structure excited in the $\alpha + ^{12}C \rightarrow ^{18}O$ reaction, where a strong dipole transition has been observed between the 1^--state and the $^+0$-state (D.A. Bromley, M. Gai, Rex Keddy).

Also from *Yale* comes the observation of an 8^+-state in the $^{12}C(^{12}C,\alpha)^{20}Ne, ^8Be$ reactions fitting together with previously observed 6^+- and 4^+-state nicely into a rotational band pattern (Fig. 35). The molecular structure of this 8p-4h-configuration in ^{20}Ne seems to have analogy to the water molecule, i.e. it is of the type

A similar configuration was identified by *Gersch* et al. from Desden for the 14^+-state in the $^{12}C+^{16}O$-reaction. They reported this at the Hvar-meeting last May. Their identified structure looked like

Let me now jump to fission isomers in Uranium. They also are a kind of molecular structure. The Heidelberg-group (*Metag, Habs, Specht* and others) have excited a 0^+-state in the second minimum by $^{238}U(d,pn)^{238m}U$ reaction. They looked at the

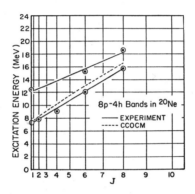

Fig. 35. Peter *Parker* investigated the $^{12}C(^{12}C,\alpha)Ne^{20}~^8Be$ - reaction. It is suggested that the $K = 0_6^+$ rotational band in ^{20}Ne, including the newly observed 8^+-state at 18.534 MeV indicates a nuclear molecule of the type C_α^α. The 0^+-state lies at 12.44 MeV (Balamuth et al., Tokyo Conference J. Phys. Soc. Japan 44(1978)244.) and the 6^+-state at 15.16 MeV (Young, Nucl. Phys. A330 (1979) 452).

conversion electrons and found strong peaks (see Fig. 36). This seems to be the discovery of the β-vibration in the second minimum; thus confirming again the richness of molecular structure also in heavy nuclei.
Other very exciting news comes from the positron front. *Greenberg, Schwalm*, et al. observed resonances in the positron spectrum of U+U-collisions (see Fig. 37). It seems to signal not only the theoretically predicted decay of the vacuum, but also the existence of rather long-living superheavy nuclear systems (molecule, nucleus). The lifetime of the super-systems seems to be considerably longer than 10^{-20} sec. Also, the fact that more positron lines are seen at various energies and at rather sharp angles suggests that we are dealing here with a superheavy molecular nuclear system (which probably cannot be distinguished from a superheavy nucleus) with several

isomeric configurations. Due to the double-resonance mechanism a whole variety of nuclear states of rather welldefined, low angular momenta ($\ell \leq 30$) can be found in the vicinity of the Coulomb barrier. Therefore, pronounced angular structure (due to $|P_\ell (\cos \theta)|^2$) and also fluctuations (due to many resonances) may appear. Moreover, supercritical conversion (pair conversion of excited states of the superheavy nucleus) will give rise to additional structure in the positron spectrum. We thus see that nuclear molecules are a rather general phenomenon, even existing in the largest systems we can possibly make on earth. If the positron signals turn out to be correct, we certainly will move into an extremely interesting future, not only allowing the establishment of extreme superheavy nuclear systems, but also allowing their spectroscopy through supercritical conversion.

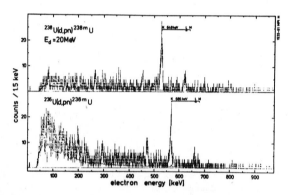

Fig. 36. Conversion electron spectrum in coincidence with fission decay of the shape isomer U. Goerlach, D. Habs, V. Metag, H.J. Specht, B. Schwartz, Heidelberg 1981.

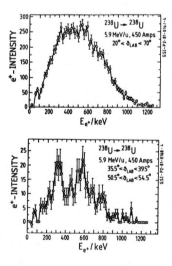

Fig. 37. The total positron spectrum (upper figure) and the positron spectrum for a angular window around 37° and 52° for U+U. Greenberg, Schwalm, Bokemeyer, Bethge, Schweppe and Gruber seem to observe resonance structure.

I believe that there lies an exciting future ahead of us:

1) The extension of the present work to higher spins (lower cross sections). I think e.g. on the continuation of the Yale elastic $^{12}C-^{12}C$ data as done by the Oak Ridge group some years ago. This becomes simpler in the future. Betts has indicated this.

2) The investigation of nuclear molecules in medium heavy and very heavy systems.

3) The better quantitative understanding of the molecular window and thus the imaginary nucleus-nucleus potential. Out of these studies we shall probably also learn completely new features of the level-densities and on the average coupling matrix elements between specific molecular configurations and precompound states (see the discussion in connection with thealignment results).

4) The coexistence of molecular and compound configurations seems to be a challenging problem to theoretical physicists and to experimentalists. Into this category belong also the effects of effective masses depending on the distance between the nuclei. *Fink and Scheid* have shown years ago this may introduce isomeric structures in the potential. (J. of Physics G (1975) 685).

5) The study of the influence of odd particles on nuclear molecules needs to be clarified. The odd particle should introduce new structures, but plays otherwise not a t-o great role. The Landau-Zener effect for nucleons in two center orbitals would prove the existence of molecular orbitals for nucleons. The investigations of *Imanishi and von Oertzen*, who measured and calculated the influence of the two-center-orbitals in the system $^{12}C + ^{13}C$ seems to be promising and interesting.

6) The investigation of molecular transfer; in particular the resonant mass transfer predicted by us within the fragmentation theory appli-d to molecular channels gives a simple explanation for the preference of the α-particle channels and transfers in the molecular reactions. The observation of Betts already supports these ideas.

7) The generator coordinate method must be further developed to yield cross sections and excitation functions. Simple band structure predictions as explained to us in the talks by *Dr. Baye* are not enough. Due to the coupling to other channels these simple predictions may get destroyed. I also won't hide some difficulties I have in the understanding of these calculations. On the one side one performs a very accurate theory starting with a basic Hamiltonian. On the other side one has to leave out a great number of channels, which have to be treated outside the Generator-Coordinate-Scheme by absoroptive potentials. I wonder whether an effective Hamiltonian wouldn't be more appropriate from the beginning.

8) Nuclear molecules seem to play an important role also in deep inelastic reactions. A kind of resonating mass transfer should exist and be identified. As I indicated, this seems already to be the case in the mass-transfer observed by Betts for the Si+Si-system. We can learn from the study of lighter systems about

mechanisms which may have relevance for the heavier systems. In light systems
explicit calculations can be done to a certain extent, while in heavy systems
classical methods have to be introduced at a very early stage. The hydrodynamical calculations of Buchwald, Cusson, Stoecker and Maruhn for Kr-Er (Fig. 38)
show for an ordinary equation of state that quasimolecular transient structures
are indeed formed. The same authors predict under the assumption of an equation
of state with a density isomeric structure (Fig. 39) that rather long living and
extremely deformed nuclear molecules can be formed. If such behaviour would be
found, this indeed, would be a most fundamental discovery.

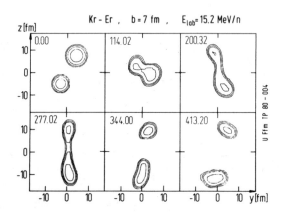

Fig. 38. The formation of a transient nuclear quasimolecule in a Kr-Er collision
with an impact parameter of 7 fm according to the hydrodynamical calculation of Buchwald, Cusson, Stoecker and Maruhn. In this case a normal equation of state has been used. The numbers in the individual figures indicate the time in fm/c (1 fm/c = 1/3 10^{-23}s).

9) Nuclear and atomic quasi-molecules can be connected with each other. I mention
the <u>atomic clock</u>, where due to the time-delay obtained in a nuclear molecule,
the emission of δ-electrons, the K-vacancy-production and in supercritical systems
also the emission of positrons should show periodic structures in their excitation functions (Fig. 40). From them the sticking- or orbiting time can be deduced. This seems to introduce an absolute clock in the time range between
10^{-19}-10^{-22} sec into nuclear collisions. An atomic clock can also be based on
the resonance scattering of particles, as has recently been demonstrated by the
Köln and Stanford groups. I should also mention that one should continue searching for quasimolecular nuclear γ-rays stemming from two center orbital transitions.

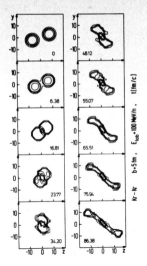

Fig. 39. The appearance of superdeformed, rather long-living nuclear quasimolecules according to hydrodynamical calculations of Stoecker, Cusson, Buchwald and Maruhn. Here an equation of state with a density isomer has been assumed. The isometric matter is formed in the collision and seen in the middle of the very elongated molecule. The numbers in the individual figures give the time-scale in fm/c.

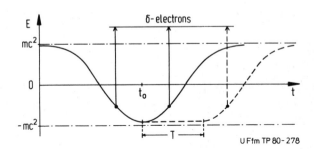

Fig. 40a. Schematic graph of the quasimolecular electronic levels in a heavy ion collision. At the time t_o the two ions are closest to each other and the electronic molecular binding energy is largest. The vertical arrows depict the interferring transition amplitudes for the emission of δ-electrons. The dashed arrow shows the time-delayed (T) transition amplitude in case there is a sticking or nuclear molecular binding.

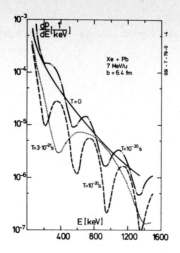

Fig. 40b. Calculated cross section for δ-electron emission as a function of the δ-electron energy E. For Coulomb trajectories (full curve in Fig. a)) an exponential decrease appears. If there is a time-delay T, the incoming and outgoing amplitudes (Fig. a) are changed and the cross section oscillates. From those oscillations the time-delay $T = h/\Delta E$ can be inferred.

10) Nuclear Molecules are interesting in themselves. Their discovery and existence alone, the knowledge we have and still shall obtain about them is an advancement of nuclear science and an interesting contribution to science as a whole. As I indicated by mentioning the recent positron-experiments, nuclear molecular structures seem also to occur in U+U-collisions. This then means that nuclear molecules are a rather general phenomenon.

We have now come to the end of our meeting. It was a week full of discussions. I found a number of exciting new informations and developments. Professor Eberhard deserves our warmest thanks for organizing this conference so nicely for us.

R. Bass
Nuclear Reactions with Heavy Ions

1980. 176 figures, 31 tables. VIII, 410 pages. (Texts and Monographs in Physics). ISBN 3-540-09611-6

Contents:
Introduction. – Light Scattering Systems. – Quasi-Elastic Scattering from Heavier Target Nuclei. – General Aspects of Nucleon Transfer. – Quasi-Elastic Transfer Reactions. – Deep-Inelastic Scattering and Transfer. – Complete Fusion. – Compound-Nucleus Decay. – Appendices. – Subject Index.

I. Lindgren, J. Morrison
Atomic Many-Body Theory

1982. 96 figures. Approx. 488 pages. (Springer Series in Chemical Physics, Volume 13) ISBN 3-540-10504-2

Contents:
Angular-Momentum Diagrams and the Independent-Particle Model: Introduction. Angular-Momentum and Spherical Tensor Operators. Angular-Momentum Graphs. Further Developments of Angular-Momentum Graphs. Applications to Physical Problems. The Independent-Particle Model. The Central-Field Model. The Hartree-Fock Model. Many-Electron Wave Functions. – Perturbation Theory and the Treatment of Atomic Many-Body Effects: Perturbation Theory. First-Order Perturbation for Closed-Shell Atoms. Second Quantization and the Particle-Hole Formalism. Application of Perturbation Theory to Closed-Shell Systems. Application of Perturbation Theory to Open-Shell Systems. The Hyperfine Interaction. The Pair-Correlation Problem and the Coupled-Cluster Approach. – Appendices A–C. – References. – Author Index. – Subject Index.

I. I. Sobelman
Atomic Spectra and Radiative Transitions

1979. 21 figures, 46 tables. XII, 306 pages. (Springer Series in Chemical Physics, Volume 1) ISBN 3-540-09082-7

Contents:
Elementary Information on Atomic Spectra: The Hydrogen Spectrum. Systematics of the Spectra of Multielectron Atoms. Spectra of Multielectron Atoms. – Theory of Atomic Spectra: Angular Momenta. Systematics of the Levels of Multielectron Atoms. – Hyperfine Structure of Spectral Lines. – The Atom in an External Electric Field. The Atom in an External Magnetic Field. Radiative Transitions. – References. – List of Symbols. – Subject Index.

This volume is continued by *Excitation of Atoms and Broadening of Spectral Lines* published as Volume 7 in this series.

"...this book presents a wealth of information about the quantum mechanics of free atoms... it is nearly a must."

Applied Optics

Structure and Collisions of Ions and Atoms

Editor: **I. A. Sellin**

1978. 157 figures, 17 tables. XI, 350 pages. (Topics in Current Physics, Volume 5). ISBN 3-540-08576-9

Contents:
I. A. Sellin: Introduction. – *S. J. Brodsky, P. J. Mohr:* Quantum Electrodynamics in Strong and Supercritical Fields. – *L. Armstrong jr.:* Relativistic Effects in Highly Ionized Atoms. – *J. S. Briggs, K. Taulbjerg:* Theory of Inelastic Atom-Atom Collisions. – *N. Stolterfoht:* Excitation in Energetic Ion-Atom Collisions Accompanied by Electron Emission. – *P. H. Mokler, F. Folkmann:* X-Ray Production in Heavy Ion-Atom-Collisions. – *I. A. Sellin:* Extensions of Beam Foil Spectroscopy. – *S. Datz:* Atomic Collisions in Solids.

Springer-Verlag Berlin Heidelberg New York

Lecture Notes in Physics

Vol. 114: Stellar Turbulence. Proceedings, 1979. Edited by D. F. Gray and J. L. Linsky. IX, 308 pages. 1980.

Vol. 115: Modern Trends in the Theory of Condensed Matter. Proceedings, 1979. Edited by A. Pekalski and J. A. Przystawa. IX, 597 pages. 1980.

Vol. 116: Mathematical Problems in Theoretical Physics. Proceedings, 1979. Edited by K. Osterwalder. VIII, 412 pages. 1980.

Vol. 117: Deep-Inelastic and Fusion Reactions with Heavy Ions. Proceedings, 1979. Edited by W. von Oertzen. XIII, 394 pages. 1980.

Vol. 118: Quantum Chromodynamics. Proceedings, 1979. Edited by J. L. Alonso and R. Tarrach. IX, 424 pages. 1980.

Vol. 119: Nuclear Spectroscopy. Proceedings, 1979. Edited by G. F. Bertsch and D. Kurath. VII, 250 pages. 1980.

Vol. 120: Nonlinear Evolution Equations and Dynamical Systems. Proceedings, 1979. Edited by M. Boiti, F. Pempinelli and G. Soliani. VI, 368 pages. 1980.

Vol. 121: F. W. Wiegel, Fluid Flow Through Porous Macromolecular Systems. V, 102 pages. 1980.

Vol. 122: New Developments in Semiconductor Physics. Proceedings, 1979. Edited by F. Beleznay et al. V, 276 pages. 1980.

Vol. 123: D. H. Mayer, The Ruelle-Araki Transfer Operator in Classical Statistical Mechanics. VIII, 154 pages. 1980.

Vol. 124: Gravitational Radiation, Collapsed Objects and Exact Solutions. Proceedings, 1979. Edited by C. Edwards. VI, 487 pages. 1980.

Vol. 125: Nonradial and Nonlinear Stellar Pulsation. Proceedings, 1980. Edited by H. A. Hill and W. A. Dziembowski. VIII, 497 pages. 1980.

Vol. 126: Complex Analysis, Microlocal Calculus and Relativistic Quantum Theory. Proceedings, 1979. Edited by D. Iagolnitzer. VIII, 502 pages. 1980.

Vol. 127: E. Sanchez-Palencia, Non-Homogeneous Media and Vibration Theory. IX, 398 pages. 1980.

Vol. 128: Neutron Spin Echo. Proceedings, 1979. Edited by F. Mezei. VI, 253 pages. 1980.

Vol. 129: Geometrical and Topological Methods in Gauge Theories. Proceedings, 1979. Edited by J. Harnad and S. Shnider. VIII, 155 pages. 1980.

Vol. 130: Mathematical Methods and Applications of Scattering Theory. Proceedings, 1979. Edited by J. A. DeSanto, A. W. Sáenz and W. W. Zachary. XIII, 331 pages. 1980.

Vol. 131: H. C. Fogedby, Theoretical Aspects of Mainly Low Dimensional Magnetic Systems. XI, 163 pages. 1980.

Vol. 132: Systems Far from Equilibrium. Proceedings, 1980. Edited by L. Garrido. XV, 403 pages. 1980.

Vol. 133: Narrow Gap Semiconductors Physics and Applications. Proceedings, 1979. Edited by W. Zawadzki. X, 572 pages. 1980.

Vol. 134: γγ Collisions. Proceedings, 1980. Edited by G. Cochard and P. Kessler. XIII, 400 pages. 1980.

Vol. 135: Group Theoretical Methods in Physics. Proceedings, 1980. Edited by K. B. Wolf. XXVI, 629 pages. 1980.

Vol. 136: The Role of Coherent Structures in Modelling Turbulence and Mixing. Proceedings 1980. Edited by J. Jimenez. XIII, 393 pages. 1981.

Vol. 137: From Collective States to Quarks in Nuclei. Edited by H. Arenhövel and A. M. Saruis. VII, 414 pages. 1981.

Vol. 138: The Many-Body Problem. Proceedings 1980. Edited by R. Guardiola and J. Ros. V, 374 pages. 1981.

Vol. 139: H. D. Doebner, Differential Geometric Methods in Mathematical Physics. Proceedings 1981. VII, 329 pages. 1981.

Vol. 140: P. Kramer, M. Saraceno, Geometry of the Time-Dependent Variational Principle in Quantum Mechanics. IV, 98 pages. 1981.

Vol. 141: Seventh International Conference on Numerical Methods in Fluid Dynamics. Proceedings. Edited by W. C. Reynolds and R. W. MacCormack. VIII, 485 pages. 1981.

Vol. 142: Recent Progress in Many-Body Theories. Proceedings. Edited by J. G. Zabolitzky, M. de Llano, M. Fortes and J. W. Clark. VIII, 479 pages. 1981.

Vol. 143: Present Status and Aims of Quantum Electrodynamics. Proceedings, 1980. Edited by G. Gräff, E. Klempt and G. Werth. VI, 302 pages. 1981.

Vol. 144: Topics in Nuclear Physics I. A Comprehensive Review of Recent Developments. Edited by T.T.S. Kuo and S.S.M. Wong. XX, 567 pages. 1981.

Vol. 145: Topics in Nuclear Physics II. A Comprehensive Review of Recent Developments. Proceedings 1980/81. Edited by T. T. S. Kuo and S. S. M. Wong. VIII, 571-1.082 pages. 1981.

Vol. 146: B. J. West, On the Simpler Aspects of Nonlinear Fluctuating. Deep Gravity Waves. VI, 341 pages. 1981.

Vol. 147: J. Messer, Temperature Dependent Thomas-Fermi Theory. IX, 131 pages. 1981.

Vol. 148: Advances in Fluid Mechanics. Proceedings, 1980. Edited by E. Krause. VII, 361 pages. 1981.

Vol. 149: Disordered Systems and Localization. Proceedings, 1981. Edited by C. Castellani, C. Castro, and L. Peliti. XII, 308 pages. 1981.

Vol. 150: N. Straumann, Allgemeine Relativitätstheorie und relativistische Astrophysik. VII, 418 Seiten. 1981.

Vol. 151: Integrable Quantum Field Theory. Proceedings, 1981. Edited by J. Hietarinta and C. Montonen. V, 251 pages. 1982.

Vol. 152: Physics of Narrow Gap Semiconductors. Proceedings, 1981. Edited by E. Gornik, H. Heinrich and L. Palmetshofer. XIII, 485 pages. 1982.

Vol. 153: Mathematical Problems in Theoretical Physics. Proceedings, 1981. Edited by R. Schrader, R. Seiler, and D.A. Uhlenbrock. XII, 429 pages. 1982.

Vol. 154: Macroscopic Properties of Disordered Media. Proceedings, 1981. Edited by R. Burridge, S. Childress, and G. Papanicolaou. VII, 307 pages. 1982.

Vol. 156: Resonances in Heavy Ion Reactions. Proceedings, 1981. Edited by K.A. Eberhard. XII, 448 pages. 1982.